THE FIVE-MILLION-YEAR ODYSSEY

The Five-Million-Year Odyssey

THE HUMAN JOURNEY FROM APE TO AGRICULTURE

PETER BELLWOOD

PRINCETON UNIVERSITY PRESS

PRINCETON & OXFORD

Published by Princeton University Press
41 William Street, Princeton, New Jersey 08540
99 Banbury Road, Oxford OX2 6JX

press.princeton.edu

All Rights Reserved

Library of Congress Cataloging-in-Publication Data

Names: Bellwood, Peter S., author.
Title: The five-million-year odyssey : the human journey from ape to agriculture / Peter Bellwood.
Description: Princeton, New Jersey : Princeton University Press, 2022. | Includes bibliographical references and index.
Identifiers: LCCN 2021052570 (print) | LCCN 2021052571 (ebook) | ISBN 9780691197579 (hardback) | ISBN 9780691236339 (ebook)
Subjects: LCSH: Human evolution. | Human beings—Origin. | Human beings—Migrations. | Social evolution. | Civilization—History. | Language and languages—Origin. | Language and culture—History. | BISAC: SCIENCE / Life Sciences / Evolution | SOCIAL SCIENCE / Anthropology / Physical
Classification: LCC GN281 .B438 2022 (print) | LCC GN281 (ebook) | DDC 599.93/8—dc23/eng/20211201
LC record available at https://lccn.loc.gov/2021052570
LC ebook record available at https://lccn.loc.gov/2021052571

British Library Cataloging-in-Publication Data is available

Editorial: Alison Kalett, Hallie Schaeffer
Production: Danielle Amatucci
Publicity: Sara Henning-Stout, Kate Farquhar-Thomson
Copyeditor: Vickie West

Jacket art: From *Animal Biology,* by Lorande Loss Woodruff (1879–1947), 1938.

This book has been composed in Arno

Printed on acid-free paper. ∞

Printed in the United States of America

10 9 8 7 6 5 4 3 2 1

This book is dedicated to my grandchildren—

Ethan, Hamish, Leo, Isla, and Eleanor—

and to the future of humanity.

CONTENTS

PREFACE

FOR SOME YEARS NOW, my family and friends have been telling me that I need to write an account of what I am here calling the Five-Million-Year Odyssey, told in a way that can be understood by nonspecialists. By way of background, I have spent most of my life as a specialist, a person who writes technical reports about archaeology that can only be understood by a few colleagues. This book is a new challenge for me, even though some of my previous books have been widely read by members of the general public.

That said, I do not wish to write a book that is simplistic, that demeans the intellect of my readers. Some of the topics that I discuss are fairly complicated, as befitting the complexity of human behavior, but I try to express them in simple language. There are times when I need to take the bull by the horns, and it is at these times that the contents of this book, I hope, will also be of interest to some of my colleagues, especially in those fields that study the human past through information sources apart from archaeology.

I am now a retired professor of archaeology, having spent my adult life teaching undergraduate and graduate students about the achievements of ancient humans around the world. I have undertaken many archaeological research projects in Southeast Asia and the Pacific Islands, and I have also been lucky enough to visit many of the archaeological wonders of the world in other regions. Should the current pandemic allow, I hope in the future to be able to visit some more.

As a result of all this research and travel, what do I have to say that will be of interest to the general reader, and that has not been presented already by other authors? The answer, I hope, will be a long-term perspective on human populations, past, recent, and in many cases still existing today. I discuss their origins, their migrations, and in some cases their ultimate fates. I commence with ancient apes, and I finish over much of the world, beyond the reach of the ancient civilizations, close to the year 1492 CE, after which the world changed in unprecedented ways that extend beyond my narrative. This book is about the world as it was, before the impacts of the Columbian exchange and the subsequent Colonial Era.

This book is also a personal account that reflects my own career and interests, as well as my conviction that the human past belongs to everyone. I write not just from the perspective of an archaeologist but as someone who has also discovered that archaeology alone, despite its undoubted merits, will not take us very far in terms of a broad understanding of the human past. We also need the bones, the genes, and, during the later part of the record, the reconstructed speech of our ancestors. I do not claim to be an expert in every field of research that I discuss, but I firmly believe that there is still space within human knowledge for a single author to present an opinion on that perennial question—Where do we all come from?

How an Archaeologist Discovered Languages and Genes

As an undergraduate student at Cambridge University during the mid-1960s, I determined many future developments in my life by studying archaeology, taught at that time as a freestanding and practical field of study within the general milieu of history and anthropology, with its own body of theory and interpretation. Nowadays, archaeology has become an integral part of a much broader multidisciplinary network of scientific approaches toward the past, a research network that tracks the histories of human populations in terms of their archaeology, their languages, and their DNA. The current boom in DNA analysis from both the living and the dead that underpins so much current knowledge was little more than a science fiction dream of the future in the 1960s, as were personal computers and online journals.

In 1966, I made a fateful decision that determined my perspective on the world of human prehistory ever after. After a stint as an archaeological supervisor on the excavation of a *tepe* (ancient city mound) in Lorestan Province, western Iran, organized by Institute of Archaeology (London) archaeologist Clare Goff,[1] I accepted a post in 1967 as Lecturer in Archaeology at the University of Auckland in New Zealand. Here, and working out of the Australian National University (ANU) in Canberra after 1973, I discovered the rest of the world—at least outside Europe, North Africa, and the Middle East, the focal regions of my Cambridge studies. In both Auckland and Canberra I had colleagues who were social anthropologists, linguists, and biological anthropologists, and they were always happy to discuss matters of common interest. I also had the good fortune to be teaching archaeology rather than learning it, which is essential if one wishes to find out what one really thinks about topics that matter.

In New Zealand, I discovered the indigenous peoples of Southeast Asia and the Pacific, rich in languages, social anthropology, and biological variation. While living there I continued doing archaeology from the ground, both in New Zealand itself and in various tropical Polynesian archipelagos, especially the Marquesas and Cook Islands. But my attention soon shifted away from the ancient artifacts toward the real people who once existed behind them. I wanted to know who these people were and where their ancestors came from.

The Pacific Islands were an exciting region in the study of the human past during the middle and later decades of the twentieth century. The continued existence to the present of many Pacific Island peoples as healthy functioning societies, despite past colonial burdens, meant that this region of the world offered a field of study focused on their origins and migrations that could mingle findings from ethnology (the comparative study of ethnographic peoples and societies), archaeology, language history, and human evolution (then usually called "physical anthropology").

Indeed, Pacific Island societies had spawned a rich tradition of observation and recording by explorers and anthropologists alike, from sixteenth-century Spanish navigators to twentieth-century anthropologists like Bronislaw Malinowski and Margaret Mead. In addition, many

Pacific Island societies kept long genealogies and oral traditions about their ancestors, sometimes with intriguing levels of accuracy that have been supported by archaeological research.[2] The whole region gave one a sense of being in a "laboratory," where people had arrived on empty islands and their descendants had adapted and diverged into different societies over fairly short periods of time. Polynesia and Micronesia still have this allure today, although the larger islands to the west, around New Guinea and in Indonesia, reveal a far greater time depth and complexity of human settlement.

After my move to Australia in 1973, I spread my research from Polynesia into Island Southeast Asia (Indonesia, Malaysian Borneo, and the Philippines), intent on searching for the origins of the Polynesians and other Pacific peoples, both within Southeast Asian archaeology and within the history of the much greater Austronesian language family to which the Polynesian languages belong. I also extended my research interests in population history into neighboring regions in Mainland Southeast Asia, India, and China. I am currently working with ANU and Vietnamese colleagues on archaeology in Vietnam, examining the activities of ancestral Austroasiatic (including modern Khmer and Vietnamese) and Kra-Dai (including modern Thai) peoples within southern China and Mainland Southeast Asia between 5,000 and 3,000 years ago. I deal with this topic in more detail in chapter 11.

As well as my research in Polynesia and Southeast Asia, my interests have also focused during the past forty years on early farming populations across the whole of the world, asking how they might have spread their food-producing lifestyles, languages, and genes. By the late 1980s, I was starting to perceive an operational linkage between the prehistory of the Austronesian-speaking peoples (including the Polynesians) and a more fundamental theory. The Austronesians were heirs to a great oceanic migration made possible by the demographic growth in population consequent upon the adoption of techniques of food production, as opposed to hunting and gathering from the wild. Of course, the ancient Austronesians had boats with outriggers and sails as well, and they were adept at extracting food from the sea, but a transportable food-producing repertoire of domesticated plants and animals was the essential support for those founding populations who braved the un-known during so many ancient voyages over the horizon. The broader and more fundamental theory that I could perceive went beyond the Austronesians and focused upon the acquisition of techniques of food production and the spread of human populations and their language families in many other parts of the world.

Nevertheless, the contents of this book are much broader than just the origins and dispersals of food-producing populations within the past 10,000 years, even if those dispersals determined the human bio-logical and linguistic tapestry that covers much of the world today. In this Odyssey, I investigate the whole five million years of humanity. I investigate our roots among older and now extinct hominin (human-like) species as well the prehistory of humans just like us, our more immediate ancestors. We begin with apelike creatures who lived more than five million years ago and finish with the precolonial populations whose descendants still form the underlying structure of our twenty-first-century world.

Reconstructing the Past from Multiple Sources

What can we really know about the past, especially its deeper layers, before written history came into existence? When we admire a majestic ruin—for example, the Great Pyramid at Giza or the Colosseum in Rome—we might try to imagine what the reality was like so long ago. Much of what we imagine nowadays is supplied for us by the media, particularly the enormous number of documentary films that relive the past in colorful detail, with narrators, actors, fabulous costumes, and exciting action. The Egyptians and Romans, of course, also had written history. But how much can we really know about a day in the life of an average human in, say, 50,000 BCE? Nothing written has survived from that time.

The answers will vary depending on the time depth. We know far more about life in ancient Rome than we do about life in the deeper recesses of prehistory, before writing was invented, for which most in-formation has to be extracted piece by piece from the ground. Back at five million years ago, when our apelike ancestors roamed tropical Africa, we have almost no direct information at all and must reconstruct possibilities through skeletal and genetic comparisons with our living chimpanzee and bonobo cousins. As we progress through the Odyssey, so the methods of extracting relevant information change. We begin with fossils and stone tools, and we end with the populations and languages that form the human tapestry of our modern world.

All of this means that reconstruction of the human past must be a multidisciplinary exercise. Four core fields of study—archaeology, paleoanthropology, genetics, and comparative linguistics—provide the central bodies of data, supported by inputs from the earth sciences, plant and animal sciences, and the human social sciences grouped within anthropology.

What, exactly, do these four core disciplines study? *Archaeologists* read the past through excavating and recording the surviving traces of the cultural and economic activities of ancient humans. In practice, this means excavating buried archaeological sites and recording surviving aboveground monuments and other traces of human activity. Archaeologists use artifacts to define past cultures, and they recover many of the materials dealt with by other specialists, such as human and animal bones, plant remains, soil samples, and dating samples. They pay deep attention to chronology using a range of geophysical dating methods and specialize in the scientific examination of artifacts to determine their compositions, sources, and usages.

By definition, the archaeological record can only be fragmentary, a survivor from the ravages of time, erosion, and deposition. As a result, many archaeologists also use the anthropological record of recent or living societies to reconstruct, through comparison, some of the materially invisible characteristics of ancient societies.

Paleoanthropologists analyze skeletons and fossils (fossils are geologically mineralized bones) and express their opinions in terms of a large array of named hominin species, including our own species, *Homo sapiens*. Many of these species are extinct, such as *Australopithecus africanus*, *Homo erectus*, and *Homo neanderthalensis*, but many genes from them also survive in us today, either through direct descent or through hybridization, as I discuss in coming chapters. Concepts such as "species" and "extinction" are not always hard and fast in hominin paleoanthropology.

Allied with paleoanthropologists are *forensic anthropologists,* who record observations from ancient bones that relate to lifestyle, pathology, and the demographic profiles (e.g., birth rates, age-at-death distributions) of ancient populations. Paleoanthropology and forensic anthropology are usually grouped together within the larger field of *biological anthropology.*

Geneticists derive their samples from blood, saliva, or hair in living human populations, and, if preservation conditions are good, from ancient bones and teeth, and even skin or hair in waterlogged or extremely dry conditions. Nowadays, they express their opinions mostly in terms of whole genome (nuclear) DNA ancestry, reconstructed from the DNA profiles of specific modern and ancient populations. They pay close attention to where ancestral genomic configurations—as expressed through the plotting of mutations among the millions of nucleotides that create our genes—are or once were located in space and time. They also identify the mixtures that can be detected between such ancestral configurations because population mixture, of course, illuminates population history.

Linguists classify languages into families defined by commonly inherited cognate sounds, words, and grammatical features ("cognate" means descended from a common ancestor). They then study the internal histories of those families by careful comparison to identify subgroups of closely related languages within them. Linguists can also suggest where the ancestors of languages and language families were located in time and space, and reconstruct ancient societies and environments through the identification of ancestral cognate words and their meanings.

On "Prehistory"

Readers will soon discover that the subject matter of this book draws little from history as recorded through the world's various writing systems during the past 5,000 years. It is about something far more fundamental: our preliterate past. I am comfortable about referring to this prewriting phase of human existence as "prehistory," and the ancestors of everyone alive in the world today lived through it.

Although I sometimes use the term "history" to refer in a general sense to the whole time span of humanity, this book focuses on human populations in prehistoric time—that is, before the use of written language by ancient civilizations. Prehistoric time encompassed 99.9 percent of hominin history on earth if we go back five million years to when hominins were emerging as a biological lineage separate from the ancestors of the living great apes, especially those of the chimpanzees and bonobos of Africa. Only the last 5,000 years have been historic in the sense of having the benefit of written documents, and large parts of the world were still prehistoric as recently as 2,000 or even 200 years ago.

I know the term "prehistoric" is sometimes a bête noir for those seeking the ethical high ground. Many of us will be aware of those amusing cartoons that illustrate prehistoric life in terms of caves, grunts, just-invented stone wheels, and large wooden clubs. But "prehistory" is not a derogatory concept. Being prehistoric does not mean being primitive, or even particularly ancient; nor does it mean that people kept no oral traditions about their past—let us not forget that Homer's *Odyssey* began life as an oral tradition, long before Homer committed it to writing. Everyone living in the world today has prehistoric ancestors, and the more recent ones were every bit as intelligent as we are. Prehistory has a variable end date, depending on the region of interest; written records are oldest in Egypt and the Middle East, and youngest in the remote regions revealed to the rest of the world only in the later part of the Colonial Era—for instance, in the New Guinea Highlands and Amazonia.

Human populations have been interested in their collective pasts, including their prehistory, for as long as written history has existed, and presumably long before. The Greek historian Herodotus (fifth century BCE) is often credited with being the first author to write what we would describe today as true history, consciously constructed as such and in many cases also referring to events that occurred before the existence of writing. A great many of the world's ancient historical and religious documents draw upon aspects of history and imagined prehistory for the simple reason that history in the general sense has always mattered. It dictated why some people lived here and others there, or why one particular religion was dominant here and another there. Sometimes, history in the broad sense (including prehistory) has glorified misery, dispossession, and war. Other times, it celebrates achievement and can be not only explanatory but also stimulating.

Early human history from written records during the past 5,000 years has focused mainly on the roles of rulers and their kingdoms. But a prehistoric hunter-gatherer who migrated from Africa into the Middle East 50,000 years ago might have played just as important a role in determining the future of humanity as any Egyptian pharaoh or Roman emperor simply by being in a significant place at a significant time. The importance of any individual in human population history is not just a reflection of political status or military success. All of our ancestors, wherever we might live on earth, have played their part in creating what exists now, even if the record is susceptible to being masked or even erased by time and circumstances.

1

The Odyssey Revealed

Five Million Years of Hominin Achievement

During the past five million years, humans and their hominin ancestors have evolved from a bipedal (two-legged) ape into the globally dominant species that we call *Homo sapiens*. We are now eight billion people rather than a few thousand; mobile phones rather than stone tools dominate the lives of many of those billions; and, by the start of the Colonial Era (1492 CE), our ancestors spoke at least 8,000 different languages, of which about 6,500 survive today. Our evolution has taken us from an African ape, through many intermediate hominin species, to *Homo sapiens* and the dizzying heights of the modern technological revolution. Indeed, the success of our global domination is currently causing many of us great concern.

How did all of this happen? The events of the past five (or more) million years have been immense in detail, and much of that detail will forever be lost to us. But there are guiding threads. Two essential processes, *evolution* and *migration*, have underpinned the histories of all species of life on earth, from viruses to whales, including *Homo sapiens* and its ancestors. Evolution creates new species out of existing ones, and migration carries the members of those new species into new environmental conditions, thus encouraging evolution to continue in new directions.

The never-ceasing production lines created by evolution, migration, and further evolution have left continuous traces of their passage, silent witnesses scattered through space and time waiting patiently for those who can find and interpret them. Those traces are the plot for a saga on a cosmic scale.

The traces are not only biological; they include two major nonbiological categories of human achievement, these being the archaeological cultures that record ancient human lifestyles and the families of related languages that record how humans communicated in the past. Our cultures and languages evolved and traveled with their human creators to far-flung corners of the world during the course of prehistory. Together with the fossils and the genes, they add to the basic conceptual scaffold around which this book is constructed.

The human Odyssey, from ape to agriculture, is thus our main field of concern. I will examine how the different hominin populations that have existed during the past five million years, including our own modern human species and its immediate ancestors, have been identified by paleoanthropologists, archaeologists, and geneticists. One ultimate goal is to show how these ancestral populations have contributed to the creation of our own place in the world, although it is not my intention to put *Homo sapiens* on a pedestal of ultimate perfection. Many might say that we deserve no such accolade.

However, we might still ask: Where does *Homo sapiens* actually fit within the Odyssey? We did not exist five million years ago as a recognizable species separate from other hominins, except perhaps in nascent form; we were an undifferentiated glimmer in the genetic cosmos of archaic humanity, waiting for the eventual chance to make an appearance and then migrate into the world to become a new species. I describe the details, such as they are known to us, of this appearance later, but the main point to be stressed in this introduction is that we are a very young species compared with the five-million-year hominin Odyssey as a whole. The oldest fossil skulls recognized as approaching a modern human status in terms of brain size and shape are only about 300,000 years old. All of us alive today descend from a common genetic ancestry of similar antiquity, at least in terms of DNA comparisons between the living human populations of the world.

Yet the genus *Homo*, within which *Homo sapiens* is the sole survivor of what were once several species, including *Homo erectus* and the Neanderthals (*Homo neanderthalensis*), has existed for at least two million years, and hominins in general for more than five million years. This recency for *Homo sapiens* as a distinct species means that we can interbreed freely, if age and health permit, with a partner from anywhere in the world. The differences we perceive in individual bodily characteristics, such as skin or hair color, are superficial.

Furthermore, the recency of our origin in Sub-Saharan Africa means that all living humans carry the same basic ability to create languages, cultures, and societies at a global level of complexity that has been recorded by linguists, historians, anthropologists, and ethnographers for well over a century. We can each learn, speak, and understand the language of anyone else in the world, if we wish to. The shared features of basic behavior and intelligence that we see across the human population today must also have characterized our ancestors since the African emergence and expansion of our species throughout the Old World, from South Africa to Australia, by at least 50,000 years ago.

Right now, therefore, humans across the whole world are a biological unity at the species level. However, it was not always so. Before the main spread of *Homo sapiens* out of Africa, many different hominin species roamed the Old World continents at any one time. There were even several distinct hominin genera (groups of related species) in Africa before one million years ago. These genera and species had been differentiating from each other for many times longer than the modern human time span, so they expressed far more diversity than we see across our own species now. All eventually became extinct, except for the still rather obscure line of genetic descent that led eventually to us, *Homo sapiens*. Some of those pre-*sapiens* species, especially the Neanderthals and Denisovans of Europe and Asia (to be discussed in chapter 4), also hybridized with our own *Homo sapiens* ancestors, in the process transferring genes that still survive among us today.

From a five-million-year perspective, one point cannot be denied. As *Homo sapiens*, we have ridden hard on the achievements of our remote ancestors to become the most successful, and now unique, heirs to those five million years of hominin biological and cultural evolution. As Charles Darwin noted over 150 years ago, "Man still bears in his bodily frame the indelible stamp of his lowly origin."[1] That five-million-year time span postdates our evolutionary separation from the ancestors of the living great apes, especially the panins (chimpanzees and bonobos, members of the genus *Pan*) of equatorial Africa.[2] After that separation, hominins forged their own unique identities as upright bipedal and increasingly large-brained primate life-forms. The panins forged their own identities in another direction to become the knuckle-walking chimpanzees and bonobos that exist in tropical Africa today.

We come from an ape heritage, as of course do our closest cousins in the natural world, the great apes themselves. Jared Diamond once referred to us as the "Third Chimpanzee,"[3] but our brains are huge by ape standards, and our cultural creations astonishing. Our ancestors spread eventually across the whole world, while those of the living great apes (panins, gorillas, orangutans) remained in tropical Africa and Southeast Asia, where they suffer threatened conditions of survival today. Our current human population numbers cause many of us concern, as does our ongoing impact on Planet Earth. In evolutionary terms we have been enormously successful, at least so far.

Brains, Cultural Creations, and Population Numbers

Let me illustrate the overall evolutionary success of humans with an impressionistic illustration of two aspects of the hominin achievement plotted against time. The first is the increase in the volume of the brain, from a chimpanzee (average 380 cubic centimeters) to a modern human (average 1,350 cubic centimeters), as recorded from fossils during the last 3.5 million years for which such brain size records exist (figure 1.1). A brain volume increase on this scale—by a factor of three or four through such a relatively brief period of evolutionary time—is unprecedented in the rest of the mammalian world.

The second aspect lies in human behavior, in the rising complexity of cultures and societies. Figure 1.2 is schematic and selective, but it

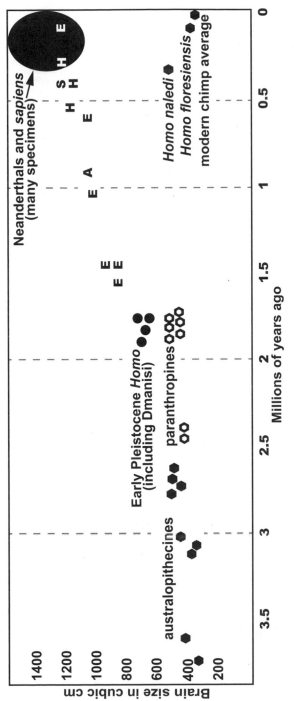

Figure 1.1. The evolution of hominin brain volume (cubic capacity) through time. A = *Homo antecessor*; E = *Homo erectus*; H = *Homo heidelbergensis*; S = Sima de los Huesos (Spain) average. Data partially from Dean Falk, "Hominin brain evolution," in S. Reynolds and A. Gallagher, eds., *African Genesis* (Cambridge University Press, 2012), 145–162.

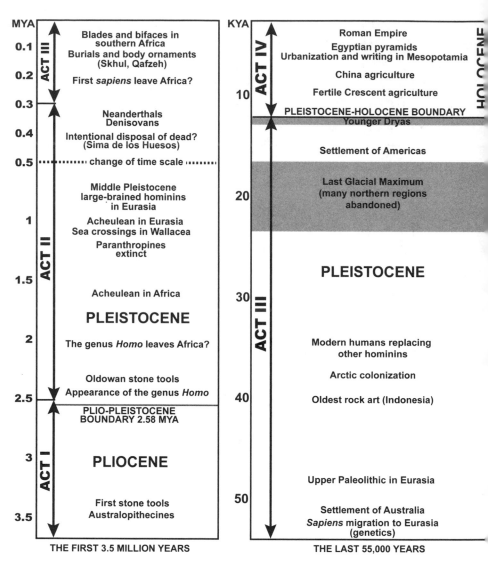

Figure 1.2. The evolution of hominin culture since 3.5 million years ago, with a time line for the four acts described in this chapter. Act I commenced six million years ago, but its early phases reveal no definite signs of cultural activity. The figure has two registers and starts at the bottom left. Note the changes in chronological scale in the vertical axes. KYA = thousands of years ago; MYA = millions of years ago.

focuses on some of the major developments in social and economic organization as recognized in the record of archaeology. These include developments in technology (e.g., stone to metal), provision of food (e.g., hunting and gathering to food production), and social organization from small nuclear family groups to the state-level empires of early history. The increasing tempo of development with the rise of food production after 10,000 years ago is evident.

There is a third aspect of the human career that is more difficult to illustrate: the increase in the estimated size of the human population. Prehistoric population sizes can be inferred from indirect sources of information, such as comparable ethnographic population densities, and areas and numbers of archaeological sites at different points in time. They can also be estimated from genetic comparisons of mutation frequency between the DNA sequences of different ancient and living populations. The larger the population, the more frequently one might expect mutation events to have occurred in its genome, and such mutation events can be dated using molecular clocks. However, I do not attempt here to create a graphical guestimate of hominin population numbers through time because there are too many uncertainties. The key point is that our numbers have grown dramatically during the course of our Odyssey.

The oldest hominin populations were small, and there were perhaps still fewer than two million humans in the world at 12,000 years ago. With the widespread establishment of food production, starting around 12,000 years ago, human populations began to increase with unprecedented speed. By 2,000 years ago we had reached an estimated 300 million people worldwide. Since 1 CE our numbers have skyrocketed, to one billion by 1800 CE and to almost eight billion now.

Of course, the trends through time in these three examples of human achievement are not identical. Our modern human brain volume was achieved by some hominin species more than 50,000 years ago, including our own *Homo sapiens* ancestors and our extinct Neanderthal cousins. There was a major development of cultural complexity (e.g., art, body ornamentation, and purposeful burial of the dead) at about the same time. But our population size only really began to explode with the beginnings of agriculture after 12,000 years ago. States, cities, and writing only became prominent in certain regions of the world after about 5,000 years ago. Before this, most of humanity lived in small egalitarian kinship-based communities.

In short, our evolution over the past five million years of our Odyssey has impacted our world on a scale equivalent to that of a major epochal change, like the coming of the Ice Ages or the appearance of mammals. As earth scientists Simon Lewis and Mark Maslin point out, for the first time in the earth's 4.5-billion-year history, a single species is increasingly dictating its future.[4]

Hominin Evolution as a Four-Act Drama

The events that have taken place in hominin prehistory can be visualized as a sequence of four Acts, which can be succinctly described as follows (figure 1.2):

- Act I: hominins before the genus *Homo* (6 to 2.5 million years ago).
- Act II: the genus *Homo* onward to the fossil appearance of *Homo sapiens* (2.5 million to 300,000 years ago).
- Act III: *Homo sapiens* onward to the appearance of food production (300,000 to 12,000 years ago).
- Act IV: the age of food production (12,000 years ago to the present).

Act I (discussed in chapter 2) was played out in Africa by the hominins who existed after the split from the panins but before the appearance of the genus *Homo* (to which we all belong today). It featured the emergence of a prototype hominin, most probably in Sub-Saharan Africa (although not everyone agrees on an ultimate African origin—see chapter 2), out of a small-brained and apelike ancestor who most likely had an upright body posture, and who later developed the ability to make and use stone tools. No definite fossils of this prototype are known just yet. The subsequent cast of Act I, as known to us today, included the extinct African hominin genera *Australopithecus* and *Paranthropus* (discussed together in chapter 2 as the "australopithecines").

Act I ended with the gradual appearance of traits that defined the emerging genus *Homo*, especially an increase in brain size. However, ancient hominins did not evolve from one genus or species into another overnight—such processes required hundreds of millennia and occurred at different rates for different traits. Bipedalism, for instance, began to develop long before any marked increase in brain size or the use of stone tools. There is no single date at which the genus *Homo* suddenly sprang forth, fully formed.

Between 2.5 million and 300,000 years ago, Act II (chapters 3 and 4) was played out on the earth's stage by a cast of species within the new genus *Homo*. The other australopithecine species that had existed in Africa during Act I gradually became extinct. As far as we know, one or more species of *Homo*, especially *Homo erectus*, left Africa early in this second Act, around two million years ago, to migrate into accessible regions of Eurasia.

Another significant migration out of Africa appears to have occurred during the later part of Act II. This gave rise eventually to new hominin species across Eurasia, one being the well-known Neanderthals, another being the recently discovered Denisovans of Siberia and eastern Asia. Back in Africa, some of the large-brained hominins who remained behind continued the separate evolution there of *Homo sapiens*, although we must wait for Act III to meet actual members of this species through their bones.

Act III (chapters 5 and 6) revolves around a new cast of ancestral modern human hunters and gatherers—*Homo sapiens*—the oldest fossil specimens of whom so far date between 300,000 and 200,000 years ago. Comparisons of skull morphology and ancient DNA suggest an earlier genetic initiation for this species in Sub-Saharan Africa, perhaps about 700,000 years ago, but no definite *sapiens* fossils yet exist from this earlier time period.

The definitive entry of *Homo sapiens* into Eurasia, in terms of founding the ancestry of living human populations outside Africa, occurred much later, between 70,000 and 50,000 years ago in terms of archaeology and genetic dating. Much mystery swirls around this topic, as I discuss in chapter 5, but we do know that ancestral *Homo sapiens* overlapped in time and sometimes interbred with the other *Homo* species that also coexisted during Act III, both inside and outside Africa. These included the Eurasian Neanderthals and Denisovans.

All of these non-*sapiens* species of *Homo* became extinct toward the end of Act III, perhaps because of cultural and demographic competition as well as interbreeding with the clever and more fecund *Homo sapiens*, "the wise human." Act III also witnessed the colonization by *Homo sapiens* of the remainder of the habitable world beyond Africa and Eurasia (except for distant oceanic islands), including Australia, New Guinea, and eventually the Americas.

Ancestral modern human populations during Act III, and especially since 50,000 years ago, also left behind a cultural record with far more detail than that left by previous hominins. Archaeologists recover aspects of these developing cultural traditions as art on cave walls and on portable objects, as red ocher used to decorate burials, and as body ornaments, including beads and pendants of stone, bone, and shell. *Homo sapiens* populations buried their dead according to rituals that sometimes involved placement of manufactured ornaments or other items for the afterlife in purposefully dug graves. These cultural traits, plus the sustained ability to travel farther at sea and farther into extreme cold than the more archaic hominins, were convincing and identifiable achievements unequaled by any pre-*sapiens* species of *Homo*.

Act IV (chapter 7 and onward), the final act that still occupies the global stage today, commenced around 12,000 years ago in the Middle East, and more recently in several other key homelands of agriculture. It emerged within the pronounced episode of global warming that followed the last Ice Age, after 18,000 years ago, and it involved only *Homo sapiens*—all other hominins were extinct as separate and independent species by this time.

The key development in Act IV was food production through the domestication of plants and animals. This created transportable food-producing economies that could underpin migration into new territories, leading in turn to dramatic population growth in latitudes where agriculture was possible. The eventual outcomes of this population growth were developing social orders that yielded cities and states in many parts of the world, and eventually a series of scientific and industrial revolutions, all leading into our current world situation. The world's largest language families underwent their expansions during this phase, many of them carried with the migrations of early farming and pastoralist populations.

Anyone who thinks further about this succession of four acts within human prehistory will quickly draw one obvious conclusion: each act was shorter in duration than its predecessor yet was increasingly dramatic in terms of human dispersal and population size. Human evolution has been like a snowball rolling downhill, gathering size and speed on steeper sections, losing momentum on flatter ones, but certainly never stopping altogether. Climate scientists even debate the possibility of an Act V which they term the Anthropocene, the Age of Humankind, although its date of commencement is not unanimously decided. The beginnings of agriculture, the Industrial Revolution, and the invention of the atomic bomb are currently all candidates.

Population Growth and Migration: Why They Mattered

Of all the hominin genera and species known to have existed, *Homo sapiens* survived the evolutionary fray as an eventual winner, alone in the world since perhaps 30,000 years ago. We are now burdened with a growing population that shares dangerously unequal access to our earth's subsistence resources.

In modern human affairs, a growing population has long been a fundamental factor in driving history, for good or ill, as stressed for recent human history by demographer Paul Morland in his book *The Human Tide*.[5] Increasing population size encouraged migration, and migration into new and productive landscapes in turn encouraged increasing population size—a powerful mutualism that must have driven a great deal of the Five-Million-Year Odyssey, especially in the case of *Homo sapiens*. Migration, after all, was one of the major factors that divorced the course of hominin evolution from that of the great apes. Our ancestors left home for good.

Prior to the start of the Colonial Era in 1492 CE, modern humans, *Homo sapiens*, experienced two unprecedented worldwide episodes of migration that led to significant population growth. The first was the successful movement beyond Africa into the rest of the world during the latter part of Act III. This unfolded as a series of consecutive migrations that commenced over 50,000 years ago, initially spreading out of Africa through Eurasia to reach Australia and New Guinea, and culminating in the settlement of the Americas from northeastern Asia by 15,000 years ago. The total human population increased greatly in overall numbers during this period, partly due to the enormous extent and resource potential of the newly colonized land masses.

The second major episode of growth resulted from the migrations of populations with transportable economies of food production during the past 12,000 years. Some of these farmer and herder migrations achieved enormous extents, even if they required centuries or millennia to unfold completely. These Act IV migrations in the Odyssey are of direct interest to billions of people in the world today because of their association with the origins and histories of many existing ethnic populations and language families. If we ever pause to wonder why English is spoken in England and Australia, Turkish in Turkey, Maori in New Zealand, and Navajo in Arizona, we will soon discover that the answers involve plentiful human migration.

Having introduced the actors, there are still two important matters that require some exposure in this introduction: the stage and the clock.

Our World as the Stage

Human evolution did not occur against an unchanging environmental background. Behind it lay the earth's surface and atmosphere, subject during the past 2.6 million years of the Pleistocene and Holocene geological epochs (defined further in chapter 3) to regular cycles of climatic change. These gyrated, in cycles of approximately 100,000 years, between glacial ice ages at one extreme and interglacial warm intervals at the other, the latter similar to the climate that our world enjoys, and fears, today. Each cycle was associated with substantial swings of temperature, rainfall distribution, and global sea level, the last varying by up to 130 meters, similar to the height of a thirty-five-story building.

Remembering that global sea level is close to a 120,000-year peak at the moment, let us imagine what the world's coastlines would have looked like whenever a 130-meter depth of seawater became locked away in the massive ice sheets that extended from the North Pole almost as far south as New York and London. Because sea level change in the open ocean is a worldwide phenomenon, all of the world's great river deltas would have been narrow incised river channels when the sea surface was so low and continental shelves would have been exposed as flat coastal plains around the edges of the continents. One would have been able to walk almost entirely on dry land, except for river crossings, from the Cape of Good Hope to Cape Horn. Asia was then joined to Alaska across a dry Bering Strait; Borneo and Bali were joined to the Malay Peninsula; and New Guinea was joined to Australia. The outlines of these land bridges can be seen in figures 3.1, 5.1, and 6.1.

One can also imagine, of course, what would have happened as rising sea levels flooded back over 130 meters in the reverse direction during warm interglacial conditions, when the ice sheets melted. Coastal impacts would have attended both directions of movement, especially when they were relatively rapid. Our ancestors lived through such glacial to interglacial cycles many times (probably more than twenty) during the 2.6 million years of the Pleistocene epoch, surviving through adaptation and movement as glaciers and sea levels waxed and waned in alternation.

Today, human activity is prolonging the current warm interglacial climate toward uncertain outcomes, causing many of us to question our future. The great climatic cycles of the Pleistocene enable us to see the results of our current actions from a long-term perspective. I do not profess to be a climate scientist or a politician, but I must state that I share the concerns of many people about the current climatic trends as they increasingly move the earth toward one of the warmest phases in its history during the past million years.

How Old Is It? Dating the Past

There is one final matter to explain before we launch into the Odyssey. To understand our past, we need a precise chronology for the many ancient populations and events that we wish to study. Obviously, it matters greatly if a given fossil from an australopithecine, a Neanderthal, or a modern human is two million, 200,000, or only 20,000 years old. The same applies to assemblages of stone tools, and indeed to all elements that survive from the human past. We need to know the real age if we are to avoid confusion in our interpretations.

So, where do "absolute" dates (i.e., dates counted in solar years ago) come from, bearing in mind that even the most precise will always have a statistical range of laboratory error? I am going to forego the temptation to explain here all of the dating methods used by investigators of the deeper levels of the human past, before written records existed, in terms of their laboratory techniques and statistical calculations. Readers who wish to know how scientists calculate dates using changes in the earth's paleomagnetism recorded in sediments, or how they measure the changing states of the various atomic particles used in radiocarbon, potassium-argon, uranium series, electron spin resonance, optically stimulated luminescence, and cosmogenic nuclide dating (to name some of the major techniques currently dominating the literature), and over what periods of time these various methods work should research the answers themselves. I can only deal in this book with the actual results.

Furthermore, archaeology is not the only source of absolute dates in prehistory. Geneticists have access to many different molecular clocks that can calculate the spans of time that have elapsed since periods of common origin between related populations and species. Linguists can calculate approximate dates for periods of common origin between related languages, based on observations about how quickly individual languages and words have changed within the historical record. As with archaeological and geophysical dating, however, many of these methods are complex and highly statistical, and this is not the place to go into them in detail.

My main interest here is to discuss how we might "trust" the dates that scientists return to prehistorians, allowing that the issue is not just one of potential laboratory or calculation error but of ancient *context*. Error ranges and variations in laboratory competence are nowadays relatively minor contributors to uncertainty. But ancient context is absolutely fundamental, and it has two aspects—the context of deposition, and whether the date is direct or indirect in terms of the material being dated.

The first aspect concerns the context of deposition, or how the object of interest reached the resting place from which it emerged into the scientific light of day. Deposition can be either primary or secondary. An undisturbed human skeleton in a grave, with all its bones in articulation, is in a primary context. If it is under 50,000 years old it can probably be dated directly by radiocarbon dating, as long as the bones still contain sufficient carbon-bearing collagen. But a piece of charcoal in the grave next to that skeleton will not necessarily be in a primary context, unless it can be shown that the burial party deliberately lit a fire during the funeral ceremony. Otherwise, the charcoal could have been dug by the gravediggers out of deeper layers laid down many thousands of years before the death of the person buried, and then thrown back with the grave fill.

As another example, a stone tool or a fossil skull found in a layer of Pleistocene riverine sediment might be in a primary context if it was incorporated directly from its user or owner into an actively accumulating flood plain. But it is also necessary to consider whether the sediment was secondarily redeposited by forces of nature long after its original deposition. Thus, tool or skull and sediment might be of the same age, or they might be thousands, even millions, of years apart in age. Only informed research of a geomorphological and stratigraphic nature will give the answer.

The second issue is that of direct versus indirect dating. For example, the bones of a human skeleton subjected to radiocarbon dating are clearly being dated directly, even if the bones come from a secondary disturbed context. If the laboratory calculation of the date is correct, then that date applies automatically to the death of the human who once carried the skeleton, wherever the bones might have been found. But a date derived from an adjacent piece of charcoal is, of course, a secondary date when applied to the skeleton, as described above—correct for the charcoal, for which it is a direct date, but not necessarily for the skeleton.

The materials that are being dated can also produce problems. Usually, dates for artifacts of stone, pottery, or metal are indirect because these nonorganic substances are difficult to date directly in terms of their actual manufacture by human artisans, as opposed to their geological ages as raw materials. However, direct dating methods can be applied to organic materials that contain carbon, such as bones or charcoal, as well to sediments that contain other radioactive minerals. Debates over the correctness of the chronologies claimed by those who have recovered ancient dating samples have peppered the literature about human prehistory for many decades.

Sometimes scientists can be led astray for long periods if wrong dates masquerade as right ones because of contextual ambiguity, especially if no further corroborating discoveries are made. More often, however, when discoveries around the studied topic are frequent, once-claimed but incorrect chronologies can be revealed as isolated outliers from the main distribution, hence unconvincing. Caution about absolute dates claimed for ancient hominins and their cultural products is always wise, as long as it is also well informed.

2

The Odyssey Begins

IN THIS CHAPTER I examine the several million years of hominin evolution following the separation between the ancestors of the panins (chimpanzees and bonobos) on the one hand and the ancestors of hominins on the other. This was apparently a drawn-out separation that probably occurred, according to one recent estimate, between 9.6 and 6.5 million years ago.[1] We thus enter Act I in the Odyssey, defined in chapter 1 as lasting from approximately 6 to 2.5 million years ago. I describe the australopithecines and continue onward to the appearance in Africa of early species within our own genus, *Homo*.

Two intriguing issues arise that currently defy simple answers: how did ancestral hominins eventually become species separate from ancestral chimpanzees and bonobos, and where and from which australopithecine lineages did the first large-brained *Homo* populations originate?

How Did Hominins Come into Existence?

To answer this question, I suppose we could begin with the first life on earth more than three billion years ago and then work our way toward the present, life form by life form, fossil by fossil. This would make for a long book. Instead, I commence a little more recently, during the final stages of the Miocene epoch of geological time, which lasted from 23 to 5.3 million years ago.

By the end of the Miocene, our uniquely human Odyssey was just beginning, as our ancestors gradually ceased to live as quadrupedal apes and began to develop unique human characteristics. Somewhere in tropical Africa at that time lived an ape that was the common ancestor of both hominins and panins, the latter being our closest surviving cousins in the natural world. Gorillas had already separated from the panin-hominin lineage from a shared ancestral gene pool located farther back in time.

Today, of the two panin species, the chimpanzees (*Pan troglodytes*) have a discontinuous distribution in central Africa north of the Congo River and in western Africa. There is no fossil record to suggest that they ever lived very far beyond where they exist now. They are adept tree climbers and knuckle-walkers, with stooping postures when on the ground. As indicated in figure 2.1, the closely related bonobos (*Pan paniscus*) inhabit a more restricted area south of the Congo River, which was apparently the barrier that originally caused the separation into these two species from a common ancestor about 1.7 million years ago.[2] Both species are highly intelligent, social animals; chimpanzees have a male-dominance hierarchy in sexual relations and bonobos have more egalitarian and promiscuous behavior.

Unlike the panins, the early hominins, of whom the australopithecines are the best-known members, combined apelike characteristics in brain size with a gradual emergence over time of hominin characteristics in bipedal posture, hand grip, and shape of the dentition (jaws) and face. Act I also witnessed some remarkable geographical achievements by ancient hominins. They appear to have colonized all habitable parts of Africa by at least 3.7 million years ago, at least south of the Sahara. Conversely, their panin and gorilla cousins stayed closer to home in the tropical woodlands of western and central Africa.

What Was an Early Hominin?

To answer the question of how an ape became a hominin, we should think first about our own physical characteristics as living modern humans, especially when compared with our closest relatives, the panins and gorillas. Five fairly obvious characteristics that define humans apart from apes come to mind.

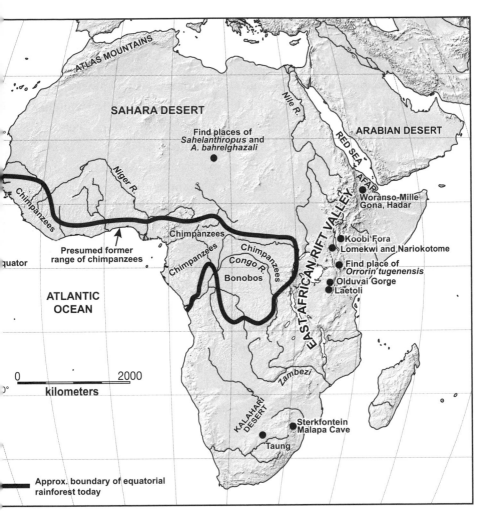

Figure 2.1. Map of Africa to show find places associated with hominin populations before the appearance of *Homo*, and the current distribution of panins.

First, humans are bipedal. We can stand upright and walk, stride, and run with our hands free. Panins and gorillas can stand upright too, but they habitually use their knuckles with a stooped posture for movement on the ground—long-distance bipedal running is not one of their skills. Interestingly, hominin bipedality evolved long before any significant growth in hominin brain size—it appears to have been the first major bodily change to prophesy the hominin arrival on earth. A femur fragment of the oldest potential fossil hominin, *Orrorin tugenensis*, found in Kenya and dated to about six million years ago, indicates a bipedal posture, although its brain was probably no larger than that of a chimpanzee.

The 3.5-million-year-old australopithecine footprints found in hardened volcanic ash at Laetoli in Tanzania, and the partial skeleton of "Lucy," a 3.2-million-year-old female australopithecine found at Hadar in the Afar region of Ethiopia (figure 2.2), also had definite bipedal characteristics, as did the oldest australopithecine remains from South Africa. Lucy walked erect with a humanlike pelvis, but had a brain volume of only 400 cubic centimeters, slightly larger than that of an average panin. Australopithecines were still only child-sized by modern human standards: Lucy was 1.1 meters tall and weighed an estimated 29 kilograms (somewhat less than a chimpanzee). Figure 2.2 shows Lucy walking alongside an early *Homo* female, who lived about 1.5 million years later.

Second, compared with panins and indeed all other mammals, modern humans have large brains (figures 1.1 and 2.3), not only in absolute terms but also when compared to our body weights. Panins have about 75 percent of the average human body weight but only about 25 percent of the modern human brain volume (roughly 350 cubic centimeters for panins versus 1,350 for humans). Most of the increase in brain size in humans, however, has occurred within the past two million years of existence of the genus *Homo*. The older hominin genera *Australopithecus* and *Paranthropus* retained brains much closer in size to those of panins. Brain size thus lagged behind bipedalism in the timing of its evolution.

Third, we are able to oppose our thumbs against each of the four fingers of our hands, a skill that enables us to craft wonderful objects, ranging from stone tools to fabulous works of art, and also to play complex musical instruments. Apes, including panins, lack this degree of precision in their grip, but they are expert at grasping tree branches with both their hands and their feet. Like the large brain, the precision grip of hominin hands also appears to have evolved well after the first appearance of bipedal posture, although it was present in nascent form during the time span of the australopithecines, as can be seen in the reconstruction of Lucy.

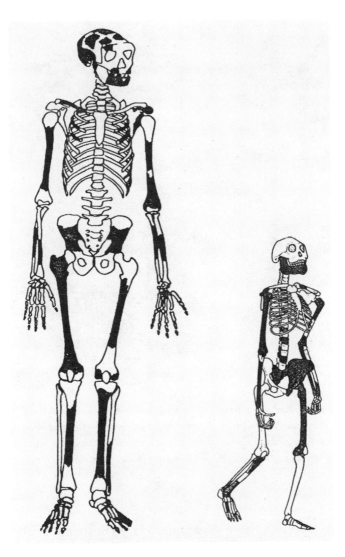

Figure 2.2. The reconstructed skeleton of "Lucy," a 3.2-million-year-old female australopithecine from Hadar in the Afar region of Ethiopia (*right*), alongside the reconstructed skeleton of a female Early Pleistocene *Homo* from Koobi Fora, Lake Turkana, Kenya, dating from 1.7 million years ago (*left*). The dark portions are the bones actually preserved. Courtesy Milford Wolpoff and McGraw-Hill. This figure was published originally as figure 136 in Milford Wolpoff, *Paleoanthropology*, 2nd ed. (McGraw-Hill College, 1999).

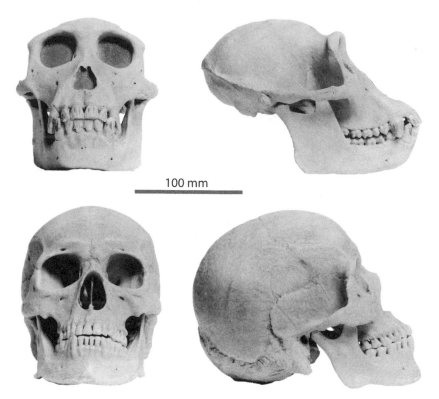

100 mm

Figure 2.3. The skull of a chimpanzee compared with that of a modern human. The modern human has a larger, higher, and more globular braincase, no brow ridges, a flatter retracted face, smaller canine teeth, and a prominent chin. Casts from the collection of the School of Archaeology and Anthropology, Australian National University. Photos by Maggie Otto.

Fourth, hominins have developed a larynx capable of uttering the complex sequences of sound that we term speech. For Karl Marx's famous collaborator, the political and social theorist Frederick Engels, speech and labor, together with erect posture and opposable digits ("free hands" in Engels's terms), were the emancipators of humanity from an apelike condition.[3] As Engels expressed it in 1884, free hands led to labor, and labor as a communicative activity meant that humans in the making had something to say to one another using speech. As a result, we have another fundamental characteristic of humans, especially modern ones. Beyond our bodies, big brains, and hands, we cooperate through speech with thousands or even millions of our fellows to create societies and civilizations, or *cultures* in the parlance of anthropologists. Australopithecines were perhaps the first hominins to create culture, given their overlap in time with the oldest deliberately flaked stone tools found in Africa.

Finally, modern humans are remarkably hairless when compared with other primates. The separate evolutionary histories of our pubic and head lice indicate that some hominins, or at least those directly ancestral to the genus *Homo*, began to lose a continuous, thick body-hair cover about three million years ago, well after the separation time from panins. It has been suggested that this loss of body hair allowed more efficient heat loss through sweating in tropical African daytime temperatures. Bipedal hominins could have moved around the landscape with their hairless bodies carried vertically to the sun but with their hair-covered scalps protected from solar radiation. This in turn would have allowed them to hunt and scavenge meat safely and efficiently in the heat of the daytime, when many other large mammal predators rested.[4]

How did all these hominin characteristics develop, and out of what?

The "Missing Link" and the Elusive Common Ancestor of Hominins and Panins

In the popular mind, the idea of a "missing link" has always been one of the most exciting and mysterious aspects of human evolution, at least for those who accept the reality of Darwin's observation that we descend from an apelike ancestor. The identification of the australopithecines and their ancestors as potential missing links, to the satisfaction of most paleoanthropologists, was a remarkable achievement, albeit one tinged with scandal.

The first serious applicant for the role of missing link appeared in 1891, when Dutch anatomist Eugène Dubois found a skull cap of what he eventually termed *Pithecanthropus erectus* (now *Homo erectus*) while excavating into the bank of the Solo River at Trinil in central Java. Small-brained "Java Man" led to much inconclusive discussion at the time, and then a new discovery was announced in 1912. That year a large-brained skull and apelike jaw were found in a gravel pit at Piltdown in southern England, and this find promptly was declared a competitor for missing link status. From this "Piltdown Man" perspective, early man was thought to have had a large brain, unlike the Javan *Pithecanthropus*. But little did people know at the time that Piltdown Man was a forgery—a modern human skull with a modified orangutan jaw.

Surprisingly, it took scientists forty-one years to lift the lid on the fraud (in 1953) through careful forensic analysis. When they did so, there must have been an audible sigh of relief from the world's paleoanthropologists. Ever since the first australopithecine skull had been found in 1924 at Taung in South Africa, it was becoming more and more apparent that the Piltdown combination of a large brain and a primitive jaw was problematic. The australopithecines, like the *Pithecanthropus* of Dubois, had small brains but recognizably hominin jaws and teeth. They were quite unlike the masquerading Piltdown skull and jaw. And so Piltdown Man left the stage.

The australopithecines, however, provide for us only the later part of the hominin story during Act I. Unfortunately, no one has ever found the bones of the immediate common ancestor of both panins and hominins, and neither are there any definite fossils of ancestral panins. We can only imagine the common ancestor through consideration of the characteristics of living and fossil apes and hominins. Any convincing candidate must have had bodily features that could have evolved into both knuckle-walking panins and bipedal hominins. Two decades ago, paleoanthropologist Milford Wolpoff proposed a small African tree-climbing primate as a likely common ancestor for hominins and panins, with a body mass around 35 kilograms, somewhat lighter than an average modern chimpanzee and perhaps similar in size to an australopithecine like Lucy.[5] This primate would have moved along the tops of branches as panins and gorillas do now, rather than by swinging (brachiating) beneath them like gibbons and orangutans.

By a remarkable coincidence, just as I was writing this chapter in 2020, the journal *Nature* published an article about fossils of a small ape found in a clay pit in southern Germany, dating from the middle

Miocene, around 11.6 million years ago.[6] This ancient primate, *Danuvius guggenmosi*, weighed between 17 and 31 kilograms and, according to its finders, had both an upright bipedal posture with relatively flat feet suitable for walking on tree limbs as well as good grasping and climbing abilities in both its feet and its hands. It therefore had both potential hominin and panin characteristics. The authors referred to the abilities of this species as "extended limb clambering," and raised the possibility that it could represent the kind of ancestor from which both hominins and panins eventually descended. Indeed, German paleoanthropologist Madelaine Böhme, the excavator of *Danuvius*, has even suggested that the hominin evolution of bipedalism prior to seven million years ago took place mainly in Eurasia rather than in Africa.[7]

Danuvius is thought provoking, but other authorities have disputed its claimed bipedal status, suggesting that it was just another species of arboreal ape.[8] An actual shared ancestor for both panins and hominins in the "right" time period—9.6 to 6.5 million years ago—still eludes researchers.

On the Panin/Hominin Split

In what follows I will assume, with the majority of paleoanthropologists, that hominins evolved in Africa. After all, the concentration of evidence in that continent, especially since five million years ago, is rather overwhelming. So how did ancestral panins and ancestral hominins eventually separate into two no-longer-communicating gene pools, thus undergoing independent processes of new species formation?

The simplest answer would be that a group of ancestral panins/hominins in the late Miocene forest of tropical Africa became physically separated from their relatives. Perhaps some of them managed to cross a wide river, just as some ancestral bonobos apparently managed to cross the Congo River during a prolonged dry period more than 1.7 million years ago to form a new species separate from chimpanzees.[9] Panins are not frequent swimmers, and if the first hominins were not either, this could be a convincing enough explanation. The evolutionary processes of mutation, selection, and genetic drift[10] would then have ensured that, eventually, reproduction of fertile offspring between the separated populations would no longer have been genetically possible, even if some of their descendants should one day have met again. Two reproductively bounded species would have come into existence out of one common ancestral species.

We might ask how much time would have been required for two medium-sized primate species that had split from a common origin to become reproductively incompatible, either completely, or via first-generation sterility (as with mules and hinnies—the sterile offspring of horses and donkeys). One to two million years would be a fair estimate according to a recent compilation of comparative mammalian fossil and molecular clock data.[11] Yet chimpanzees and bonobos are still fully interfertile in captivity, despite such a potential separation time in their case.

In fact, we have no exact idea how much time would have been needed to achieve complete speciation for hominins, because all other species of *Homo* except for *sapiens* are now extinct. We have no living hominin cousins (apart from the panins, who are not hominins) with whom comparisons can be made. In this regard, however, it is known from ancient DNA research that at least 700,000 years, perhaps even a million years, of *Homo sapiens* and Neanderthal separation out of a common ancestral population (to be discussed in chapter 4) did not eradicate the ability for successful reproduction between these two species. Some Neanderthal genes were transmitted into modern human populations through mixed matings in Eurasia as recently as 45,000 years ago, and these Neanderthal genes still survive in small percentages in the modern human population today, indicating that hybrid offspring between the two species were fertile. The story of human evolution does not deal in hard and fast species boundaries, at least not as far as we can tell, although some tiny hominins that we will meet later could have become exceptions in this regard vis-à-vis other hominin species due to their long isolation.

As it happens, a simple split between ancestral hominins and ancestral panins is generally considered less likely than a long-term succession of episodes of gene pool separation alternating with periodic remixing, as environmental barriers ebbed and flowed across the tropical African landscape. In such a "sympatric" speciation process within a continuous territory, the split from a single into two or several reproductively isolated gene pools would have occurred intermittently, perhaps over millions of years of successive separation and rehybridization, culminating in no-return separation only if and when the mixing ceased for long enough for reproductive barriers to harden. Hence, we have the estimate of roughly three million years (9.6 to 6.5 million years ago) for the hominin-panin separation given above, which is favored by recent genetic and statistical modeling.[12]

Beyond the Nest: The First Hominins Emerge

According to the record from fossils, the ancestral hominin population had definitely separated from its great ape cousins by at least five million years ago, at the beginning of the Pliocene epoch of earth history (the Pliocene is dated from 5.3 to 2.6 million years ago). Our most fundamental knowledge about these oldest identified hominins comes from tropical East Africa, especially for the period between 5 and 3.5 million years ago. By 3.7 million years ago, there were also hominins in South Africa.[13]

The East African fossils come from exposed sediments in the East African Rift Valley (shown in figure 2.1), a 3,000-kilometer-long series of deep fault lines created by the tectonic activity that has resulted from the movement of continental plates. The Rift Valley commences in the Jordan Valley in the southern Levant, and runs under the Red Sea into the Afar region (Afar Triangle), which flanks the Bab el Mandeb strait between Africa and the Arabian Peninsula. It continues south through the East African lakes region toward Zambia and Malawi. The East African portion of the Rift Valley contains several large lakes whose ancient food-rich shorelines were frequented by hominins. Of all landscape features with potential to attract hominins, the presence of permanent water was certainly one of the most important.

The Rift Valley also has a rich history of volcanic activity, meaning that its fossil-bearing geological layers are interstratified with volcanic deposits that can be dated by geologists, using potassium-argon and other geophysical methods of dating. It is this volcanic chronology that provides a basic time line for the earlier phases of African hominin prehistory. In this landscape we find the fossil remains of several increasingly bipedal hominins in the stratified sediments laid down by ancient rivers and lakes. The first potential members of the ancestral hominin club belonged to two Late Miocene species, *Sahelanthropus tchadensis* and *Orrorin tugenensis*, who lived more than five million years ago in Chad and Kenya, respectively (at least, that is where their fossil remains have so far been found). The remains are fragmentary and their roles in hominin evolution rather disputed, but they are considered by many paleoanthropologists to be closer in surviving bone morphology to ancestral hominins than to ancestral panins.

The next actors on the stage belonged to two successive late Miocene and Pliocene genera—*Ardipithecus* (the older of the two) followed by *Australopithecus*. The ardipithecines (species *kadabba* and *ramidus*), who appeared first, were adept tree-climbers whose remains are reported from sites in Ethiopia. Not all paleoanthropologists agree that the ardipithecines were hominins; others regard them as the likely ancestors of the East African australopithecines, a genus better understood because of the greater number of its fossils. The australopithecines have become central to all debates about the subsequent origins of the genus *Homo*.

Pliocene Ancestors: The Australopithecines

Australopithecine fossils have been discovered in large numbers, with the oldest in East Africa dating to between 4.5 and 4 million years ago. They had spread by at least 3.7 million years ago into South Africa as well as into what is now the Sahara Desert. The 3.6- to 3-million-year-old teeth and jaw fragments of *Australopithecus bahrelghazali* from Koro Toro, located just south of the Tibesti Mountains in northern Chad (and not far from where *Sahelanthropus* was found), are particularly interesting because this area is within the Sahara today. Most probably australopithecines roamed there with access to a vegetated landscape during a warm and wet Pliocene climatic interval, when the limit of summer monsoon rainfall in Africa was located farther north than it is now.

The widespread distribution of *Australopithecus* is particularly striking because it implies that these hominins were living across a large range of both open and wooded environments from Ethiopia to South Africa, extending west into what nowadays has become the central Sahara (but so far with no clear evidence for a presence north of the Sahara). Their bipedal posture and stamina for long-distance walking would have enabled them to roam far in search of essential water sources, especially during droughts.[14] If they had been meat eaters (which is not certain), their posture would also have enabled them to move far afield following herds of animals for hunting and scavenging opportunities.

Despite widespread fossil occurrences in eastern and southern Africa, many parts of the continent, especially in the west and center, have so far yielded no hominin fossils at all. This may be because of preservation problems. Obviously, some environments are better than others for the preservation and visibility of fossils, with wet tropical rainforests being among the most difficult, and drier unforested landscapes with widespread sediment exposure, such as those in the East African Rift Zone, being among the most productive.

Landscapes also change with time, particularly through erosion and sedimentary deposition. This can affect the visibility of caves because they can fill with sediment or collapse, eventually becoming buried deep beneath the modern ground surface and so subject to luck in discovery. Many famous fossil-bearing ancient caves of the paleoanthropological record—for instance, those near Johannesburg in South Africa, at Zhoukoudian near Beijing in China, and at Atapuerca in northern Spain—were sediment-filled with stone and soil deposits known as "breccia" long before archaeologists discovered them. The Atapuerca caves only emerged when a mining company excavated a railway cutting through the limestone hill that contained them (to be discussed in chapter 3).

Taking these factors into account, existing discoveries imply that virtually the whole African continent south of the Sahara was inhabited by hominins by at least 3.7 million years ago, except possibly for canopied rainforests and waterless deserts. This is a striking observation because it indicates that hominins by this time had expanded far beyond the ranges of any of their ape cousins. They appear already to have been a migratory success. Why?

The answers to this question surely lie in the gradually developing hominin body and brain characteristics revealed by australopithecine fossils, and in the demonstration of a gradual shift away from apelike into increasingly hominin-like behavior. As examples, the newly reported *Australopithecus anamensis* skull from Woranso-Mille in the Afar region of Ethiopia, dated to 3.8 million years ago, still had a brain volume of only 370 cubic centimeters.[15] By 3.6 million years ago, however, an australopithecine skull from Sterkfontein in South Africa had reached 408 cubic centimeters.[16] By 2.5 million years ago, some late australopithecine (or early *Homo*) brain volumes had attained 450 cubic centimeters, which is well above the 350-cubic-centimeter average for modern panins (see figure 1.1). Although not yet a rapid increase, the growth in brain volume was at least consistently upward as some late australopithecines evolved to become the first members of the genus *Homo*.

Australopithecines were also fully bipedal, with legs for striding that were longer than their arms, although there was much variation in foot morphology.[17] Their hands were shaped much like ours, with a developing ability that was gradually approaching our modern precision grip.[18] However, they still retained an apelike grasping ability in both hands and feet that would have continued to assist them in climbing trees— this is especially obvious in the reconstruction of Lucy (figure 2.2).

Another apelike feature that continued with the australopithecines is that male bodies continued to weigh rather more than female bodies, indicating a degree of sexual difference in body size roughly as great as (and perhaps even greater than) that between male and female chimpanzees today.[19] Australopithecine females were mostly under 1.2 meters tall, whereas large males might have reached 1.5 meters. With the rise of the genus *Homo* after two million years ago, these male-female body size differences began to decrease, a trend thought by many paleoanthropologists to reflect increasing levels of male-female cooperation, especially in order to reproduce and raise infants.

"Man the Tool-Maker"?

One specific humanlike activity among the later australopithecines was probably the manufacture of stone tools, although there is still uncertainty over whether the oldest ones were made by the last australopithecines, the first *Homo*, or both. A site called Lomekwi, near Lake Turkana in Kenya, has produced some of the oldest tools, which have been claimed to date from about 3.3 million years ago; they are almost certainly of australopithecine origin if the date is correct.

The first regular use of stone tools entered the record about 2.6 million years ago, for instance at the sites of Bokol Dora 1 and Gona in Ethiopia.[20] Some of the animal bones from Gona show cut marks from stone flakes, suggesting removal of meat. Interestingly, there are sites, such as Lokalelei in Kenya, from which the flakes recovered can actually be fitted back together again to re-create the original cores from which they were struck. Recent research into such "refits" suggests that demonstration and copying by teachers and learners enabled the transmission of the required technology.[21]

As for stone tool usage, stone flakes used as sharp-edged knives would have been one fairly obvious requirement for a hominin wishing to cut some meat off a bone. But it has also been suggested by Jessica Thompson and colleagues that the first stone tools could have been hammers and pounders rather than knives, manipulated for smashing animal bones and skulls to extract servings of fat- and protein-rich marrow.[22]

Whatever the reasons for the initial hominin use of stone tools, we should remember that chimpanzees have been observed to use tools of stone and organic materials taken from nature, including stone pebbles as hammers and pounders, to help in the extraction of food or other desired materials.[23] Panins and orangutans have also been taught in laboratories how to flake stone in order to access rewards. However, no apes have been observed to make stone tools by flaking regularly in the wild. Only hominins did this, so the hoary old concept of "man the tool-maker" might still have a point.[24] We return to it later because hominin tools and bigger brains might have been linked in development.

Big Strides after 2.5 Million Years Ago: Early *Homo*

After two million years ago, the australopithecines gradually disappeared as a paleoanthropological category, with some evolving onward to become the genus *Homo*, and others edging toward extinction. Among the most recent australopithecines might have been the partial skeletons of a possible family (two males, a female, and three infants) found together in Malapa Cave in South Africa. They lived just under two million years ago and are known to paleoanthropologists as *Australopithecus sediba*.

Perhaps the best-recorded extinctions involved the several species of "robust" australopithecine, some placed in a separate genus called *Paranthropus*, who finally disappeared in both East and South Africa by about one million years ago. These robust and heavily built species differed in their larger faces and teeth compared with the more lightly built australopithecines that probably gave rise to *Homo*, and also in the projecting bone crests (sagittal crests) for muscle attachment along the tops of their skulls, strong muscles being necessary for working their large crushing jaws and molar teeth. Presumably they lived alongside and perhaps in competition with their less robust australopithecine cousins as well as early *Homo*, remaining as separate species because they were different in body size and appearance, diet, and possibly even sexual behavior and reproductive compatibility. Climate change might also have played a role in their demise—there has been a recent suggestion that changes in plant food species due to increasing aridity in southeastern Africa could have led to the extinction of the South African robust species *Paranthropus robustus*.[25]

By two million years ago, hominin characteristics were sufficiently represented among late australopithecines that paleoanthropologists agree, with a comforting level of consensus, that we are witnessing the emergence of our own genus *Homo* out of a more apelike australopithecine ancestor. Although there was no sudden, sharp evolutionary change involved in the emergence of *Homo*, the changes that occurred around two million years ago were more rapid than those of the previous three million years. As for where this occurred, both East and South Africa have their champions.

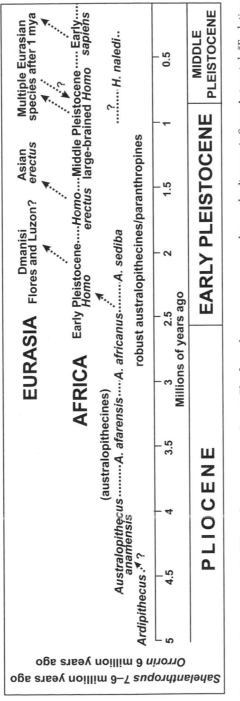

Figure 2.4. Hominin species in Africa and movements into Eurasia. This figure draws on many sources, but see the diagrams in Susan Anton et al., "Evolution of early *Homo*," *Science* 345 (2014): 1236828; and Xijun Ni et al., "Massive cranium from Harbin in northeastern China," *The Innovation* 2 (2021): 100130.

Paleoanthropologists currently recognize several species of early *Homo* living between roughly 2.6 and 1.6 million years ago in both East and South Africa, including *Homo rudolfensis, Homo habilis, Homo ergaster,* and *Homo erectus.* Exactly how all of these were related to each other remains uncertain, and in what follows I refer to them all as "Early Pleistocene *Homo.*" Figure 2.4 presents a visualization for the evolution of Early Pleistocene *Homo,* and figure 2.5 presents a selection of hominin crania from frontal and side views to show some of the major size and contour differences that identify them.

In figure 2.4, the earliest members of *Homo* overlap with and gradually replace the australopithecines before two million years ago, but without precisely identified ancestral connections. Early Pleistocene *Homo* was differentiated from australopithecines by having larger average body and brain sizes, by factors of about 30 percent. By around two million years ago, members of the species termed *Homo habilis* and *Homo rudolfensis* had undergone increases in brain volume to between 510 and 800 cubic centimeters, far beyond the late australopithecine average. We average about 1,350 cubic centimeters now, with a wide range of individual variation.

Currently, the oldest claim for a potential *Homo* fossil centers on a mandible dated to 2.8 million years ago found in Ethiopia, so we could, at a pinch, be dealing with a Late Pliocene rather than Early Pleistocene appearance of the genus.[26] This find is older than the upward trend in brain size around two million years ago that appears to mark the emergence of *Homo,* but it could support the view of some paleoanthropologists that the characters that defined the new genus did not all appear one fine day in a single population in one place, as a result of some massive mutation, and then spread. On the contrary, they could have

---

Figure 2.5. A panoply of hominin evolution, from *Australopithecus* to us. The skulls are labeled with location, species, and approximate age, and the arrangement does not represent a family tree. Mostly casts in the collection of the School of Archaeology and Anthropology, Australian National University. Photos by Maggie Otto. *Homo floresensis* is the original specimen, reproduced courtesy of the Liang Bua team (Matt Tocheri and Thomas Sutikna). The casts of Dmanisi 3 and Atapuerca 5 are reproduced courtesy of © Bone Clones, www.boneclones.com. The cast of Amud 1 is reproduced courtesy of the Australian Museum, photos by Abram Powell.

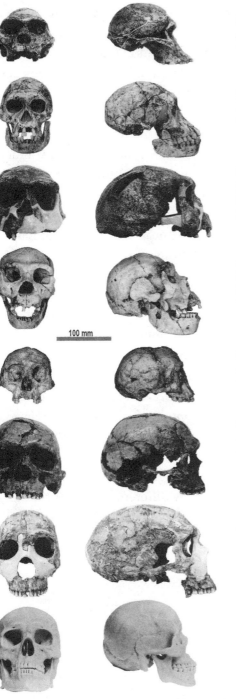

Sterkfontein 5, South Africa
Australopithecus africanus
~2 million years old

Dmanisi 3, Georgia
Early Pleistocene *Homo*
~1.7 million years old

Sangiran 17, Java
Homo erectus
500,000 – 1,000,000
years old?

Atapuerca 5, Spain
Sima de Los Huesos
Early Neanderthal
~430,000 years old

Liang Bua 1, Flores, Indonesia
Homo floresiensis
~100,000 years old

Qafzeh 6, Israel
Early *Homo sapiens*
~100,000 years old

Amud 1, Israel
Late Neanderthal
~55,000 years old

Modern *Homo sapiens*
(of East Asian origin)

100 mm

evolved in what is often called a mosaic fashion—an incipient trait here, an incipient trait there, in a sequence that commenced among the later australopithecines and that could have spanned half a million years or more across a very wide region of Africa. As discussed above, hominins and panins had perhaps separated in a similar mosaic fashion several million years beforehand. All we can say for sure is that hominins with brains large enough to be classified as non-australopithecine had evolved by at least two million years ago, and that the remaining australopithecines, especially the robust ones, continued to exist independently until their eventual extinction by one million years ago.

Perhaps the best surviving example of an Early Pleistocene *Homo* from Africa is the almost complete skeleton of a young boy or early teenager discovered at Nariokotome, near the shoreline of Lake Turkana in Kenya. Known as "Nariokotome Boy" or "Turkana Boy," and dated to about 1.6 million years ago, this individual might have grown to an adult stature of around 1.65 meters, similar to a smallish modern male human. He was fully bipedal and would probably have had an adult brain volume, had he lived, roughly double that of the australopithecines.[27] The ratio of brain mass to body mass was also increasing, from an estimated average of about 1.2 percent in australopithecines to over 1.5 percent in Nariokotome Boy. That ratio is about 2.75 percent today.[28] Figure 2.2 shows a reconstruction of another Early Pleistocene *Homo* skeleton from Lake Turkana, a female contemporary of Nariokotome Boy.

Nariokotome Boy was thus a relatively large-brained and large-bodied young member of the emergent genus *Homo*, variously classified as *Homo ergaster* or *Homo erectus* by different paleoanthropologists. Epitomized by this young terrestrial biped, and having shed many of the apelike skeletal features that had formerly characterized the australopithecines, Early Pleistocene *Homo* was coming of age.

The Origins of Human Behavior

Why did brain sizes begin to increase among Early Pleistocene *Homo* populations? One point demonstrated by Leslie Aiello and Peter Wheeler in 1995 was that the eating of meat or bone marrow would have provided more direct energy from fats and proteins to fuel brain expansion than would the eating of fruits, nuts, tubers, and grass seeds alone.[29] Others had also thought of this possibility long beforehand—let me quote Engels again, writing in 1884: "The most essential effect, however, of a meat diet was on the brain, which now received a far richer flow of the materials necessary for its nourishment and development than formerly."[30]

Alternatively (or in combination), did the increasing size of the hominin brain reflect increasing levels of communication and socialization within hominin groups? Did more brain volume equate with larger alliance networks between individuals and social communities of increasing size?[31] Larger cooperating groups would have provided greater defense against predators and enemies, which would have been useful when a bipedal life on the ground in open country would arguably have exposed early hominins to more dangerous predator species than they would have faced if living entirely in trees.[32] Cooperating groups can also assist in child-rearing, a major factor that could greatly have increased sociability among the early members of the genus *Homo*. We might expect that increasing needs for social interaction led to increased levels of interpersonal communication, with recognition of kinship between individuals and perhaps even rudimentary language.

This is not all. The intriguing idea that material culture, especially the use of tools, might also have helped to create the first large-brained members of the genus *Homo* also has keen adherents.[33] This might have occurred not just through the use of stone tools (and possibly fire) to extract and prepare meat and marrow, but also perhaps through the use of fiber slings to carry helpless babies, thus allowing their infant brains to grow safely outside the mother's womb. A human infant must be born through the mother's pelvis early in its developmental career because of its large brain, given that the modern human brain increases in volume from approximately 350 to 1,400 cubic centimeters between birth and maturity—in other words, by a factor of four. Human infants therefore need longer and more intensive maternal care than do the infants of great apes.

There are other technological advances that, in theory, must have mattered, even though we have little direct evidence for anything at this time depth apart from simple stone tools. Projectiles such as wooden spears, which the human hand, arm, and shoulder can propel with considerable accuracy, together with digging sticks for grubbing out tubers and other tasty morsels located in the ground would have assisted in providing more food for the group. So also would animal skin containers for foods and liquids. Easier access to food would have encouraged an increasing birth rate by improving maternal health, thus raising the success rate and frequency of conception.

I have already mentioned the ability to make fire, a relatively easy process once the necessary raw materials and their manipulations are understood. Fire can also be kept alive easily, if one knows how. Unfortunately charcoal disintegrates and dissipates rather easily in many sediments, so the remains of actual hearths can be difficult to recognize in ancient archaeological deposits. Burned bone can provide a clue to the presence of fire, but cooking of meat to make it palatable would not necessarily lead to the burning of bone.

Evolutionary biologist Richard Wrangham has suggested that fire was used by Early Pleistocene *Homo* populations, but many archaeologists only accept evidence for it after one million years ago,[34] suggesting that earlier hominins pounded and sliced tubers and raw meat to make them palatable without cooking. However, as Wrangham points out, our relatively small and even-surfaced human teeth, without large canines, and our small intestines are adapted to eating and digesting cooked foods, including cooked meat and starchy vegetables. We are not raw meat carnivores, and neither, according to their dentitions, were any of our hominin ancestors. Cooking makes food easier to digest, in turn releasing protein and carbohydrate energy for brain growth. It also encourages eating after dark around a potential campfire with predators kept at bay—hence more socialization. And a good fire could keep hominins warm on a chilly night.

As for actual group sizes on the ground, hunter-gatherer populations in the ethnographic record sometimes congregated in local groups of up to 50 or 100 people. Although we cannot apply the ethnographic record directly to the interpretation of human behavior two million years ago, at least it can give some idea of the possibilities. Archaeologist John Gowlett has estimated that Early Pleistocene *Homo* group territories extended over perhaps 80 to 150 square kilometers, based on the movement of stone raw materials used for tools in archaeological sites.[35] Many archaeologists believe that locations of habitual congregation were present within these territories, recognizable through accumulations of artifacts and discarded animal bones.

A suggested example of one such location is the archaeological site of Koobi Fora FxJj50 on the eastern shore of Lake Turkana in Kenya, which has produced refitting stone artifacts and refitting segments of an antelope humerus, presumably broken open by hominins.[36] The animal bones in the site were mostly of medium to large game animals. If humans had hunted them directly, they would have needed to share the fresh meat quickly between several individuals to avoid having to consume it in a rotting state. Cooperation via food sharing would have had its rewards.

Indeed, there has been much debate about whether or not early humans could hunt directly by using projectiles or if they were obliged to scavenge for meat after carnivore kills, approaching carcasses around the middle of the day when the carnivores themselves were hopefully taking a siesta. The surviving evidence does not allow an easy separation between these two modes of behavior, and there is no obvious reason why both should not have occurred. I am still inclined to agree with archaeologist Glynn Isaac, who offered a hypothesis back in 1981 about base camps and the behavior associated with them from his experiences at Koobi Fora:

> The food sharing hypothesis [for early hominins] would predict the following as having been important: tools, transport of food, meat eating, gathered plant foods, division of labour and the existence of places at which members of a social group would reconvene at least every day or so and at which discarded artifacts and food refuse would accumulate. The archaeological configuration observed at Olduvai [Tanzania] and at Koobi Fora [Kenya] fits many of these predictions.[37]

From this perspective, the nonbiological essence behind hominin success, especially within the genus *Homo*, was cooperation and sharing at the group level, as well as a growing significance for the use of artifacts to assist their daily activities. Another skill that paid off for Early Pleistocene *Homo* was its success in migration over remarkable distances.

3

Out of Africa

TO SUMMARIZE SO FAR, some australopithecines evolved into the larger brained genus *Homo* by two million years ago in Africa. Those that did not, including the robust australopithecine (or paranthropine) species, eventually became extinct as independent gene pools. A new genus was on the launchpad—the genus *Homo*—and takeoff was imminent. With its increased brain size, fully bipedal posture, precision grip, tool-using culture, and impending ability to migrate to new and far away environments—moving literally to the other side of the world— early *Homo* was rapidly becoming unequaled (figures 2.4, 3.1, and 3.2). What happened next?

In this chapter I discuss the first part of Act II in the Odyssey, when Early Pleistocene *Homo* left Africa and traveled overland to Georgia, China, and Java, possibly leaving Africa on more than one occasion. Small hominins at some point continued onward across sea gaps to reach Flores in eastern Indonesia and Luzon in the Philippines, where they survived in isolation until the eventual arrival of *Homo sapiens* about 50,000 years ago. I terminate this chapter at around one million years ago when larger-brained Middle Pleistocene hominins, including the ancestors of *Homo sapiens* and the Neanderthals, began to spread through Africa and Eurasia.

In order to appreciate what follows, it is necessary first to introduce the chronological divisions that define the Pleistocene.

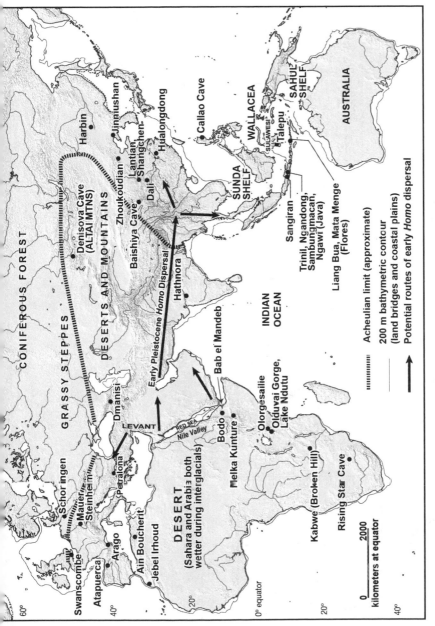

Figure 3.1. Map of Africa and Eurasia to show archaeological sites and hominin find places associated with the genus *Homo*. Acheulean limits are estimated from A. P. Derevianko, *Three Global Human Migrations in Eurasia*, vol. 4: *The Acheulean and Bifacial Lithic Industries* (Russian Academy of Sciences, 2019), 769.

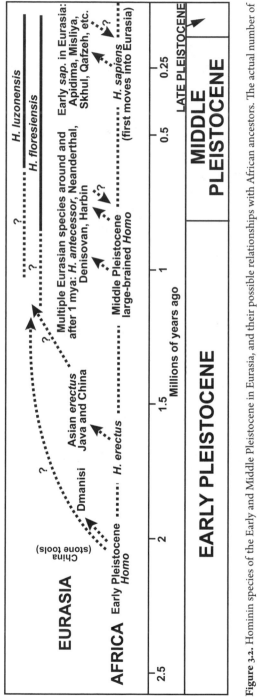

Figure 3.2. Hominin species of the Early and Middle Pleistocene in Eurasia, and their possible relationships with African ancestors. The actual number of movements out of or into Africa during this more than two-million-year time span remains unclear, especially for the large-brained species of the Middle Pleistocene.

Pleistocene Chronology: The Basics

As I stressed in chapter 1, accurate chronology is of the utmost significance in ordering our knowledge of the hominin past. Equally important is acknowledging the long-term changes in earthly environments
that occurred during the Pleistocene, especially those that were cyclical
in occurrence and caused by natural phenomena. Such cycles determined the Pleistocene ice ages and the interglacials that occurred between them.

Since the nineteenth century, scientists have divided earth history
into epochs defined by different geological layers and lifeforms, the two
most recent being the Pleistocene and the Holocene. We met the preceding Miocene and Pliocene epochs in chapter 2. The Pleistocene is
agreed by geologists to have commenced 2.58 million years ago, and it
was marked by a shift in the species compositions of Mediterranean
plankton faunas as well as by more frequent glacial cycles. It continued
through many cycles of high-latitude glaciation separated by warm interglacials until 11,700 years ago, a date that marks the termination of a
final and brief resurgence of Ice Age conditions known to climate scientists as the Younger Dryas. The Younger Dryas was followed by the
Holocene, our current interglacial, although in reality the Holocene is
merely another Pleistocene interglacial, the most recent of many, that
is being actively extended through global warming that is the result of
human activity.

For our evolutionary and archaeological purposes, the Pleistocene
has three main chronological divisions of unequal length—Early,
Middle, and Late—into which we can set the record of hominin fossils
and artifacts (figures 2.4 and 3.2).

The *Early Pleistocene* lasted from 2.58 million to 780,000 years ago, the
latter date associated with a reversal of the orientation of the earth's
magnetic field, which can be traced in many rocks and sediments. The
Early Pleistocene witnessed the eventual disappearance of the australopithecines and the emergence and expansion of the genus *Homo*.

The *Middle Pleistocene* lasted from 780,000 to 129,000 years ago, the
latter date marking the beginning of the penultimate interglacial. The

Middle Pleistocene was the age of several major large-brained *Homo* species (to be discussed in chapter 4) such as the Neanderthals and the Denisovans. It also saw the rise of early *Homo sapiens* in Africa (to be discussed in chapter 5).

The *Late Pleistocene*, when *Homo sapiens* rose to prominence, lasted from 129,000 to 11,700 years ago, the latter date marking the end of the Younger Dryas cold interval. The Late Pleistocene included the penultimate interglacial, which lasted for about 11,000 years from 129,000 until 118,000 years ago. At that time, average global temperatures were 2°C above those of the present, and sea levels up to 9 meters higher. Then followed the long cooling trend toward the last glaciation (the "Ice Age" of popular media), which sank to its maximum cold between 25,000 and 18,000 years ago. Under what are termed *Last Glacial Maximum* conditions, and compared with today, mountain snow lines descended about one vertical kilometer, sea level fell about 130 meters, and tropical ocean surfaces were on average 3–5°C colder.[1]

The Pleistocene Glacial-Interglacial Cycles, and Hominin Migration to Asia

Since the beginning of the Pleistocene around 2.6 million years ago, and continuing until now, the earth has undergone many gyrations through "Milankovitch cycles" of movement in its orbit around the sun and in the tilt of its axis. These Milankovitch cycles reflect the gravitational effects of other planets within the solar system, and they cause climatic swings between cold and dry glacial conditions (ice ages) on the one hand and warm and wet interglacial conditions, such as those that we enjoy today in the Holocene, on the other.

Scientists plot these temperature swings from the changing ratios through time between two isotopes of oxygen. The oxygen samples needed to do this are taken from the calcium carbonate shells of marine plankton cored from marine sediments, from calcium carbonate stalactites in limestone caves, and from carbon dioxide trapped in ancient air pockets in ice sheets in Greenland, the Himalayas, and Antarctica.

These paleoclimate records reveal that there were many cycles, at least twenty or more, from ice age to interglacial and back again during the two-million-year life span of the genus *Homo*. Furthermore, the cycles gradually increased in intensity with the passage of the Pleistocene. During the last 900,000 years they have each lasted around 100,000 years, but beforehand they occurred in 40,000-year cycles of lesser magnitude.

As opposites, ice ages and interglacials provided different conditions for the ancient hominin societies that had to exist within them and migrate through them. Ancient hominins reacted accordingly, pulling back from expanding glaciers and deserts during the ice ages, and spreading forward during the interglacials through the inviting grasslands and wooded parklands that recolonized the former glacial-era landscapes. Both extremes encouraged movement but for different reasons and in different directions.

For the Early Pleistocene time span with which we are concerned in this chapter, the precise details of these glacial to interglacial swings are difficult to relate to specific events within hominin prehistory. The record is simply too thin. All we can state is that the first hominins to leave Africa probably did so during an Early Pleistocene period with an inviting climate that allowed them to traverse the Sahara Desert or Arabian Peninsula from the south and to move into a previously unoccupied Levant, along the eastern Mediterranean coastline.

However, climate change by itself, in terms just of rainfall and temperature, was not the only natural factor prompting human migration. As I noted in chapter 1, during each successive climatic cycle of the Pleistocene the worldwide sea level changed too, in cycles of up to 130 meters. During glaciations, the shelves that surround continents and large islands were exposed as dry land.

Thus, the large islands of western Indonesia (Java, Bali, Borneo, and Sumatra) were joined to the mainland of Southeast Asia by the exposed Sunda Continental Shelf that today lies beneath the South China Sea. Australia, New Guinea, and Tasmania were joined into one Greater Australian continent, across the Sahul Continental Shelf (beneath the

Arafura Sea) in the north and Bass Strait in the south. Alaska and the Chukotka Peninsula of northeastern Siberia were joined across the Beringian land bridge, which was inhabited during ice ages by large herds of tundra-loving caribou, musk oxen, and mammoths.

Not surprisingly, early hominins sometimes took advantage of these land bridges to migrate—they moved, for instance, across dry land from the Southeast Asian mainland to Java and Borneo, from Siberia to Alaska, and from the European mainland to the British Isles. Some early hominins were also capable of crossing narrow but permanent sea gaps to reach islands, especially in eastern Indonesia and the Philippines. Whether early hominins reached Sahul (New Guinea and Australia) prior to the arrival of *Homo sapiens* is an interesting question that I will discuss briefly in chapter 5. So far, there is no evidence in support of a pre-*sapiens* arrival in the Americas.

Escaping the Homeland

Most paleoanthropologists believe that in order to reach Asia, the Early Pleistocene *Homo* migrants from Africa either entered the southern Levant through the Sinai Peninsula, accessible from the Nile Valley, or swam/floated to what is now Yemen across the narrow sea passage called the Bab el Mandeb at the southern end of the Red Sea (figure 3.1). This passage is 29 kilometers wide now, although it might have been much narrower, even briefly dry land, during periods of minimum glacial sea level.[2]

Paradoxically, when sea levels were low during the Pleistocene ice ages, and the Bab el Mandeb was conveniently narrow, the adjacent climates on land were extremely dry due to the trapping of global moisture in the ice sheets. The Saharan and Arabian deserts, as well as Sinai, would have been hostile to hominin passage at such times. Conversely, these deserts became green and inviting during the higher rainfall conditions of interglacial periods, although the Bab el Mandeb sea passage then became too wide for easy crossing.

A high sea level in the Bab el Mandeb may have been no problem if one had a raft of some kind, even just a floating log. This idea might not be as silly as it sounds given the arrival of the ancestor of *Homo floresiensis* on the permanently sea-girded island of Flores in Indonesia by one million years ago (I discuss this later in the chapter). But, as with Flores, we can only guess about the precise circumstances behind such events linking Africa and Asia at two million years ago.

Early Exits from Africa: How Many?

Because there were multiple Pleistocene cycles of Milankovitch-enforced climate change within the prehistory of our genus, potentially there could have been multiple environmental episodes during which Early and Middle Pleistocene hominins could have migrated from Africa into the Levant and other regions of Eurasia, as well as vice versa. So, we might ask: How many times did early hominins actually leave Africa or even migrate back by the same route?[3]

Neither the fossil nor the molecular records of hominin and mammalian evolution indicate a particularly high frequency of migration, although perhaps we have too many gaps in the record to know for sure. If we stick to fossil hominins alone, we will be told by many paleoanthropologists that hominins only left Africa a few times, not the twenty times or more throughout the whole of the Pleistocene that the Milankovitch cycles might suggest. However, is it possible that the fossil record, being small and scattered, is leveling out a record of much greater complexity? Was hominin movement from Africa into Asia, and even vice versa, more frequent during Pleistocene interglacials than is currently apparent?

There are two opposing potential answers to this question, and at this stage I doubt that anyone knows which is closest to being correct. On the one hand, but without direct evidence, we can simply accept that hominin pioneers could have migrated from Africa into tropical and temperate regions of Asia multiple times during the Early Pleistocene, particularly during wet interglacials, following the route down the Nile through the Sahara and across the Sinai Peninsula, or across the Bab el Mandeb. Middle and Late Pleistocene fossils of hyenas, hippos, antelopes, and buffalos of African origin have been found with stone tools made by fairly recent hominin species near ancient lakes in Israel and

Arabia; they indicate that such animal migrations sometimes occurred, although they might not have penetrated far into Asia.[4] For early hominins, however, the fossil record might never be detailed enough to give certainty, and each interval between migrations might perhaps have been too short for any significant evolutionary changes in the skulls and faces of hominins to be visible to paleoanthropologists.

Against this optimistic view of frequent migration out of Africa there are two opposing observations that urge caution. First, Early Pleistocene hominins often appear to have migrated alone.[5] For instance, the 1.7 million-year-old hominins of Dmanisi in Georgia, a crucial location to be discussed soon, coexisted with Asian rather than co-immigrant African animals. A second important issue is that of competition with other species (competitive exclusion). Mammals, including hominins, would only have colonized new regions if potential competitor species allowed them to do so, or if the migrants had certain genetic or behavioral advantages. The Sinai Peninsula land route from Africa into Asia was also a narrow bottleneck and not always attractive to hominin would-be migrants, especially during dry glacial periods.

The factor of competition might not have been a problem for the first hominin migrants to leave Africa because they would have had no other hominin competitors in Asia. Furthermore, the animals in Asia would initially have been naive (i.e., unadapted to flight when facing an unfamiliar bipedal predator) and hence easily killed for food. However, these advantages would have been compromised if later migrants arrived who had evolved genetic advantages prior to their arrival, as happened when African *Homo sapiens* migrants into Eurasia eventually replaced the Eurasian Neanderthals during the Late Pleistocene. For the Early Pleistocene this possibility seems less likely. But what does the archaeological evidence tell us?

Early Pleistocene *Homo* Reaches North Africa and Asia

We still have a great deal to learn about the movements of the earliest members of the genus *Homo* within and out of Sub-Saharan Africa. But we know they left, in no uncertain terms. A few crucial discoveries make this clear.

First, someone crossed the Sahara from the south. Riverine deposits at Ain Boucherit, across the Sahara and on the northern edge of the Atlas Mountains in Algeria (figure 3.1 shows the location), have produced stone tools and animal bones with cut marks left by removal of meat in layers dating between 2.4 and 1.9 million years ago.[6] The stone tools belong to the same "Oldowan" archaeological stone tool industry—named after the many locations that have been found in the archaeological landscape of Olduvai Gorge, Tanzania—that occurs at contemporary sites in Sub-Saharan Africa. Oldowan tools consist basically of large implements made by flaking a stone core to create a straight or pointed sharp working edge at or around one end. These pebble tools are found with large numbers of the usable sharp-edged flakes that were also struck off the cores during the manufacturing process.

As primary evidence of hominin technology and meat removal from bones, these stone tools indicate a passage by hominins of some kind (no hominin fossils have yet been found at Ain Boucherit) through what is now the Sahara Desert. The movement occurred during the transition from late australopithecines into the first *Homo*. However, there is no evidence at this time for any movement by hominins directly from Africa across the Mediterranean Sea into Europe. The Strait of Gibraltar was always at least 10 kilometers wide during the Pleistocene, and it is 14 kilometers wide now.

Asia was another story. Before two million years ago, it appears that hominins had reached virtually the latitude of Beijing in northern China, a rather mind-boggling 8,000 kilometers from Sinai on the other side of some fairly hostile desert and mountain terrain in central Asia. The evidence comes from the site of Shangchen, located in Lantian County, Shaanxi Province, close to the Yellow River, and it is dated to between 2.4 and 2.1 million years ago. Like Ain Boucherit, this site also contained Oldowan-like stone tools, but again no hominin fossils.[7]

Beijing nowadays is hardly a warm and snug place in the middle of winter, although the excavators of Shangchen suggested that these early hominins lived there during a period of relatively warm climate. Perhaps they headed south when they wanted to escape the cold, or wrapped themselves in animal pelts (fur to the inside) and toughed it out, possibly using fire and caves for shelter, and accessing enough fats, marrow, carbohydrates, and vitamins in their diet to keep themselves alive.[8] I suspect they did both, perhaps tracking movements to the south by their preferred animal prey species.

As noted above, if the first Eurasian migrants were already hunting for meat, they would have had the additional advantage of the native Eurasian animal prey species being naive.[9] The first settlers could have taken advantage of this situation and moved quickly through new landscapes, their numbers fueled by a bonanza in easily available sources of meat. Later arrivals might not have fared so well, and mammals in both Sub-Saharan Africa and Eurasia would have become more wary in this regard as they adapted to the presence of hungry hominins. The first Asian hominins probably hit a relatively short-lived jackpot.

These discoveries at Ain Boucherit and in the Yellow River valley of China have opened our eyes to just how early in hominin prehistory the initial movements out of Africa might have been. But neither of these two sites has produced hominin skeletal remains. For the makers, we must move forward in time roughly half a million years to the site of Dmanisi in Georgia, situated evocatively within the walls of a ruined monastery in a green, tree-clad Caucasian landscape.

I was able to visit this site in 2017 with my wife Claudia in order to see the exact location where five hominin skulls were discovered in what the excavators regard as a possible carnivore accumulation of bones. The group might have been scavenging for meat, perhaps a little unwisely given that they lived alongside saber-toothed cats, wolves, and hyenas.[10] The skulls were found in volcanic ash soil above the uneven surface of a basalt lava flow that was laid down about 1.8 million years ago. It is therefore assumed that they are about 1.7 million years old.

The five Dmanisi skulls are remarkable. They are small-brained, at only 550 to 730 cubic centimeters; in this regard they are similar to the oldest *Homo* fossils that date from about two million years ago in East and South Africa (see figure 2.5, second row). With large projecting faces and brow ridges, they were regarded by their excavators as belonging to a single species of Early Pleistocene *Homo*, and they have been variously termed *Homo erectus* or *Homo georgicus* by different paleoanthropologists. It is thought from their postcranial (nonskull) bones that the Dmanisi people stood about 1.5 meters tall, perhaps a little shorter than the adult male relatives of African Turkana Boy of a similar date. Like their contemporaries in Africa and China, they also used Oldowan stone tools to cut meat from animal bones. However, their dental health was not always impressive—one individual was completely toothless and so presumably could only have eaten soft (cooked?) food.

Because the Dmanisi skulls are perhaps half a million years younger than the oldest stone tools claimed from Algeria and China, it might be unwise to regard this population as directly representing the first hominins to leave Sub-Saharan Africa. Yet they strongly suggest that the initial emigrants had rather small brains and bodies. They represent a rich fossil record that is only exceeded at this time by the record from Sub-Saharan Africa itself.

Homo erectus: Getting to China and Java

The oldest hominin fossils beyond Dmanisi currently occur in China, a region remarkably rich in hominin remains. The most famous *Homo erectus* remains, popularly known as "Peking Man," date back as far as 800,000 years ago and come from the breccia-filled caves at Zhoukoudian near Beijing. These have cranial capacities that average 1,000 cubic centimeters, approaching those of the large-brained Middle Pleistocene hominins (which I describe in chapter 4).

However, China has much older hominins, including an Early Pleistocene skull dating to possibly 1.6 million years ago from Lantian in Shaanxi Province, close to the middle Yellow River and not far from the site of Shangchen that I discussed above. This is normally recognized as a *Homo erectus*, but with a cranial capacity of only 780 cubic centimeters it fits better with the Dmanisi hominins, which are only slightly older. Were the Lantian and Zhoukoudian hominins, separated by almost one million years, simply consecutive versions of *Homo erectus*? Or were they different species? I return to this question below.

The first *Homo erectus* groups to reach southern hemisphere Java in Indonesia ("Java Man," the *Pithecanthropus* of Eugène Dubois) could only get there by crossing the Equator, with its dense rainforests.[11]

These are not easy environments for hunters and gatherers to exploit because most resources exist high above the ground, although some recent human populations have adapted to these conditions and even thrived if they have good animal-trapping technology. Early hominins seem to have preferred more open country.

Java is an island now, but it was joined during glacial periods to Asia via Sumatra and the exposed Sunda Shelf. If the first migrants waited for glacial-era low sea levels in order to cross Sundaland on foot, they might have been able to take advantage of passageways of open grassland and parkland opened through the rainforest by the correspondingly dry climatic conditions.[12] Such landscapes would have offered opportunities for hunting large animals such as wild cattle, elephant-like creatures such as stegodons, and deer, although the Javan archaeological record reveals few direct associations between hominin fossils and those of hunted fauna because of poor survival conditions.

Just when hominins first reached Java is unclear. The most recent dating scheme for the fossil-bearing location of Sangiran in central Java places the hominin arrival there at only about 1.3 million years ago,[13] but there have been claims of older dates going back to as much as 1.8 million years ago from other sites. There are problems with *Homo erectus* in Java that relate to chronology and to the imprecise stratigraphic contexts of the many discoveries made there.

In figure 3.1 I have indicated schematically what I regard as the most likely general directions of movement of Early Pleistocene *Homo*, especially *Homo erectus*, through Asia. However, the question still remains: how many hominin movements occurred out of Africa before one million years ago? One, two, or lots? Marcia Ponce de Leon and colleagues suggest, based on the contours of brain cavities inside skulls, that the Dmanisi population had a more primitive frontal lobe organization than *Homo erectus* in Java.[14] They ask if two migrations out of Africa are required to account for this situation, one happening perhaps two million years ago and another about 1.6 million years ago. I have indicated this possibility in figure 2.4, and I expand on the repercussions in figure 3.2 because they involve not just *Homo erectus* but also the small hominins of Flores and Luzon, to which I now turn.

The Enigma of Flores Island

Early hominins did not stop when they reached the eastern edge of Sundaland and faced the unbridged ocean beyond the islands of Bali and Borneo, especially when they could see the tops of volcanoes rising above the horizon on distant islands. Nevertheless, getting beyond Java to the volcanic island of Flores, located in the Nusa Tenggara island chain of southeastern Indonesia, must have been more difficult than crossing the exposed Sunda Shelf. This is because reaching Flores would always have required more than one sea crossing from the Asian mainland, even during periods of low glacial sea level.

Yet somehow these passages were made, potentially more than one million years ago. Flores lies in a biogeographical region of islands between Asia and Australia known as Wallacea, named after the nineteenth-century naturalist Alfred Russel Wallace, a contemporary of Charles Darwin. Wallacea was never traversed by glacial-era land bridges, and it consists entirely of islands and island chains thrust up by volcanic arc tectonics from the sea bed, each flanked by deep sea.

In 2004, the first report was published on the finding of a fairly complete but tiny hominin skeleton in a cave in western Flores called Liang Bua. The skeleton was of a female, who apparently died face down in a pool of water. She was found with stone tools of Oldowan affinity and the bones of mainly juvenile pygmy stegodons, with cut marks left by the removal of meat. Stegodons were trunked elephant-like creatures that became extinct in Indonesia during the Late Pleistocene. On Flores, they appear to have become dwarfed in the small island environment, a process known to have affected large mammals on many small islands around the world with limited food resources (although small mammals such as rats got bigger).

When first discovered, *Homo floresiensis* was thought to have survived until about 12,000 years ago, according to radiocarbon dating in the cave. The excavators now think that they initially misinterpreted the Liang Bua stratigraphy, which was deposited discontinuously in the cave through periods of erosion followed by redeposition. A more recent program of dating by other geophysical methods apart from radiocarbon has made it clear that *floresiensis*, since found to be represented by more than one individual (and possibly as many as eight according to the other hominin bones found in the cave), lived in Liang Bua between 195,000 and 50,000 years ago, after which it became extinct.[15]

Recent research in open-air sites such as Mata Menge in central Flores has produced yet more small hominin bones (including a small mandible, but no cranial parts), perhaps of an ancestor of the Liang Bua population, together with stone tools and more bones of stegodons, Komodo dragons, and giant storks. These come from geological deposits dating to between 1.3 and 0.7 million years ago, before which time there are, as yet, no signs of a hominin presence on the island.[16] *Homo floresiensis* and its potential ancestor presumably lived on Flores for more than a million years, in a high degree of isolation from all other hominin life. This is apparent from its tiny brain size, which would have become larger had there been genetic mixing with larger-brained hominins such as *Homo erectus*.

Two explanations for the existence of *floresiensis* come to mind, given its extremely small stature, body weight, and brain size (circa 106 centimeters, 28 kilograms, and 420 cubic centimeters, respectively), and the australopithecine characteristics surviving in its pelvis, hands (especially wrist bones), and elongated feet, all rather out of place at such a relatively recent date in the case of Liang Bua. One explanation favored by paleoanthropologist Debbie Argue and colleagues is that its ancestors were derived from a small-brained Early Pleistocene *Homo* source population, distinct from *erectus*, which simply retained small brains for a million years or more in the isolation of Flores Island.[17]

This explanation seems possible according to a recent study of hominin foot bones, which suggests that those of *Homo floresiensis* were intermediate in shape between those of *Australopithecus afarensis* of Ethiopia (e.g., "Lucy") and Early Pleistocene *Homo*, including the Dmanisi population.[18] However, if the current dating evidence for both Java and Flores is correct, suggesting that the first hominins only arrived on both islands after 1.3 million years ago, then the chronology is too recent for direct transmission from an *Australopithecus* ancestor because that genus was probably extinct by this time.

The second explanation is that *floresiensis* became dwarfed in body and brain entirely in Flores, like the stegodons, after its ancestors landed as somewhat larger-bodied *Homo erectus* settlers from somewhere in Sundaland. This explanation would fit a date of only 1.3 million years ago for the hominin crossing of the equator into the Southern Hemisphere, but it would not explain the elongated feet, unless these can somehow be explained as evolutionary reversals toward a prior australopithecine state.

As we will see soon, there is new evidence from the Philippines that supports the first hypothesis, that *floresiensis* represented an early arrival of a tiny hominin with residual australopithecine characteristics, followed by an immense time span of isolation within Wallacea. If correct, this would make a hominin arrival date in Java and Wallacea well before 1.3 million years ago quite likely. I also referred above to a recent suggestion based on brain endocasts that there might have been two movements from Africa into Eurasia, the first giving rise to the Dmanisi population and the second to Javan *Homo erectus*. The first might then also account for *floresiensis*, as hinted in figure 3.2. However, certainty on this issue is elusive.

Another Flores mystery concerns the pygmy stegodons. Their ancestors on the island were larger in size than the ones that coexisted with *floresiensis*, so their dwarfing was definitely local to the sea-girded islands of eastern Indonesia. But why did *floresiensis*, who was presumably eating stegodon meat, not quickly exterminate them, given the small size of the island and the one million years of hominin-stegodon coexistence? *Homo sapiens* arguably would not have been so considerate.

Interestingly, recent studies of large herbivore extinctions in Africa and Indonesia indicate that early hominins were not necessarily always responsible for such events and that climatic changes, as they affected plant food resources for grazers and browsers, were more important.[19] Perhaps *floresiensis* was a timid hunter or smart enough not to kill off what could have been a major source of meat, whether hunted directly or scavenged from kills by Komodo dragons.

The greatest mystery of all is the question of how these hominins reached Flores. This island was never joined by a direct land bridge to

Java or the Asian mainland, as indicated by the absence of Javan land mammals such as deer, pigs, cattle, rhinos, and tigers in the animal bones found in Flores archaeological sites dating from the Pleistocene. The stegodons were able to swim there using their trunks as snorkels, like true elephants (which incidentally did not make it to Flores). Paleoanthropologist Madelaine Böhme has even suggested that the tiny hominins perhaps rode on the backs of stegodons swimming between islands—an imaginative scenario, but alas not one for easy testing.[20] The Komodo dragons arrived much earlier in geological time through the continental drifting of pieces of the Australian continent into the volcanic island arc of eastern Indonesia.

So we must conclude that no one ever walked to Flores from Java or Bali during the Early Pleistocene, or for that matter at any other time, because there was no land bridge. They had to float there somehow. The north-to-south sea passages between the Lesser Sunda (Nusa Tengg-ara) islands themselves have strong currents, and archaeologists today regard *Homo floresiensis* as having arrived most likely via the central Indonesian island of Sulawesi to the north, reaching Flores by crossing sea passages between intermediate islands exposed by Pleistocene low sea levels.

However, the story becomes even more mysterious when we realize that Sulawesi itself is also within Wallacea and likewise was never joined by dry land to the Sunda Shelf. To reach this island, the ancestors of *floresiensis* would have needed to cross the Strait of Makassar, a perma-nent sea channel that separates Sulawesi from eastern Borneo. Cur-rently, Sulawesi has evidence for a human presence from stone tools dating between 200,000 and 100,000 years ago at a site called Talepu, but no hominin fossils have yet been found there.

The *floresiensis* saga is still much of a mystery. Perhaps these tiny hominins could swim, and their low body weights would have made flotation on logs or mats of vegetation more likely than for a heavier hominin such as *Homo erectus*. Whatever the answers, the arrival of *flo-resiensis* in some form on Flores before one million years ago stands as one of the most remarkable events in the early Pleistocene prehistory of the genus *Homo*.[21]

Luzon, the Philippines

This is still not the end of the story for the tiny hominins of Island Southeast Asia. More recently, the discovery was announced in the Philippines, also a part of Wallacea, of a few bones and teeth from another tiny hominin with residual australopithecine characteristics. These remains date from before 60,000 years ago, hence they are Late Pleistocene in age like the Liang Bua skeleton in Flores. The bones were found in Callao Cave in the Cagayan Valley of northern Luzon Island, a site excavated by my former student Armand Mijares of the University of the Philippines.[22] Like Flores and Sulawesi, the Philippine Islands (except for Palawan) were never joined by a land bridge to the rest of Southeast Asia. Luzon could only have been reached by crossing sea.

The Callao hominin bones were found without stone tools in waterlaid deposits in the base of the cave, but some associated bones of deer, pig, and a species of buffalo carried cut marks from stone tools. These bones had been washed into the cave from an external location that has not yet been located. Interestingly, the surviving foot bones of *Homo luzonensis*, like those of *Homo floresiensis*, also appear to have retained some degree of australopithecine-like grasping and climbing ability, so perhaps this species also descended from an early movement out of Africa, maybe the same one that gave rise to the Flores hominin.

Luzon, like Flores, also has traces of a much older Middle Pleistocene hominin presence, dating from about 700,000 years ago in other sites in the Cagayan Valley and in association with animal (but not hominin) bones with cut marks, including those of extinct Middle Pleistocene species of pig, deer, buffalo, and rhinoceros.[23] These large mammal species never reached Flores, but all can swim short distances, albeit not as strongly as elephants. It is possible that they reached Luzon across narrow sea channels from eastern Borneo via the island of Palawan.

The ancestors of the tiny Wallacean hominins drive home an important message. They were capable of crossing short sea passages, even if we do not know exactly how they did so. Such events apparently allowed them to reach Flores, Sulawesi, and Luzon by around one million years ago or before.[24] However, the small stature of these populations, apparently constant through perhaps a million years, suggests that they did not make such sea crossings often. They appear to have been remarkably isolated from any direct interaction with their larger hominin cousins in Sundaland.

The Handiwork of *Homo erectus* and Its Contemporaries

One point about all the Eurasian Early Pleistocene hominin species discussed in this chapter is that their bones are associated with Oldowan stone tool industries, as described above. The bones of *Homo luzonensis* have not yet been reported with stone tools, but archaeological finds from elsewhere in the Cagayan Valley on Luzon suggest an Oldowan affinity for them.

Such Oldowan core and flake tools (or "Dmanisian" tools, after some Russian archaeologists[25]) still dominated production as recently as 900,000 years ago in many archaeological sites in both Africa and Eurasia. Production of Oldowan tools never really stopped, and they continued to be made alongside younger and more complex forms until the end of the Pleistocene and the arrival of farming populations in many parts of the world. Clearly, the Oldowan was a perfectly functional stone tool industry, capable of doing the tasks that hominins needed to do. Bipedal hominins with patience and stamina had the ability to move huge distances on foot, hunting and collecting with these tools as they went. Was the massive increase in territorial range that occurred when hominins entered Eurasia—far beyond the ranges achieved by other primate species—linked to Oldowan stone tool use? I suspect that it was.

Despite this, something fundamental was about to change in the technological manipulations of African and Eurasian hominins at around one million years ago. In terms of stone tool manufacturing technology, we come in the next chapter to the first major pattern shift in the archaeological record to have occurred since Early Pleistocene *Homo* left Africa one million years beforehand, a change found across Africa and large parts of Eurasia. Archaeologists term this new technology the Acheulean, and it became highly significant during the next major stage in the evolution of modern humanity. It is time to approach the Middle Pleistocene and to bring to the stage a new cast of large-brained hominin species. These include the large-brained Eurasian ancestors of the Neanderthals, the Denisovans, the newly-announced Harbin population of northern China, and, not least, the presumed African ancestors of modern humans like us.

4

New Species Emerge

THIS CHAPTER CONSIDERS the second part of Act II of the Odyssey, the stage for the large-brained hominins who dominated the Middle Pleistocene of Africa and Eurasia. Genetic evidence makes it clear that a major set of expansions must have occurred within the genus *Homo* to explain the existence of what ultimately were to become *Homo sapiens*, the Neanderthals of western Eurasia, and the Denisovans of eastern Asia. The most recent chronological calculations, based on comparisons of DNA and morphological characters in dated crania, push back the common ancestor of these three major species to at least 600,000 years ago, and perhaps to as much as 1.25 million years ago (figure 4.1).[1]

But these three species are not the end of the story. Even as I write this chapter, it is becoming ever clearer that *Homo sapiens*, the Neanderthals, and the Denisovans were only the tip of a poorly understood iceberg. Paleoanthropologists also discuss several other hominin species from this time period, known mainly from scarce and fragmentary remains. They include *Homo antecessor* in Spain, a large and widespread group in Africa and Eurasia termed *Homo heidelbergensis*, the newly announced "Harbin human group" from northern China, and possibly another species from Israel (which I discuss in chapter 5). The last two literally appeared in the literature as I was putting the finishing touches to this chapter, illustrating that what we might see as a nice neat picture one day can rapidly be obscured by new findings the next.

The later part of the Early Pleistocene and the following Middle Pleistocene combine to form a difficult and complex period in the study

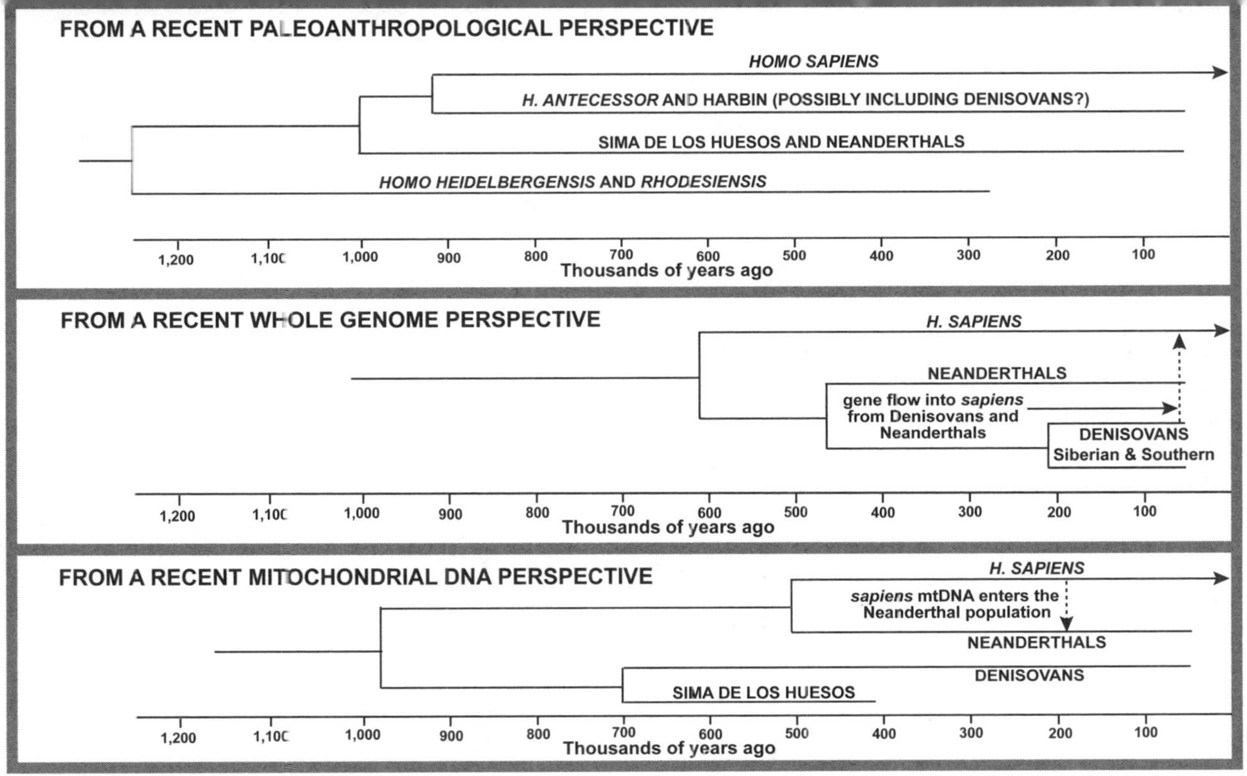

Figure 4.1. Evolutionary relationships between early *Homo sapiens* and contemporary large-brained Middle and Late Pleistocene species in Eurasia. (*Top*) From the paleoanthropological perspective (skull morphology), *Homo sapiens* clusters with *Homo antecessor* and the Harbin/Denisovans. (*Middle*) From the whole genome perspective, Neanderthals cluster with Denisovans. (*Bottom*) From the mitochondrial DNA (mtDNA) perspective, Neanderthals cluster with *Homo sapiens*. For the sources of these diagrams, see note 1 in this chapter. They are presented here without ranges of error, so the dates should be read as approximate.

of human evolution, made more difficult by the lack of complete skeletons, the fragmentary nature of many of the relevant fossils, and, not least, some remarkably variable calculations concerning chronology. For those who might wish, ever hopefully, that the above list of seven potential species provides the last word on the matter, Southeast Asia also continued to harbor populations of *Homo erectus*, *Homo floresiensis*, and *Homo luzonensis* until well within the Late Pleistocene, perhaps until as recently as 50,000 years ago, when *Homo sapiens* commenced in earnest to dominate all of the hominin world occupied to that time. Indeed, there was also yet another small-brained species termed *Homo naledi* in southern Africa. We are currently facing the possibility of at least eleven hominin species, some better established than others, all living at some point during the Middle Pleistocene!

Who gave rise to whom? That question stands as the Gordian knot of modern paleoanthropology (admirers of Alexander the Great will understand the metaphor). One point I must make from the start is that, although impressive, the time span of ~1 million years of shared ancestry for *Homo sapiens*, the Neanderthals, and the Denisovans is much less than the roughly 2.5-million-year time span of the genus *Homo* as a whole. This tells us clearly that the age of *Homo erectus* was brought to an end over much of Africa and Eurasia by the expansions of new hominin species.

I have tried to give some idea of the current situation in figure 4.1, which puts the main performers into a time and space framework and indicates their internal relationships, insofar as they are known at present. The figure is informed by three sources of data manipulated by complex statistical techniques—morphological variations in cranial shape, whole genomic comparisons between *Homo sapiens* and extinct hominins, and corresponding comparisons of mitochondrial DNA haplogroups.

These three sources give similar but by no means identical results. One complication is that Neanderthals appear to have altered their mitochondrial DNA haplotypes more than 170,000 years ago as a result of interbreeding with early populations of *Homo sapiens*, presumably in

NEW SPECIES EMERGE   63

Eurasia. This means that Neanderthals cluster with *sapiens* in terms of their mitochondrial DNA, but with Denisovans in terms of their whole genomes. It is also apparent that the ages derived from morphological comparisons are somewhat older than those derived from genetics. Human ancestry was never meant to be simple!

Before we move to examine all the actors on the stage during the later part of Act II, I wish to introduce an issue that currently muddies the water for those who like nice clean family trees of human evolution. What, exactly, was a "species" in terms of the named categories created from ancient bones by paleoanthropologists? Was a species of hominin an exclusive kind of club, breeding purely within its own boundaries? Or were those boundaries porous?

Understanding the Course of Human Evolution

Before the recovery and analysis of ancient DNA, a development that generally postdated the year 2000 as far as human prehistory is concerned, paleoanthropologists and archaeologists tended to fall into two camps with respect to how one should interpret human evolution. The split concerned a major question. How can we best understand the human past—by lumping fossil populations at any one time into one single species that encompassed the globe, or by splitting them apart into many?

Thus, can the five-million-year hominin Odyssey be best described by a model of regional continuity throughout the whole inhabited world, such that only one species existed at any one time? Or, conversely, does a model that involved reproductive isolation between populations, leading to frequent speciation and, for some, eventual extinction, fit the evidence better?

The regional continuity (or multiregional) model, as clearly expressed by University of Michigan paleoanthropologist Milford Wolpoff and his Australian National University colleague Alan Thorne in the 1990s, claimed that only a single species existed at any one time within the genus *Homo* across the inhabited portions of Africa and Eurasia, and that regional populations were linked into a network by interbreeding, more commonly referred to as "gene flow" by paleoanthropologists. They believed that no regional populations speciated, and none became extinct.[2] *Homo erectus*, the Neanderthals, and *Homo sapiens* formed an unbroken line of evolutionary descent from one to another through time, albeit with regional variation at a subspecies level.

Today, most paleoanthropologists would probably agree that the gene flow that was essential for the regional continuity model to operate exclusively could never have been continuous all the way from South Africa to Java and Flores.[3] Had it been so, we would not expect to have up to eleven separate hominin species in the Middle Pleistocene, as opposed to just one. But we might still expect that linkages between neighboring populations would eventually have carried biological developments quite far afield, especially if they carried advantages in terms of reproductive success that could be acted upon by natural selection.

The regional continuity model also meets obstacles with small-brained hominins in Indonesia, the Philippines, and South Africa (the last to be described below), who were quite unknown and unguessable back in the 1990s. These species had extremely small brains, yet they existed into relatively recent times. Each of them implies genetic isolation, perhaps for over a million years or even two, certainly not gene flow. Had there been frequent gene flow with larger-brained populations, the brains of these hominins could never have remained so small.

The opposing model, which has dominated much recent discussion about the origins of *Homo sapiens*, suggests that successful populations spread periodically and physically replaced their predecessors. In the specific case of *Homo sapiens*, the replacement model is often associated with the concept of an "African Eve," the name given to a female *Homo sapiens* who lived in Africa about 150,000 years ago and from whom all living humans have inherited their mitochondrial DNA.[4] This relatively young date attached by molecular clocks to African Eve convinced many paleoanthropologists, before the arrival of ancient DNA and genomic (whole nuclear genome) analysis, that the carriers of her mitochondrial heritage replaced all earlier hominin species, without genetic mixing, especially during a burst of *Homo sapiens* expansion beyond Africa that is currently dated between 60,000 and 40,000 years ago.

The replacement model, however, is also undermined by the relatively new observations from ancient DNA about interbreeding between hominin species, especially during the Middle and Late Pleistocene. Gene flow was certainly happening across species boundaries, especially between Neanderthals, Denisovans, and *Homo sapiens*, as I will again describe in more detail below. These were not just regional populations of a single species in the way that living Indigenous East Asians, West Eurasians, Sub-Saharan Africans, Americans, and Australians are populations of *Homo sapiens* today. They were far more diversified than living humans. The differences between these archaic species reflected time depths of separation on scales of hundreds of thousands rather than tens of thousands of years. Despite this, the history of interspecies mixing between them indicates that both gene flow and extinction could operate side by side.

We must therefore abandon polarized models of regional continuity versus wholesale replacement. Gene flow and interaction linked populations in corridors of communication, such as the African Rift Valley, the Levant, around coastlines, and along major rivers. However, it did not do so in situations of geographical or behavioral isolation, as in the cases of the Southeast Asian or South African small-brained hominins, for whom clean speciation and ultimate extinction provide the best explanation.

In my view, forged as an archaeologist with an interest in classification, species names such as *Homo sapiens* and *Homo neanderthalensis* do assist us to explain human evolution with a greater level of coherence than would be the case if every hominin fossil was simply classified as *Homo sapiens* from the Early Pleistocene onward. We just have to remember that ancient hominin species were somewhat negotiable concepts, sometimes with porous and sometimes with sealed boundaries.

The time has now come to examine the long list of named species and potential species that are about to enter the Act II stage. Some are known from their actual bones. Others, especially the Denisovans of Asia, are known mainly from their ancient DNA.

Figure 4.2. Excavation in the former cave of Gran Dolina, Atapuerca, northern Spain. Part of the railway cutting that exposed the sediments can be seen at the far right, as can the original limestone cave roof above the infilling sediments. Photo by the author.

Homo antecessor in Europe

One of the first new species to make an appearance in Eurasia was the aptly-named *Homo antecessor,* dated to about 850,000 years ago from fragmentary remains found in the caves of Sima del Elefante and Gran Dolina, both in the rather amazing cave complex of Atapuerca, near Burgos in northern Spain (figure 4.2). *Homo antecessor* was about 1.6 meters tall, the oldest to qualify as a European hominin, with a 1,000-cubic-centimeter brain volume—considerably larger than the Dmanisi brains of almost a million years before. This hominin was not a *Homo erectus* but a herald for something new.

As with Dmanisi (see chapter 3), I was also able to visit Atapuerca, in company with Spanish archaeologist Robert Sala, to look at the excavations underway during 2011 in a number of breccia-filled caves in a large limestone massif. These caves with their sedimentary fillings had been cut through and exposed in the sides of an abandoned railway cutting excavated by a former mining company in the late nineteenth century, offering to science an archaeological gold mine that must contain large numbers of still-undiscovered Pleistocene living places. The records from Sima del Elefante and Gran Dolina make it clear that the hominin inhabitants were hunting relatively warm climate animals such as horses, wild cattle, macaque monkeys, and rhinos by one million years ago, leaving cut marks on animal bones from stone tools of Oldowan tradition, and possibly also engaging in cannibalism (some of the hominin bones themselves are cut marked). However, these sites, despite the richness of their remains, still have no trace of any use of fire.

Most interestingly, the *antecessor* remains from Atapuerca have facial characteristics similar to those that ultimately developed in *Homo sapiens*, as likewise did several newly analyzed Middle Pleistocene hominin fossils from China to which I return below (the Harbin group; see figure 4.1, top).[5] But *antecessor*, it seems, did not evolve directly into *Homo sapiens*. Recent analysis of proteins from an *antecessor* molar tooth found in Gran Dolina places this species in a sister lineage, rather than one directly ancestral to us.[6]

Despite this, the facial similarities between *antecessor* and *sapiens*, who is generally thought to be of African origin, might lead us to wonder if hominins sometimes also migrated back into Africa carrying new biological characteristics that had evolved in Eurasia. Paleoanthropologists are becoming increasingly aware of such possibilities, although precise understanding still eludes us.[7]

The Mysterious *Homo heidelbergensis*

The name *Homo heidelbergensis* has been used by some paleoanthropologists to refer to a number of Middle Pleistocene hominin remains from Africa and Eurasia, most rather poorly dated, with the first to be found (in 1907) coming from Mauer, near Heidelberg in Germany. Paradoxically, Heidelberg would have been one of the last places to be reached by these hominins if they migrated from Africa, but I am afraid we must live with the incongruity.

Other potential *Homo heidelbergensis* candidates (several are located in figure 3.1), unfortunately without complete skeletons, come from Bodo in Ethiopia, Kabwe (Broken Hill) in Zambia, Lake Ndutu in Tanzania, Arago Cave in southern France, Petralona Cave in Greece, Swanscombe and Boxgrove in southern England, and, with less certainty, Hathnora in central India and Dali in central China. The species in Africa has also been called *Homo rhodesiensis*, after the Kabwe skull from the former Northern Rhodesia, which has recently been dated to 300,000 years ago.[8] Overall dates for the whole group perhaps fall between one million and 300,000 years ago, but there is considerable uncertainty.

What do we know about the Heidelberg group? Beyond sharing long and low brain cases, and continuous projecting brow ridges, there is much variation in their cranial and skeletal attributes, and most paleoanthropologists today question a Heidelberg unity as a single species. However, the fossils just listed, whether one species or not, reveal powerfully built individuals with male body weights up to 90 kilograms, similar to both Neanderthals and modern humans in body size and stature, with large faces and brain sizes averaging from 1,200 to 1,300 cubic centimeters—not much smaller than our own. Footprints preserved at Melka Kunture in Ethiopia, which some consider to belong to a Heidelberg individual, indicate a fully bipedal posture.

Given the wide distribution across Africa and western Eurasia of the Heidelberg group as a whole, and its wide distribution in time, it is likely that some of its members were related to, or perhaps even members of, the earliest *sapiens* populations in Africa as well as the ancestral Neanderthals and Denisovans in Eurasia. These three species are all younger than 500,000 years in terms of their fossil records, so origins for them within the broad range of Heidelberg variability are quite possible, even if there is as yet no ancient DNA from a Heidelberg source to clarify the matter. Perhaps, if ancient DNA or proteins can ever be recovered from their bones, the Heidelberg group as known to date will simply merge with the ancestors of the Neanderthals, the Denisovans, the Harbin human group (to be described soon), and *Homo sapiens*. We must wait and see.

The Acheulean

What was the clue to the apparent success of the large-brained hominins discussed so far in spreading over so much of Africa and Eurasia, evidently replacing *Homo erectus* in the process? Did the Heidelberg group have some kind of cultural advantage that would have allowed them to migrate freely? What about their stone tools?

Most of the sites listed above, except for Gran Dolina (*Homo antecessor*), have produced stone tools that belong to what is known to archaeologists as the Acheulean industry. Named after nineteenth century finds at Saint Acheul in the lower Somme Valley, France, the Acheulean was different in concept from the preceding Oldowan, even if many of the Oldowan basic pebble and flake tools continued to be made alongside the newer models. Large oval or pear-shaped nodules, technically both cores and large flakes, were now shaped by flaking from both sides (i.e., bifacially) around almost all of their edges to produce what archaeologists call "hand axes"—fist-sized tools that could be used for almost any purpose (figure 4.3A). The previous Oldowan had some bifacial flaking in its later stages, but not on the scale of the Acheulean.

Hand axes are instantly recognizable when found in large numbers, and they are characteristic of the earlier part of the Middle Pleistocene across Africa, much of Europe, and through Asia to as far east as India, Turkmenistan, Kazakhstan, and Mongolia (see the distribution in figure 3.1).[9] They occur less often in Paleolithic sites in East and Southeast Asia, although examples are reported from Java, southern China, and Korea, where the spread of the Acheulean appears to have faced some resistance from *Homo erectus* populations who continued using Oldowan technology. In the mid-twentieth century, archaeologists even wrote of a "Movius Line" (named after American archaeologist Hallam Movius) that indicated a boundary to the spread of hand axes beyond eastern India. This Movius Line now seems to have been breached in several places, although the main distribution of the Acheulean is still thought to have terminated essentially in central and southern Asia (as shown in figure 3.1). Acheulean tool makers never quite colonized the whole of the hominin world.

Figure 4.3. Paleolithic stone tools. (*A*) One side of an Acheulean bifacial hand axe of flint from the Thames Valley, England. (*B*) Replica of a Levallois "tortoise" core of English flint with a large flake removed. (*C*) Cast of a Mousterian side scraper (with the retouched scraping edge at the bottom) made on a thick flake from Combe Grenal Cave, Dordogne, France. (*D*) Upper Paleolithic prismatic blade and blade core of English flint. All from the collections of the School of Archaeology and Anthropology, Australian National University. Item *B* was made by Mark Newcomer (Institute of Archaeology, London) in 1972. Item *C* was donated by the late François Bordes. Photos by Maggie Otto.

Was the Acheulean a great breakthrough in technological efficiency? I doubt it. Acheulean hand axes were not necessarily any more efficient at cutting and chopping than Oldowan pebble tools, and both industries retained a use of similar flake tools. But Acheulean hand axes would have been easy to resharpen by flake removal from their edges, and they had a recognizable style and shape, certainly visible to the archaeologists who have discovered thousands of them during the past two centuries. In other words, they can be regarded as a coherent cultural tradition, not a result of chance.[10] Were they a badge of membership developed during a new hominin spread? Perhaps they were. The importance of identity and group awareness among early hominin populations should not be taken lightly, even if it is impervious to direct demonstration.

We do not know where the Acheulean originated, although I would bet on East Africa, where the oldest reliable dates seem to exist so far. But—and this is a big "but"—the evidence does not insist on an African origin. It could possibly have been in Eurasia, in a region such as the Levant or South Asia, where Acheulean tools are also found in large numbers.

Currently, there are claims for a presence of populations using Acheulean tools between 1.7 and 1 million years ago in East Africa (for instance, at Olduvai Gorge), the Levant, and southern India, hence solidly within the Early Pleistocene and potentially within reach of early *Homo erectus*. However, the absence of hand axes in layers as young as 850,000 years ago in the dense remains of *antecessor* occupation in Gran Dolina at Atapuerca calls for caution about chronology, at least outside Africa and the Levant. The Dmanisi people at 1.7 million years ago also used Oldowan, not Acheulean tools, as did most populations of *Homo erectus* in eastern Asia. All in all, I see little strong evidence for any widespread occurrence of Acheulean stone tool technology outside Africa before one million years ago.

Although the correlation is not perfect, the overlap in time between the appearance of the large-brained non-*erectus* hominin populations just described and Acheulean stone tools outside Africa appears not to be coincidental. However, whatever its merits during the earlier part of the Middle Pleistocene, the Acheulean did not continue to be the stone tool industry of choice for the three species we are now about to examine.

The "Big Three" Species of the
Later Middle Pleistocene

We are now approaching one of the major questions to be addressed in this book. How did *Homo sapiens* emerge? To answer it, we need to examine the three widespread and quite closely related species that dominated the later part of the Middle Pleistocene across much of Africa and Eurasia, and then ask, "Where did *they* come from?" Each of these three is of crucial importance in the debate about the genomic and paleoanthropological origins of early *Homo sapiens*.

The three major species of concern are the Neanderthals in Europe and western Asia, the enigmatic "Denisovans" in Siberia and East Asia, and early *Homo sapiens* itself, located initially in Africa according to the fossil record and DNA molecular clocks. Indeed, there might also have been a fourth species in the form of the newly announced Harbin human group in China, but I return to this issue later. For the time being, I will stick with a "big three." With them, for the first time in the course of human evolution, ancient DNA provides exciting information on ancestry and population mixture.

Molecular clock calculations based on DNA comparisons between all three make it clear that they descended from a common ancestral population that lived between 700,000 and 500,000 years ago, perhaps somewhat earlier from the perspective of cranial morphology, albeit with a wide range of chronological latitude (figure 4.1). The first genomic split in the trail of common ancestry separated *Homo sapiens* and an ancestral Neanderthal/Denisovan lineage, presumably because the latter departed from Africa and entered Eurasia, while ancestral *sapiens* stayed in Africa. Neanderthals and Denisovans then separated into two species as they spread across Eurasia.

In terms of their archaeology, this two-way genomic split between *sapiens* and the combined Neanderthals and Denisovans was associated with stone tool industries that postdated the Acheulean and presumably evolved from it. They are known as the Levalloisian and the Mousterian, and I discuss them further below.

The Neanderthals

Having mentioned Neanderthals[11] so frequently, it is now time to introduce them in more detail (see figure 2.5, numbers 4 and 7 from the top, and figure 4.4). British paleoanthropologist Chris Stringer has described them as follows:

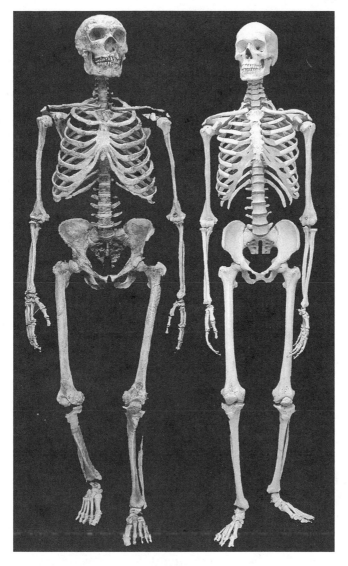

Figure 4.4. A reconstructed Neanderthal skeleton (*left*) alongside that of a modern human (*right*). Note especially the differences in shape of the rib cages and pelvic girdles. Reproduced by courtesy of the American Museum of Natural History Division of Anthropology.

Neanderthals had short, stocky bodies with large heads dominated by an enormous and projecting nose, while the cheekbones were swept back. It seems likely that the large front teeth were used as a clamp for food, toolmaking or skin processing. Above the eyesockets there was a marked brow ridge, very well developed in the centre but reduced at the sides. The cranial vault was long and low, but very broad, and almost round when viewed from behind . . . Neanderthals—and even their children—were very muscular. To judge from their body proportions . . . they may have been adapted to life in cold conditions.[12]

As stated, it is likely that the Neanderthals evolved from a hominin population that was spreading through western Eurasia around 700,000 to 500,000 years ago, splitting from the eastward-moving Denisovans in the process. The "classic" Neanderthals as described by Stringer existed from around 250,000 years ago and onward, but older members of the lineage are known from an amazing discovery.

The Atapuerca cave complex in northern Spain also contains, besides the Sima del Elefante and Gran Dolina caves discussed above, a 13-meter-deep shaft called Sima de Los Huesos, or Pit of the Bones. At its base, archaeologists have excavated more than 5,000 hominin bones, representing at least thirty individuals thrown into the shaft as whole bodies about 430,000 years ago, according to uranium-series dates on associated deposits of calcite. They belonged to a species that was clearly part of the ancestry of the Neanderthals, and their Neanderthal rather than Denisovan genomic affinities (except for their mitochondrial DNA) have been confirmed by ancient DNA analysis.[13] The Sima population is also the oldest in hominin evolution for which hyoid bones survive, a hyoid being an unattached U-shaped bone in the neck that supports the tongue and makes human speech possible. The Sima hominins could perhaps have produced a similar range of sounds to us, at least to some degree.

The Sima hominins had brain capacities that varied between 1,100 and 1,390 (average 1,250) cubic centimeters (figure 2.5, fourth row), only a little below the range for classic Neanderthals and early Homo sapiens. Their body height ranged up to 180 centimeters for large males, with weights possibly up to 100 kilograms. Lucky individuals lived to about thirty-five years. The Sima de los Huesos teeth reveal considerable wear, but no plant remains or animal bones indicative of diet were found in the site, apart from those of cave bears which would have died naturally in the cave. Neither were stone tools found, except for a single Acheulean hand axe that was dropped into the pit as the bodies were accumulating.

Sima de Los Huesos was clearly not a habitation site, and the evidence has suggested to many archaeologists a deliberate disposal of the dead 430,000 years ago in an uninhabitable (and perhaps unreachable, for them) womb-like orifice in the ground, with a deliberate hand axe offering. One individual had been killed before burial by deliberate blows to the skull, but the bones had no cut marks that might indicate cannibalism.[14] No one ever lived or dined in this cave, but the Sima hominins reveal to us some remarkable evidence for behavior that was perhaps fueled by a belief in something that lay beyond their immediate sensory world.

After 250,000 years ago, many sites across the unglaciated parts of Europe and Asia have revealed the remains of the classic Neanderthals, spread across a vast area that extended for 8,500 kilometers from west to east and 2,500 kilometers from north to south. Neanderthals lived from the British Isles to Gibraltar, and across Eurasia from France, through eastern Europe, to Israel, Iraqi Kurdistan, the Caucasus, Uzbekistan, and the Altai Mountains of Siberia, as shown in figure 5.1. Through their bones we witness the expansion of a single well-defined species across a huge geographical region, with skeletal remains from at least seventy locations, some more complete than others, excavated since the nineteenth century and representing almost 300 individuals. Partly because of the preservation of human remains in sheltered caves, whether placed there intentionally or not, the Neanderthals have become a gold mine for paleoanthropologists, archaeologists, and geneticists alike.

What bounded the Neanderthal world? To the south, the Mediterranean Sea and the arid Sinai Peninsula blocked their access to Africa, and there are no Middle Pleistocene traces of their presence in that continent, nor yet in fully tropical regions of Asia. Since the Last Glacial

Maximum, Neanderthal genes have been carried into northern Africa by modern human migrants from Eurasia, but these were a kind of "backwash" and do not imply an original presence of Neanderthals in that continent.[15] Indeed, many of the living populations of Sub-Saharan Africa lack these Neanderthal genes altogether.[16] To the north, beyond 55° latitude, cold probably kept them away, although there have been controversial suggestions from the archaeological record that some might have reached the Arctic coastline of western Russia, close to the northern end of the Ural Mountains. Elsewhere, the fossil and DNA records reveal a presence of competitors—surviving *erectus* populations in Java, and the mysterious Denisovans in much of the region from Siberia to Indonesia.

The Denisovans and the Harbin Human Group

In 2010, geneticists in the Max Planck Institute for Evolutionary Anthropology in Leipzig announced that ancient DNA had been extracted from a finger bone of a young girl excavated within Denisova Cave in the Altai Mountains of southern Siberia. The DNA analysis indicated that the girl belonged to a hitherto-unknown hominin species related to Neanderthals, but not identical to them. Subsequent analysis also revealed that some modern populations, including Philippine hunter-gatherers and Indigenous Australians and Papuans, still carry today up to 3 percent of Denisovan DNA as a result of interbreeding events that probably took place around 50,000 to 45,000 years ago.[17] Clearly, modern humans and Denisovans were interfertile, just as were modern humans and Neanderthals farther west.

More Denisovan bone fragments belonging to four separate individuals have since been identified in Denisova Cave, which has also yielded evidence for a contemporary presence of Neanderthals. Recent luminescence dating of the sediments in the cave by Zenobia Jacobs and colleagues, and radiocarbon dating by Katerina Douka and colleagues, combined with ancient DNA extraction from the sediments, indicates that the Denisovans were in occupation between 250,000 and 50,000 years ago, with the Neanderthals overlapping them from about

200,000 years ago.[18] Both groups used similar kinds of "Middle Paleo-lithic" rather than Acheulean tools, in the Levalloisian and Mousterian categories that I describe further in the next section.

The greatest surprise of all came when the DNA extracted from a bone fragment that belonged to a Denisova Cave teenage girl indicated that her mother was a Neanderthal, and her father a Denisovan with some recent Neanderthal ancestry.[19] This remarkable discovery repre-sents a catching-in-the-act of a hybridization between two species who were presumably differentiated by almost half a million years of separate evolution. Somehow, around 90,000 years ago, Neanderthals and Deniso-vans were getting together in this chilly corner of Siberia—the temperature in Denisova Cave oscillates around 0°C through much of the year.

So far, Denisovans are known mainly from the few bones recovered from Denisova Cave itself, and from the ancient Denisovan DNA that survives in its soil deposits. In addition, part of a 160,000-year-old Den-isovan mandible together with more Denisovan sediment DNA comes from Baishiya Cave on the Tibetan Plateau, in Gansu Province of China, located at an altitude of 3,280 meters above sea level. This mandible has been identified as closely related to Denisovans from amino acid se-quences in its proteins because the jawbone contained no ancient DNA.[20]

Baishiya Cave lies almost 3,000 kilometers southeast of the Altai Mountains, and its high altitude implies a Denisovan ability to inhabit low-oxygen (hypoxic) atmospheric situations. The Baishiya Denisovan might thus have been a member of the highest-altitude hominin popula-tion on earth when it was alive, and it possibly arrived there by following selected prey animals to ever-higher altitudes.

The still-tiny haul of Denisovan bones means that this species cannot yet be given a binomial (Linnaean) name with a precise skeletal defini-tion. However, there was possibly more than one Denisovan genetic population, and possibly even separate species of them. The evidence comes from new research on the surviving segments of Denisovan DNA in the chromosomes of living Asian people, including Island Southeast Asians and Melanesians. It suggests three possible instances of interbreeding between modern humans and Denisovan popula-tions, occurring separately in South Asia, the Altai region, and Island

Southeast Asia with New Guinea.[21] In the last case, it has been suggested that Denisovan mixing with *Homo sapiens* could have continued until almost the end of the Pleistocene.

The Denisovans in eastern Asia currently present us with some of the greatest mysteries in human evolution, partly because up until now they have had no skeletal definition beyond a few tiny bones and one mandible. It seems unlikely that any of the fossil remains identified as *Homo erectus* in eastern Asia could belong to Denisovans or vice versa because *erectus* and the Denisovans were apparently derived from separate hominin migrations into Eurasia around one million years apart. This means that these two species must have been deeply different by the time they met during the Middle Pleistocene.

A more likely paleoanthropological candidate for Denisovan identity walked onto the stage just as I was putting the finishing touches to this chapter. It has long been known that China harbors a large and impressive collection of Middle Pleistocene hominin remains. Access to them by Western scholars has until now been difficult, so they have rarely featured in global analyses, except for *Homo erectus* from Zhoukoudian and other Chinese sites. That picture has just changed dramatically with new publications by Chinese and Western paleoanthropologists working in combination.

A large-brained Middle Pleistocene potential new species, referred to as "the Harbin human group" in the main report—but more daringly as *Homo longi* ("Dragon Man") in an associated report with all-Chinese authors—has just hit the headlines.[22] The main exhibit for Dragon Man was buried by its finder down a well in 1933 in Harbin, Heilongjiang Province, to hide it from an invading Japanese army. It is a cranium with separate brow ridges over each eye socket, a large brain size of 1,420 cubic centimeters, and a uranium-series age that exceeds 150,000 years. Similar Chinese specimens have been reported from other sites, including Dali, Jinniushan, and Hualongdong (see locations in figure 3.1).

The authors of the report state that the Harbin human group was perhaps a sister species to the Denisovan mandible from Baishiya Cave, but certainty is elusive because the Denisovans are known mainly from their ancient DNA, whereas the Harbin group is known only from crania. Dragon Man cannot be stated definitely to have been a Denisovan, but the obvious possibility has not been lost on commentators.

At 45° north, Harbin lies at a similar latitude to the Altai Mountains and Denisova Cave. It also has winter temperatures that can drop far below 0°C. Like the Neanderthals, and the Altai and Baishiya Denisovans, the Harbin human group could either withstand extreme cold, or perhaps it had access to fire and warm skin clothing. Perhaps also it had the capacity to head south to warmer climes during the winter. Another interesting observation about the Harbin group is that it falls close in its cranial morphology to *Homo antecessor* from Spain (figure 4.1, top). Explaining why this should be so is for the future to determine.

What global paleoanthropology will decide eventually about Dragon Man is not yet clear, but my suspicion, with *Science* writer Ann Gibbons,[23] is that the Denisovans and the Harbin human group will eventually be recognized as one species. What will it be called? Chinese scholars have already put forward the name *Homo longi*. Russian scholars might prefer *Homo altaiensis*. Will that species also include *Homo antecessor* far to the west? We will have to wait and see.

Neanderthals and Denisovans: Braving the Cold and Painting the Walls?

The high density of Neanderthal fossils and archaeology dating between 250,000 and 40,000 years ago offers paleoanthropologists and archaeologists a remarkable picture of a hominin lifestyle that was clearly sentient and human, but not directly ancestral to the behavior of *Homo sapiens*. We cannot yet say much about the Denisovans or the Harbin group because of the small quantity of relevant skeletal and archaeological evidence. However, both Neanderthals and Denisovans mated successfully with *Homo sapiens* in various locations in Asia. We return to these instances again in the next chapter because they illuminate the deeper prehistory of our own species.

As far as lifestyle is concerned, we are restricted to the Neanderthals for good data. Archaeologists have for many years compared reconstructed Neanderthal behavior with the complex behavior that characterized early

Homo sapiens, especially in terms of language ability and capacity for conceptual thought and symbolic behavior. Not surprisingly, perhaps, the Neanderthals have tended to come off unfavorably, partly because modern humans replaced them. Sometimes, however, they enjoy a more sympathetic press.

In recent years, for instance, there have been suggestions that Neanderthals were capable of behaving like *Homo sapiens* and producing stone and bone artifacts similar to those of the so-called Upper Paleolithic, which I discuss in chapter 5. As examples, there is evidence (albeit debatable) from various locations in Europe that Neanderthals sometimes used blade tools of the Upper Paleolithic type, and that they could decorate cave walls with hand stencils made by blowing red ocher paint through the mouth around their hands and outspread fingers. Evidence also exists for hunting birds and using eagle feathers for symbolic activities, and for constructing circular stone and mammoth bone arrangements in both caves and open sites.[24] An eloquent plea for a high level of complexity in Neanderthal hunting, subsistence, and artistic behavior that might have approached that of *Homo sapiens* has recently been made by archaeologist Rebecca Wragg Sykes.[25]

Did the later Neanderthals sometimes adopt or copy modern human cultural habits? The instances that lie behind claims for actual copying are few in number, but if Neanderthals were capable of emulating *Homo sapiens* and breeding with them, even if only occasionally, it must mean that their patterns of behavior were not that much different from ours. There is considerable food for thought here about the complexity of Neanderthal behavior.

Indeed, all is not lost for the Neanderthals as members of a large-brained club of intelligent Middle Pleistocene hominins. Like us, they had hyoid bones in their throats for speech, and a slightly different version of a gene called *FOXP2*, which is associated with speech ability in living humans. These possessions alone do not necessarily guarantee a *sapiens* level of speech ability and sentience, but a few Neanderthal partial skeletons have been found in caves in France, Israel, Syria, and Iraqi Kurdistan with their bones still in articulated positions, suggesting deliberate burial.[26]

Neanderthals were hunters of medium to large herbivorous mammals, users of fire, and lived mainly in Eurasian middle latitudes where the food sources available would have changed markedly from glacial into interglacial periods. In Europe, reindeer, musk oxen, and woolly rhino were present during glaciations, and warmer climate species such as fallow deer, wild cattle (aurochs), and wild sheep were present during interglacials. A recent study of Neanderthal bones indicates that the males suffered frequently from injuries similar to those found on the bones of Paleolithic *Homo sapiens* male hunters, suggesting similar ways of demonstrating their hunting prowess.[27]

There are also organic remains that suggest developed hunting skills. For example, eight 2-meter-long sharpened wooden spears of spruce and pine, dating from about 300,000 years ago, have been excavated with bones of extinct horses in a lignite mine at Schöningen in Germany.[28] An early Neanderthal would have been a likely maker and user of these. Such hunting weapons could have allowed this population to accumulate larger supplies of meat, and thus to support larger cohabiting and cooperating social groups. An increasing availability of meat—especially cooked meat, because use of fire is now attested archaeologically—could also have improved maternal health and increased the birth rate, at least prior to the final phase of the Neanderthal population decline that I discuss in chapter 5.

In terms of developments in stone tool technology, the Neanderthals and Altai Mountains Denisovans were associated with a stone tool industry that archaeologists term the Levalloiso-Mousterian, which was in part continuous from the Acheulean, especially in its continuing use of small hand axes. We also see within this industry the appearance of a new kind of flaking technology that archaeologists have called the Levallois technique, named after another suburb of Paris where such tools were first recognized. Yet again, many major Paleolithic stone tool industries have French names because French scholars first realized their significance. Le Moustier is also in France, in the Dordogne region.

The Levallois technique is basically a method of predesigning a flake by first shaping a core into a humped tortoiseshell shape, then preparing a striking platform on one end, and finally removing a large flake of a preconceived shape and size. The cores are often referred to by archaeologists as "tortoise cores" or "prepared cores" (figure 4.3B). The technique works well on fine-grained stone such as flint, and many archaeologists regard it as of great significance. However, it was in reality only an advance over what was already well known to earlier hominins, this being that a sharp flake of a desired shape and size would most easily be produced if the core from which it was struck also carried the required shape.

Prepared core flaking is found widely across Africa and Eurasia after about 300,000 years ago, but the source region of the technique is unclear. It could have been in Africa, where early dates exist for similar prepared core tools at Olorgesailie in Kenya and Jebel Irhoud in Morocco (the term "Levallois" is not commonly used in Sub-Saharan Africa), or in the Levant, or even farther east in Asia. Recently, a Levallois presence has been detected in Asian locations as far apart as the Altai Mountains, China, the Caucasus, and India.[29] Like the Acheulean, however, it did not flourish in Southeast Asia, where the Oldowan techniques favored by *erectus* continued until the end of the Pleistocene.

Because prepared core techniques are found in association with Neanderthal, Denisovan, and early modern human communities, through such wide areas of Africa and Eurasia, it is perhaps pointless to ask which species, if any, actually invented them. It might be better to think of the Levallois concept as a useful idea that, once invented, would simply have spread by itself through large-brained Middle Pleistocene populations who were in contact, with occasional interbreeding.

What about the Other Middle Pleistocene Hominins?

What else was going on during the later part of the Middle Pleistocene, between 300,000 and 120,000 years ago? In Southeast Asia, the tiny hominins of Flores and Luzon continued in existence until the arrival of *Homo sapiens*, as perhaps did *Homo erectus* in Java. It is not clear when *erectus* populations on the Asian mainland gave way to Denisovans or the Harbin human group, but Indonesian *erectus* ("Solo Man") was still in existence, now with a brain size approaching 1,200 cubic centimeters, as recently as 100,000 years ago in the Solo Valley of central Java. The

Solo population was presumably replaced by *Homo sapiens* around 50,000 years ago.[30]

Whether this increase in *erectus* brain size in Java reflected internal processes or the results of interbreeding with other mainland Asian hominins is unknown. As noted, however, although traces of the Acheulean hand axe tradition occur occasionally in East Asia and Java, the long-lived *erectus* populations of that island continued to use mainly Oldowan tools.[31] This suggests that the brain size increase in Java might have occurred independently of that elsewhere in Eurasia.

On the other side of the Old World, in southern Africa, there was another mysterious species termed *Homo naledi* that also managed somehow to survive until about 300,000 years ago.[32] The remains of at least fifteen individuals of this species have been found relatively complete within two deep shafts in the Rising Star cave system near Johannesburg in South Africa, but without any stone tools. The deposition of so many individuals in deep and secluded cave shafts inaccessible to large scavengers provides plenty of food for thought. Are we witnessing another example of human burial behavior like that at Sima de los Huesos, perhaps reflecting some degree of kinship awareness?[33]

Homo naledi is also a mystery because of its small brain size (460 to 610 cubic centimeters), even smaller than the Dmanisi brains dated to over a million years before. That small brain sat atop a more normal (but still fairly small) Middle Pleistocene body averaging 1.5 meters in stature, weighing around 40 to 55 kilograms, with bipedal posture and hands like other species of *Homo*.

Homo naledi carries the hallmarks of being an unusual and long-isolated species, like *floresiensis* and *luzonensis* in Wallacea. How could this be in the apparent hominin hotspot of southern Africa? Could the dates be incorrect? This seems unlikely. Could they be hybrids between *sapiens* and a smaller-brained archaic hominin? This also seems unlikely because there is no evidence that any hominin species with a sufficiently small australopithecine-like brain still existed in Africa as recently as 300,000 years ago. The Naledis were apparently contemporary with early modern human populations in Africa that had brain sizes twice as large as they did.

I think we are left with only one conclusion. The Flores, Luzon, and Naledi small-brained but recently surviving hominins indicate that the genus *Homo* was capable of forming and maintaining reproductively isolated and morphologically diverse species when opportunities arose to do so. Exactly what those opportunities were in the case of the Naledis we may never know. The three small-brained species just mentioned presumably became extinct and contributed no genes to *Homo sapiens*, although we currently lack ancient DNA from any of them to prove this.

Perhaps we should not assume without genetic proof that ancient hominin species, even if capable of interspecies breeding, would always have done so with enthusiasm. Physical appearances, cultural and dietary differences, and other behavioral characteristics might have differentiated Pleistocene hominin species sufficiently to create a pattern that was strangely unlike that produced by the cosmopolitan situations of mixing that we witness between *sapiens* populations today. For archaic hominins, no mixing at all might sometimes have been an option, even when migration brought dispersed species back together after hundreds of millennia of intervening separation.

5

The Mysterious Newcomer

ACT III OF THE ODYSSEY takes us into the age of *Homo sapiens*. Our genetic ancestors can be tracked back in Sub-Saharan Africa to at least 250,000 years ago, in terms of the DNA that exists in living human populations. Comparisons of DNA and cranial morphology among *sapiens*, Neanderthals, and Denisovans pushes back our deep ancestry much farther, perhaps beyond 700,000 years ago, as I discussed in chapter 4 (and see figure 4.1). However, *Homo sapiens* at this much deeper time depth is not yet visible separately from other hominins in paleoanthropological or archaeological contexts.

Once on the stage, *Homo sapiens* populations were not slow to migrate. This chapter continues onward to the settlements of Eurasia and Australia by a date that is actually rather uncertain (figure 5.1). Somewhere between 70,000 and 50,000 years ago is a current best bet, or perhaps 60,000 to 50,000 years ago. Ranges of error fluctuate rather wildly at such time depths. In actuality, we will find that the observations made by archaeologists, paleoanthropologists, and geneticists do not agree on the date of initial *sapiens* migration out of Africa, or for that matter into Australia. Understanding the origins of *Homo sapiens* is one of the most debated issues in archaeology, paleoanthropology, and genetics at the present time. In this chapter I explain why that is so.

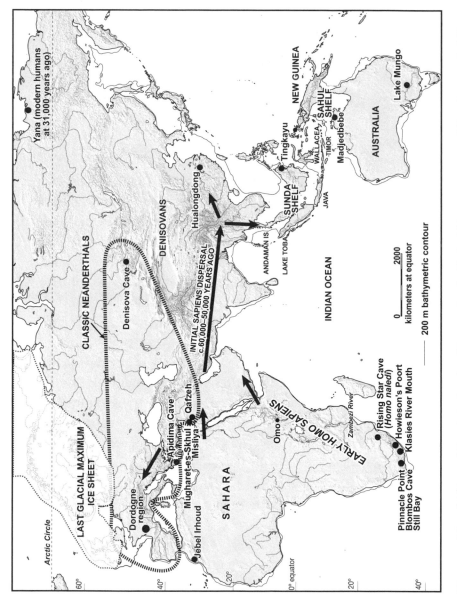

Figure 5.1. Map of Africa and Eurasia to show localities associated with the African emergence and Eurasian spread of *Homo*

Here Comes *Homo sapiens*

When I used to lecture to undergraduate students about the origin of our species, I would go to the laboratory of my late colleague Colin Groves and take out some casts of skulls, some shown in figures 2.3 and 2.5. I would set them up in a line on the table and explain to the students which was who, and why—all the way from chimpanzees and gorillas (neither of which, of course, represented direct ancestors of hominins), through australopithecines, Early Pleistocene *Homo*, *Homo erectus*, large-brained Middle Pleistocene *Homo*, Neanderthals, and onward to *Homo sapiens*.

In undertaking this exercise I stressed three points about how *Homo sapiens* differed from other hominins. These were possession of a smooth forehead rather than prominent ridges above the eyebrows, a projecting rather than a receding chin, and a high skull profile that was parallel-sided above the ears when viewed from the rear, rather than widest at the ears as in most extinct hominins. In other words, *sapiens* brain cases are high and round (globular), and this is a major definitive character of the most ancient members of the species in East Africa. Those of the other extinct hominins are long and low.

Other pointers for *Homo sapiens* included relatively small jaws and teeth, a face that did not project as far forward as it did in extinct hominins, a foramen magnum (the hole in the base of the skull that connects the brain with the spinal cord) in a position that indicates fully upright posture, and a rounded occipital region (back end) to the skull.[1] Some of these features are clearly visible in figures 2.3 and 2.5.

Such is *Homo sapiens*, the "anatomically modern human." If one sits a modern human skull alongside a Classic Neanderthal skull from a European cave, the differences are clear (figure 2.5, bottom two rows).

For another class I would lay out stone tools and other artifacts. Out would come the categories that I discussed in chapters 3 and 4 for the extinct hominins, especially from the "Lower Paleolithic" Oldowan and Acheulean industries, and the "Middle Paleolithic" Levallois and Mousterian industries associated with Neanderthals (figure 4.3C). Pebble tools worked around one end, bifacial hand axes, tortoise cores, and retouched (i.e., heavily used or deliberately sharpened) scrapers, useful for cleaning animal skins, would decorate the table.

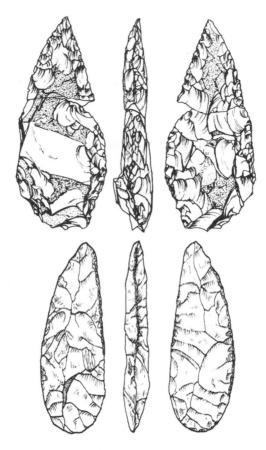

Figure 5.2. Two bifacial tools from Tingkayu, Sabah, Malaysia (see pages 110–111). The upper one (12.5 centimeters long) is unfinished because it broke during manufacture. The lower one (11 centimeters long) appears to have been used as a knife rather than as a point, according to the damage on its edges. Compare with the similar bifacial tools from Japan and North America in figure 6.4. Drawn by Lakim Kassim, Sabah Museum, Malaysia.

Then came an apparent sharp break in the procession of artifacts. *Homo sapiens*, at least in Europe, was usually hailed as much more interesting than the extinct hominins in terms of its handiwork, which included long stone blades struck from cylindrically shaped cores (figure 4.3D), delicate bifacially flaked stone knives and spearheads (figures 5.2 and 6.4), "Venus" figurines, sometimes made of mammoth ivory, and an occasional barbed bone spear point or eyed bone needle. Other creations claimed to be trademarked by *Homo sapiens* included engraved or painted art on cave walls, intentional burials with grave goods and red ocher coloring, and perforated beads and pendants of shell and bone.

Regardless of views about where and how *Homo sapiens* evolved, it was believed during much of my student and working life that this species was the sole maker and owner of a new-fangled "Upper Paleolithic" artifact assemblage that included the above items. Even though it was never clear where the Upper Paleolithic originated, its importance always seemed obvious in Europe, especially in France and Britain, where Paleolithic archaeology came of age a century or more ago.

Nowadays, new finds are turning much of this old wisdom on its head, especially in regions far beyond western Europe. *Homo sapiens* was not simply an appendage of the Eurasian Upper Paleolithic, at least not just in terms of stone tools, even though many *sapiens* populations were associated with it after 47,000 years ago. As we will soon see, the cultural roots of early *sapiens* ran much deeper. The problem is that untangling these roots is by no means an easy process. Let me outline the situation before we go further.

The Riddle of Early *Homo sapiens*

Homo sapiens evolved in Africa prior to 300,000 years ago, and at some point escaped from that continent to settle Eurasia and Australia. But when, and with what cultural attributes? This might seem like an easy question to answer, but it is not. The record from ancient skulls hints that *Homo sapiens* in some form might have left Africa as much as 200,000 years ago. Genetic molecular clocks contradict this possibility, pointing instead to successful out-of-Africa migration, ancestral to existing human populations beyond that continent, between only 70,000 and 50,000 years ago, most probably toward the later end of this time span.

The archaeological record of Upper Paleolithic stone tools could adjudicate the problem if there happened to be a 100 percent association between its appearance and the skulls of the first *Homo sapiens*, both within and beyond Africa. Alas, there is no such association. Instead, the oldest modern human cranial remains from Southeast Asia and Australia are associated with Middle Paleolithic core and flake tools similar to those made by Neanderthals, Denisovans, and other non-*sapiens* hominins. There is simply no Upper Paleolithic stone tool signature that is universally attached to the oldest modern *sapiens* skeletal remains.

I will not labor these points further here. The time has come to examine the evidence piece by piece.

The Emergence of *Homo sapiens*: Skulls and Genes

The archaeological collections of Africa, Eurasia, and Australia contain hundreds of Late Pleistocene skulls that belong to our modern species, *Homo sapiens*. All that are fully modern in their cranial anatomy are less than 60,000 years old, mostly much less. But what happened before 60,000 years ago, especially in Africa, the continent which both the fossil record and genetic comparisons identify as the *sapiens* homeland? Here, we should surely find older roots.

Paleoanthropologists have identified ten or more sets of skeletal remains from sites in Morocco, Sudan, Ethiopia, Tanzania, Zambia, and South Africa that have *sapiens*-like features in their crania and faces, and that date between 300,000 and 90,000 years ago, hence well before the appearance of full *sapiens* cranial modernity. Somewhere within this ancient and rather diverse group, presumably there existed the ancestors of all humans alive today, although the remains themselves are often fragmentary and of uncertain date. Many show signs of overlap with non-*sapiens* hominins in their cranial profiles, large teeth, and prominent brow ridges.

Current genetic research adds more to the perspective from paleoanthropology. As discussed in the previous chapter, the molecular clock dates for the evolutionary separation between *sapiens*, Neanderthals, and Denisovans go back to around 700,000 years ago. However, molecular clocks applied to the DNA diversity in living human populations give a much younger common ancestry for all of us alive today, back to only about 300,000 years ago in terms of whole genome and mitochondrial DNA comparisons between southern African populations and people elsewhere in the world.[2]

These molecular clock dates roughly match the 300,000-year chronology for *sapiens* from the ancient skulls. Prior to this time, neither paleoanthropology nor genetics as yet provide an unequivocal picture for the origin of *Homo sapiens*. An ancestral *sapiens* population must have existed somewhere in Sub-Saharan Africa before 300,000 years ago. But where? Was there even a single place of origin at all, or did *sapiens* emerge from interactions between contributor populations located across much of Sub-Saharan Africa? Nowadays, many paleontologists and geneticists think that this is what happened.

Genetic observations on living humans that could illuminate a homeland shine the spotlight on a number of deeply indigenous Sub-Saharan populations. They include the San hunter-gatherers, formerly known as Bushmen, of the Kalahari region of southwestern Africa, together with their Khoekhoen pastoralist relatives. Further north are the Sandawe and Hadza hunter-gatherers of Tanzania (located in figure 12.1), and small groups of rainforest hunter-gatherers in the Congo Basin. These groups exist nowadays as minorities among Bantu-speaking farmers, whose ancestors began to spread from West Africa through most of Sub-Saharan Africa from about 3,000 years ago (chapter 12).

A striking observation about these non-Bantu populations is that the genetic mutations in their mitochondrial DNA lineages have a deeper antiquity than those in all other human populations, both within and beyond Africa. This was the first major discovery of the modern era of DNA research into human population history, published in 1987. The mitochondrial lineages of all living people in the world trace back to a mutation that occurred with the conception of a female widely known in the popular media as "Mitochondrial Eve," or "African Eve."[3] All of us descend from this illustrious mother, who lived around 160,000 years ago, at least in terms of our mother-to-daughter-inherited mitochondrial DNA. Eve's original mitochondrial lineage group, termed Lo (L zero) still exists among the San and the Congo Basin hunter-gatherers today, and indeed in many other African populations.

Based on their mitochondrial and nuclear DNA profiles, these non-Bantu African populations represent the oldest in-place human populations in the world.[4] However, I must stress one point strongly lest there be misunderstanding. "Oldest" in this context merely means that the ancestors of these African populations have migrated the least during prehistory; to put it another way, they have stayed at home the longest and hence retained the oldest traceable mutations within their genomes.

In terms of the genetic history of humanity as a whole, the San and Hadza are no more "ancient" as a *sapiens* biological population than anyone else because all of us, including the San and Hadza, share mitochondrial lineages that mutated in and beyond Africa as a vast genealogy out of the foundation haplogroup of African Eve. Many of us had ancestors who migrated far and underwent successive and younger episodes of mitochondrial mutation. But some Africans did not—their ancestors stayed at home. Either way, indigenous Africans, Eurasians, Australians, and Americans are all equally modern human.

Since 1987, the hunt for the origin region for the most recent common ancestor for all humans alive today has proceeded apace. One recent research publication by geneticist Eva Chan and colleagues suggests a mitochondrial origin between 240,000 and 165,000 years ago in a region of lakes and wetlands to the south of the Zambezi River, in what is now northern Botswana.[5] This suggestion has received strong criticism,[6] and a separate research program by Mark Lipson and colleagues favors a separation into at least three regional *sapiens* populations in Sub-Saharan Africa by 200,000 years ago.[7] An age of 250,000 years ago has also recently been calculated for the common ancestor of all human Y-chromosome lineages, inherited through males, similar to the ages derived from whole genome and mitochondrial DNA comparisons.[8]

Despite some differences in the chronological (molecular clock) conclusions from these analyses, an identifiable genetic presence of *Homo sapiens* in Sub-Saharan Africa between 300,000 and 200,000 years ago, as suggested also by the paleoanthropology, seems assured. Exactly where that presence was located, and what happened beforehand, are topics still under debate.

Beyond Africa, with a Mystery

According to Harvard geneticist David Reich, the main *sapiens* departure from Africa occurred much later in time, between 54,000 and 49,000 years ago based on genomic molecular clock comparisons between living and ancient Eurasian populations, backed by dated occurrences of interbreeding between incoming *Homo sapiens* and Eurasian Neanderthals.[9] Some other geneticists give slightly broader date ranges for this main out-of-Africa migration, stretching back as far as 65,000 years ago, but molecular clocks generally draw the line around this time as far as movement outside Africa is concerned. Anything older than 65,000 years outside Africa should not really be *Homo sapiens*, at least not genetically. Within Africa, the time depths are greater.

The main problem, however, is that there are dates for early *Homo sapiens* crania from Eurasia that are much older than the out-of-Africa date range provided by the geneticists. One of the oldest, dated to about 210,000 years ago, comes from Apidima Cave in Greece.[10] The Levant has even richer evidence for an early *sapiens* presence, albeit currently a little younger in date than Apidima. A *sapiens* maxilla (upper face) found with Levallois stone tools in Misliya Cave in Israel is dated to between 194,000 and 177,000 years ago. By the time of the penultimate interglacial around 120,000 years ago, early *sapiens* was becoming even more firmly established in the Levant, especially in the form of the human remains preserved in the caves of Mugharet es-Skhul and Qafzeh in Israel.[11]

Among these Skhul and Qafzeh human remains are several intentional burials, of which the most remarkable is a double burial of a young woman from Qafzeh with a 6-year-old child at her feet, both buried in a flexed posture.[12] Some of these Skhul and Qafzeh human remains had red ocher dusting the bones, and some were associated with small perforated shell beads.[13] One individual from Qafzeh was buried with a set of deer antlers on its chest.

These Skhul and Qafzeh early *Homo sapiens* burials are no less than remarkable. They are thought to date collectively between about 120,000 and 90,000 years ago, according to geophysical dating of the cave sediments. Like the Neanderthals with whom they overlapped in time, they used stone tools related to the Mousterian and Levallois technologies of the Neanderthals. The Upper Paleolithic in the Levant had clearly not yet been invented or introduced. But their burial habits, associated with ocher, shell beads, and intentional grave offerings, are perhaps the oldest unarguable examples of such practices in the world. Even in Africa, the presumed homeland of *Homo sapiens*, the oldest intentional human burial known so far, of a young child from Kenya, dates only to about 78,000 years ago.[14]

Like Apidima, the crania of these Skhul and Qafzeh burials are recognizably *Homo sapiens*, but they carry certain features such as prominent brow ridges that overlap with extinct hominin species (a skull from Qafzeh can be seen in figure 2.5). These other species included Neanderthals, who were widespread in southwestern Asia when Skhul and Qafzeh caves were used for burial, as well as a newly announced 126,000-year-old hominin of hotly debated species affinity from Nesher Ramla in central Israel.[15] Were the Skhul and Qafzeh hominins interbreeding with these other Middle Pleistocene populations? It seems likely, even if no ancient DNA has yet been recovered to throw further light on the situation.

These skeletal remains of early *Homo sapiens* found beyond Africa, like Apidima, Misliya, Skhul, and Qafzeh, which are so much older than the genomic date for out-of-Africa movement by *Homo sapiens*, are puzzling. What was going on? So far, they are limited to Greece and the Levant, and older claims for early *Homo sapiens* from China are now doubted.[16] Perhaps more importantly, why are there no surviving traces of these ancient *sapiens* populations in the genomes of living humans outside Africa today?

Paradoxically, there is actually some evidence for interbreeding between Neanderthals and *sapiens* within Eurasia that could date from before 170,000 years ago, but it comes from a one-way mixture of *sapiens* genes into Neanderthals, as indicated in figure 4.1, without a reverse impact that can be traced in living *sapiens* populations.[17] It is almost as if some early *sapiens* populations migrated out of Africa before 70,000 years ago, mixed with Neanderthals, and then simply disappeared in genetic terms. Given the complexity of human behavior at Skhul and Qafzeh, this is one of the greatest mysteries in the genealogy of our ancestors.

Perhaps the emergence of *Homo sapiens* was a more complex process than is currently apparent. In this regard, a current view that accommodates many possibilities comes from paleoanthropologist Linda

Schroeder: "Although single-origin theories still persist . . . a modern human origin scenario that includes multiple episodes of introgression with contributions from Africa and Eurasia, widespread interactions between populations across Africa, and early migrations out of Africa (and possibly back into Africa), seems to be our current best-supported model."[18]

The Emergence of *Homo sapiens*: Archaeology

When archaeologists discuss the Upper Paleolithic, they tend to focus on characteristic stone tools, which bulk large in the archaeological record and can be used to define different cultures. The archaeological associations of the oldest *sapiens* remains in Africa indicate that the cultural evolution of our species was firmly rooted in the Levallois prepared core and flake tool technologies that I described in chapter 4. In Israel, the Misliya, Skhul, and Qafzeh stone tools were also of this type, similar to those made by neighboring Neanderthals and the Nesher Ramla population. More basic stone tool industries, closer to Oldowan predecessors and without the typical Levallois prepared cores, spread with the main out-of-Africa *Homo sapiens* migration into southern China, Southeast Asia, and Australia.

In temperate regions of Africa, and in western and northern Eurasia, these prepared core and flake industries appear to have changed into Upper Paleolithic counterparts, with their diagnostic blade and bifacial tools, as modern humans penetrated into higher and colder latitudes, where tailored clothing, warm shelters, and more efficient hunting equipment would have been in high demand. Not surprisingly, given the genetic evidence for a Sub-Saharan African origin of *Homo sapiens*, the world's oldest appearance of such a blade and bifacial point technology is recorded, so far, in caves in temperate South Africa.

Here, developments in stone tool making occurred well before 50,000 years ago that resembled those of the Eurasian Upper Paleolithic, although in South Africa they are dated earlier in time than in Eurasia, back toward the last interglacial. These developments confront us with our oldest archaeological evidence for the appearance of a fully *sapiens* level of technological expression in terms of stone and bone artifacts. However, in cultural and artistic terms, including a use of ocher and body ornamentation with intentional human burials, we cannot overlook the even older examples of modern human behavior from Israel that I described above. Not everything that came to characterize the behavior of early *Homo sapiens* vis-à-vis the archaic hominins necessarily appeared first in Africa.

Having said this, the South African evidence is still unique and thought-provoking. It comes in the form of blades and flakes with one edge blunted for gripping or mounting ("backed tools"), bifacially flaked spearpoints, small round ostrich shell beads, and engraved pieces of ocher used for making red pigment. Similar examples of such backed and bifacial tools, although not from Africa, can be seen in figures 5.2 and 12.2. These tools and art objects are technologically more complex than those of preceding archaeological assemblages, including those from Skhul and Qafzeh, although, as always, some earlier artifact forms continued side by side with the newer ones. Suggestions have also been made that these people knew the use of the bow and arrow.

Archaeologists in South Africa have found these stone tool assemblages in a number of celebrated caves around the South African coastline, such as Howieson's Poort, Pinnacle Point, Blombos Cave, and Klasies River Mouth (located in figure 5.1). Fragmentary human remains found in the last cave are attributed to early *Homo sapiens*.[19]

Available dates place the main period of manufacture of these stone tool industries between 75,000 and 65,000 years ago, almost 20,000 years before the beginning of the Upper Paleolithic in Europe and the Levant, and well before the main *sapiens* migration out of Africa. Industries of this type were not common in other regions of Africa at that time, although there are some examples of tanged (stemmed, for hafting) spearheads in North Africa, and some backed tools by 60,000 years ago in East Africa.[20] Nevertheless, there is no evidence for any sudden expansion of such stone tool types across the whole African continent or beyond.[21]

Why did such developments occur? The explanation that I favor is that *Homo sapiens*, a tropical creature in origin, moved southward into regions of southern Africa that had relatively cold winters, in the process perhaps meeting other resident hominin species such as *Homo naledi*.

Merely competing with other hominins might not have been a sufficient explanation for the innovation of these new kinds of artifact, but additional adaptations required by cold winters might have tipped the balance. Among the artifacts found, bone awls and needles would have been useful for making and decorating tailored skin clothing for keeping warm in winter. Shell beads and red ocher pigment would have added decorative finesse to that clothing.[22] Efficient projectile points would have been useful for hunting among herds of large mammals in the cool and dry Late Pleistocene climates of southern Africa.

Some other archaeologists also favor the idea that these South African artifact assemblages were adaptations to the requirements of cool or cold climate. For instance, Stanley Ambrose suggests that they developed during a few millennia of severe cold caused by the massive eruption of the Toba volcano in Sumatra around 74,000 years ago.[23] This was one of the largest eruptions of the past half million years, and its volcanic ash output might have led to a darkening of skies for hundreds or even thousands of years, although geomorphologists are not in agreement over the precise consequences.[24] Nevertheless, Ambrose's opinion is strengthened by an analysis of the archaeological record of southern Africa through the Late Pleistocene, which shows that the quantities of body ornaments, reflective of a use of clothes and the tools needed to make them, were greatest during the coldest recorded periods in the regional climatic record, one of which occurred around 75,000 years ago.[25]

In Eurasia, as in Africa, cold climates and hominin competition with Neanderthals and Denisovans might also have propelled populations of *Homo sapiens*, independently and perhaps on more than one occasion, toward the invention of Upper Paleolithic backed and bifacial tools, but here only after about 47,000 years ago. Such developments never occurred in the tropical climates of Southeast Asia, where cold weather was not a relevant issue.

The Upper Paleolithic in Eurasia

In Europe, the "classic" sequence of Upper Paleolithic industries that has dominated archaeological textbooks for the past century or more is named after specific archaeological sites in France, as with the preceding phases of the Paleolithic. For the Upper Paleolithic, the names come mainly from limestone caves located in the Dordogne region of south-central France. These industries are defined by differences in their stone and bone artifacts. One of the oldest to have blade tools was the Aurignacian, which commenced in Europe and the Levant around 45,000 years ago. There was also a brief preceding "Initial Upper Paleolithic," with a continuing Levallois stone tool element, recognized at about 47,000 years ago in Bulgaria.[26]

The Initial Upper Paleolithic and Aurignacian were followed by the Gravettian, which commenced across Europe with an emphasis on backed tools about 35,000 years ago. This was followed by the Solutrean with its finely made bifacial spear points, restricted to France and northern Spain during the Last Glacial Maximum, around 20,000 years ago, when much of Europe to the north was unoccupied. With the onset of postglacial warming, the Magdalenian, famous for its cave art and carved figurines, spread across much of Europe between 17,000 and 12,000 years ago. Ancient DNA evidence indicates that not all of these populations transmitted DNA through to the present, which suggests episodes of replacement as different populations moved around in response to ice sheet advances and retreats.[27]

In the higher latitudes, people making Upper Paleolithic stone tools eventually spread across most of Eurasia. These tools enabled humans to penetrate Siberia and northern China by 42,000 years ago, Japan by 38,000 years ago, and finally the Americas by 16,000 years ago. I return to these remarkable expansions in the next chapter. Such tools also reached South Asia and tropical Sri Lanka, where small backed stone artifacts made an appearance in caves soon after 40,000 years ago.[28]

There is another point to emphasize. As well as specific kinds of stone tools, there are many other aspects of life that we associate with *Homo sapiens*, such as the remarkable cave and portable art, the deliberate and sometimes highly decorated burials, and the making of body ornaments, skin clothing, and perhaps even woven textile clothing. Currently, the oldest dated cave art in the world, a hunting scene recently announced as 44,000 years old through uranium-series dating of its encasing stalagmite skin, comes from the island of Sulawesi in

Indonesia—not Africa, France, or Spain.[29] It does not occur with Upper Paleolithic stone tools. Intentional burials and body ornaments are also reported from Middle Paleolithic contexts by at least 90,000 years ago in Skhul and Qafzeh caves in Israel, as I discussed above.

Indeed, the inflated significance of Upper Paleolithic stone tool technology for identifying a presence of *Homo sapiens* in the archaeological record becomes even clearer when we consider ethnographic populations—for instance, recent New Guineans. This island lacked Upper Paleolithic tool types throughout its prehistory, right into the twentieth century in the case of the New Guinea Highlands. However, we only have to look at the incredible richness of the ethnographic record in New Guinea to understand how insignificant this absence of blade and tools was in a tropical location.

New Guinea Highland societies, as described by anthropologists at European contact, exhibited remarkable complexity in social organization, ritual activity, art and body ornamentation, agriculture, and polished stone axe production. What is more, much of this cultural repertoire was developed within New Guinea itself from indigenous Late Pleistocene roots, not introduced from outside. We need to be realistic and to see Upper Paleolithic stone tool industries as necessary for survival and perhaps the marking of group identity in certain mostly cold regions settled by early *sapiens* populations, but not as markers of a totally new level of complexity in modern human biological and cultural evolution.

Homo sapiens and the Extinction of the Neanderthals

There is one other topic that arises from the Upper Paleolithic expansion of *Homo sapiens* into the colder latitudes of Eurasia. What happened to the Neanderthals, and for that matter the Denisovans? The Neanderthals are thought to have had smaller populations than the modern humans who eventually replaced them, with Upper Paleolithic populations of *Homo sapiens* in France exceeding those of contemporary Neanderthals by a possible factor of ten.[30] This implies smaller family sizes for the later Neanderthals, and a lower birth rate—probably one reason why the percentage of Neanderthal DNA in the bones of Upper Paleolithic *Homo sapiens* in Eurasia is generally below 10 percent, and falling with time.

Science writer Tim Flannery has described Upper Paleolithic Europeans before 14,000 years ago as Neanderthal-*sapiens* hybrids.[31] Many no doubt were, with up to 10 percent Neanderthal DNA in terms of analyzed ancient bones, but there are questions concerning the overall success of this genetic mixing between the two species, especially if hybrids had relatively low survival rates.[32] For instance, one Upper Paleolithic individual from Romania had an ancestor who interbred with a Neanderthal four to six generations before its radiocarbon-dated lifetime, somewhere between 42,000 and 37,000 years ago. For some reason, this individual did not contribute DNA directly to any living human population.[33]

In this regard, it is apparent that Neanderthal genes were not always conducive to survival when transplanted into ancestral modern humans. For example, just as I was finishing this book, the astonishing news was announced that we humans might have acquired our susceptibility to COVID-19 through distant liaisons with Neanderthals.[34]

Opinions on hybrid vigor vary, however. Charles Darwin was greatly in favor of it, and there have been positive suggestions that some Neanderthal genes might have been beneficial when transplanted into modern human populations. Perhaps they assisted adaptations to low levels of ultraviolet radiation, and Neanderthals might also have carried genes to lower their metabolic rates through decreased oxygen consumption when at rest during cold nights, in order to inhibit shivering.

What specific technological advantages were available to *Homo sapiens* populations as they entered Neanderthal territories? Better hunting abilities come to mind, with more advanced projectile weapons that probably included the bow and arrow. *Homo sapiens* may have had better management of fire and control of cooking, more effective clothing and shelter, and perhaps greater social cohesion and cooperating group sizes than Neanderthals.[35] A more efficient use of language could have been another factor. Once Neanderthals began to perceive themselves as a population under predation and territorial dispossession we can expect their birth rates to have fallen, as happened in postcontact colonial situations in Australia and the Americas, where immigrants had large family sizes but dispossessed Indigenous peoples underwent severe population declines.

There is some fairly direct genetic evidence that the Neanderthal population was undergoing internal birth rate decline prior to its eventual extinction. For instance, the male-inherited Y-chromosome reveals low diversity among later Neanderthals, as if the number of fertile males in the population was slowly decreasing. In-breeding between Neanderthal half-siblings appears to have been common, and there is evidence for congenital (inherited) abnormalities in bones.[36] The interbreeding with *Homo sapiens* prompts the question: did Neanderthals eventually disappear around 40,000 years ago not through violent extinction but rather through hybridization into a much larger population of *Homo sapiens*?[37] It seems quite possible.

The Spread of *Homo sapiens* toward Eastern Eurasia

As with the first hominin migrations out of Africa into Eurasia around two million years ago, so also the Late Pleistocene migrations of *Homo sapiens* through the main bulk of Asia had to deal with the presence of inhospitable deserts and mountain ranges. The first movements were probably to the south of these forbidding landscapes, following the tropics toward Australia, although there is no definitive evidence to prove this apart from the potentially early date for *sapiens* arrival in that continent that I discuss below. However, one would imagine that *Homo sapiens* populations of recent tropical African origin would have been able to migrate into Asia north of the desert and mountain barriers *only* if they were assisted by warm clothing, which would rely on the use of Upper Paleolithic technologies. There are no signs of these technologies in northeastern Asia until about 42,000 years ago.[38]

Nevertheless, these northerly regions were eventually settled by *sapiens*, with many eventually migrating in from the west as well as the south. By the end of the Pleistocene, about 12,000 years ago, ancestral modern humans in northeastern Asia had become different in their genomes and physical appearances from their contemporaries in

Southeast Asia, Australia, and New Guinea. Populations ancestral to modern Siberians, Chinese, and Native Americans developed their distinctive physical characteristics in the colder north and east. Populations ancestral to modern Andaman Islanders, Indigenous Australians, and Papuans developed their distinctive physical characteristics in the tropical south.[39]

During the Holocene era of East Asian food production after 8,000 years ago, the northern population gradually shifted its center of gravity southward, into the mainland and islands of Southeast Asia, and eventually beyond into the distant islands of Polynesia. These movements created the human pattern that exists in eastern Asia and Oceania today. I deal with them in chapter 12, but at this point it is necessary to follow the first *sapiens* migrants onward into Australia and New Guinea.

Onward to Sahul

When sea levels were low during much of the Late Pleistocene, the now-separate landmasses of Australia, Tasmania, and New Guinea were joined as a single dry-land continent termed Sahul (figure 5.1). If this continent was first settled by Homo sapiens rather than by earlier hominins such as Denisovans, then the oldest archaeological dates for its settlement should give a minimum age for the expansion of Homo sapiens out of Africa. What are those dates?

There are two problems here. First, there is a vagueness in the chronology for initial settlement in Sahul that reflects the half-life realities of geophysical dating methods. Put simply, the archaeological sequence begins uncomfortably close to the time that radiocarbon dating ceases to work, this being when the remaining quantities of radioactive carbon-14 in organic materials become too small to measure for dating purposes. This occurs between 60,000 and 50,000 years ago, depending on the quality of the sample and the laboratory techniques used to measure its radioactivity. As a result, the oldest Australian archaeological dates come mostly from luminescence analysis of quartz and felspar grains in sediments; there are sometimes contextual difficulties with these techniques because luminescence dating does not operate directly on artifacts or bones, only on mineral grains within sediments.

Second, it is not entirely obvious that the first hominins to reach Australia and New Guinea were actually *sapiens*, the direct ancestors of modern Australian Aboriginal and Papuan populations. The oldest human burials in Australia, dating to around 42,000 years ago from Lake Mungo in New South Wales, are certainly fully modern *sapiens*. One was apparently a cremation, one of the oldest ever found anywhere in the world, the other an ocher-covered inhumation. But before this there are no well-dated skeletal remains, despite a presence of stone tools that are currently claimed to date as far back as 65,000 years ago, as I discuss below.

Unfortunately, the flaked stone tools found in most Pleistocene Southeast Asian and Sahul sites are relatively uninformative for the task of identifying hominin species. Blade tools and even the Levallois technique are generally lacking.[40] "Middle Paleolithic" flaked stone tool industries similar to those found in Late Pleistocene Australia and New Guinea were also made by *Homo erectus* in Indonesia, by Neanderthals in western Eurasia, and by early *sapiens* in Africa and Eurasia.[41] Neither is there a clear association between tool types and hominin remains identifiable to species in regions further west, such as the Arabian Peninsula and South Asia. Direct associations between dated *sapiens* crania and specific stone tool industries in these regions of southern Asia simply do not exist.

It is, therefore, possible that Sahul was reached by Denisovans or even *Homo erectus* before the ancestors of the modern Aboriginal population arrived, especially given the earlier ocean passages by *Homo floresiensis* and *luzonensis*. At present there is no fossil or genetic evidence to clinch the matter, although some ancient Australian crania have features that could suggest interbreeding with Denisovans and *Homo erectus*, possibly occurring in Asia rather than in Australia itself.[42] The record from both ancient and modern DNA also cannot rule out a possible Denisovan occupation of Australia and New Guinea before *Homo sapiens* arrived.[43] Because Australia and New Guinea were joined by dry land across Torres Strait and the Gulf of Carpentaria for most of the Late Pleistocene, hominins arriving on one land mass could presumably have reached the other quite easily.

When Was Australia Settled?

I referred above to problems in Australia with radiocarbon dating at the limits of the method, so it comes as no surprise that the date of human settlement is one of the hottest topics of debate in modern Australian archaeology. Currently, the most heated of these debates concerns the findings from a cave called Madjedbebe (formerly Malakunanja) in Arnhem Land, northern Australia, where initial human occupation without human remains has been dated to between 65,000 and 53,000 years ago by luminescence dating of sand grains from the cave sediments.[44] The dates are convincing, at least for the sediments, because the excavators state that they are in correct order, becoming older with depth, and reveal no sign of disturbance.

However, this date range has been challenged on the grounds that some of the Madjedbebe sediments could have been intruded by termite nests, leading to downward movement of artifacts within the overall stratigraphy.[45] The artifacts themselves, which cannot be dated directly, also raise questions, because the lowest layers in the site contain a number of stone axes up to 20 centimeters long with ground cutting edges. These Madjedbebe edge-ground axes are potentially the oldest of their kind in the world, and the date range claimed for them extends to almost 20,000 years older than the commencement date for the Upper Paleolithic in Eurasia. Such tools do not occur anywhere else in the world in Upper Paleolithic contexts, apart from Japan, where they appear with first human settlement in Kyushu and Honshu about 37,000 years ago, as I discuss in chapter 6. This Japanese date is almost 30,000 years younger than the Madjedbebe range.

In this instance, we can hardly claim that the Madjedbebe basal layers were occupied by a pre-*sapiens* hominin such as a Denisovan because no such hominins are known elsewhere ever to have ground the edge of a stone tool. At Madjedbebe, at least according to the stone tools, we are surely dealing with a *sapiens* presence. But when was the cave first occupied? That is the question.

A few other caves in northern Australia have fragments of edge-ground axes that are believed to date back to as much as 40,000 years, but the southern two-thirds of the continent lack such tools until the Holocene. There are many puzzling features in this situation. Are the dates claimed for the Madjedbebe edge-ground axes correct? There are significant issues of context. Was Australia really settled by *Homo sapiens* 65,000 years ago, closer to 55,000, or even 45,000 years ago, as favored by some geneticists and archaeologists?[46]

I have no further observations to make about the Madjedbebe axes, but I should make it clear that I have no problem in accepting a date of 55,000 years ago for the *sapiens* arrival in Sahul, given the large number of convincing radiocarbon dates from both Australia and New Guinea that extend back this far, close to the radiocarbon dating limit. There is also mitochondrial DNA molecular clock information that supports a modern human occupation of both Australia and New Guinea by at least 55,000 years ago, possibly by two separate groups of settlers.[47] Perhaps there is room here for some accommodation between the different views, although the debate over when exactly Australia was first settled by hominins, modern human or not, will surely continue, perhaps without easy resolution.

How Was Australia Settled?

How was Australia settled by *Homo sapiens*, given that it was always an island continent that could only be reached through the sea-girded islands of Wallacea (eastern Indonesia) with at least one journey out of sight of land, possibly over as much as 90 kilometers of open sea? The exact routes taken between the many islands of Wallacea can never be known, even if some scientists like to guess on the basis of distances and visibility between islands. Approaches via Timor or New Guinea are obviously most likely from a purely geographical perspective.[48]

As for the kinds of seacraft that were available 55,000 years ago, the ethnographic record suggests that ancestral Australians might have used rafts of logs or reed bundles. Seacraft built partly of wood at least give the Madjedbebe edge-ground axes a viable function, and cutting of holes in trees to assist in climbing to collect honey would be another.

In terms of food resources available to people migrating toward Australia, the isolated Wallacean islands of Late Pleistocene eastern Indonesia were relatively poorly provided compared with the land-bridged islands of Sumatra, Java, and Borneo on the Sunda continental shelf, especially in large land mammal faunas. As we saw in chapter 3, the pigs, deer, and cattle that inhabited Sundaland never reached the Lesser Sunda Islands, at least not until humans transported them during the Holocene, although a few Asian mammal species did reach the Philippines and Sulawesi through prehuman natural dispersal. A compensatory interest in marine fishing and shellfishing appears to have developed by at least 40,000 years ago in some islands in eastern Indonesia, where some of the world's oldest shell fishhooks have been found (figure 6.4).[49]

Once hominins/humans reached Australia and New Guinea, however, land-based resources would have reappeared in the form of large marsupial mammals and flightless land birds. Whoever arrived first, whether Denisovans or *Homo sapiens*, they would have experienced a landscape populated by naive prey with no bipedal competitors, and some of these mammals and flightless birds rapidly became extinct as a possible result of hunting, at least according to some paleontologists (opinions on the question of extinction through human activity differ, often strongly).[50] Australia had formidable marsupial carnivores, but these declined rapidly in numbers as hunters killed their more docile herbivorous prey and began to manage the seasonally dry landscapes that supported them through the use of fire.

The Australia that existed when humans arrived 50,000 or more years ago bore little resemblance to the Australia that existed on the eve of the European arrival at the end of the eighteenth century. The demise of the leaf-browsing megafauna meant that fuel loads across the landscape increased. This resulted in more frequent fires, both natural and human-lit, followed by expansions of fire-resistant eucalypt and banksia (sclerophyll) forests, none rich in edible botanical resources for humans.[51] The first *sapiens* settlers, assuming they were the first hominins to arrive, must have had it good. If they were anything like the First Americans to be discussed in chapter 6, they might have increased their population size many times over within the continent during the first few millennia of settlement.

How Many First Australians?

Speaking of fecundity, how many First Australians would have been necessary to establish a population large enough to survive the reproductive bad luck (for instance, too many boys and too few girls) and inbreeding threats that could sometimes have led to extinction in isolated conditions? As I noted in my book *First Migrants*, hunter-gatherers at 55,000 years ago were probably far fitter genetically than we are today, given their unremitting exposure to natural selection and lack of medical care. Anyone harboring a deleterious gene that produced chronic ill health or a highly visible physical defect would, in theory, have had a diminished chance of reproduction, so the offending gene might have had a short life within the population. Physical good health beyond genetic causation is another matter, and many people no doubt would have had bodies that were rather heavily used, even battered and poorly repaired, by our standards. But successful genetic procreation in circumstances of frequent inbreeding with close relatives is the issue here. A genetic profile well-honed by selection would presumably have lessened the chances of having offspring with life-threatening inherited disorders.[52]

With this in mind, it might come as a surprise that a recent exercise in demographic modeling has suggested that a founding population of between 1,300 and 1,550 individuals, both male and female, would have been required to ensure future survival after the initial settlement of the Pleistocene Sahul continent.[53] Such a large number of people can only have arrived in many seacraft (presumably hundreds of them?), rendering purposeful ocean-going behavior undeniable. Other research teams have calculated somewhat lower numbers by examining diversity in existing mitochondrial DNA lineages in the living Aboriginal population, thereby suggesting arrivals of between 72 and 400 individuals.[54] All of these estimates involve analysis of ethnographic and/or present-day genetic data, not data (such as ancient DNA) drawn directly from the time in question.

Even the smallest of these numbers, if correct, is large enough to imply that many watercraft expeditions must have reached Australia, although whether all at once or spread out over a period of time is not clear, and perhaps never will be. However, as with Polynesian migrants tens of thousands of years later, once the news was out that new land existed over the horizon, presumably meaning that some people got home again to tell the tale, we might expect a flood of interest to have followed.

We will never know exactly how many First Australians landed after such an enormous lapse of time, but I am doubtful about the highest of the above-cited numbers. As it happens, another demographic exercise long ago with an Oceanic island context in mind concluded that only three fertile couples would have been sufficient to found a viable population in total isolation, with a 50 percent chance of long-term survival.[55] Furthermore, we know from recent demographic records elsewhere in the world that small numbers of humans entering new, food-rich environments were capable of incredible rates of reproduction in total isolation, doubling or trebling per generation.

As an example, a party of 28 people, comprising 9 male Britons accompanied by 6 Tahitian men, 12 Tahitian women, and one infant girl reached Pitcairn Island in eastern Polynesia in 1790, fleeing into hiding from vengeful British authorities after the mutiny in Tahiti on the ship *HMS Bounty*. They were finally discovered in 1808, by which time they numbered 35 people, many being children. By 1830, there were 79 people on the island, and almost 200 by 1850. Thus, the population increased sevenfold in sixty years, with a few deaths but no immigration from outside. Admittedly, the Pitcairners were farmers rather than hunter-gatherers, but they do tell us that humans can be extremely fecund if they so desire, especially if they arrive in an environment that has no diseases or dangerous predators, apart perhaps from other members of the same population (there were several murders on Pitcairn in the early days!).[56]

There seems no reason why the First Australians in their healthy and fertile new hunting grounds should have been any different. I suspect relatively small numbers arrived, perhaps on more than one occasion, and reproduced rapidly once they had landed. More than sixty years ago, biological anthropologist Joseph Birdsell estimated that Australia was settled by "an immigrant handful" who doubled their population every generation to reach a "saturation" population of 300,000 hunter-gatherers after only 2,200 years.[57] I suspect that he was right.

A Beyond-Africa Scenario

How to tie all of this together? We know that footloose bands of early *Homo sapiens* began to depart Africa on a small scale starting perhaps 200,000 years ago, as discussed earlier in this chapter. They came face-to-face in Europe and the Levant with populations of Neanderthals, with some of whom they interbred. During this time period, we have no evidence to suggest that early *sapiens* populations who lived outside Africa before 70,000 years ago enjoyed any particular cultural advantages over archaic hominins, and neither can it be demonstrated that human populations living today have inherited DNA directly from them.[58]

The main out-of-Africa migration into Eurasia occurred much later, certainly after 70,000 years ago, and it appears initially to have been confined to tropical and warm temperate latitudes, convenient for a hairless and dark-skinned hominin of tropical African origin who would not habitually have needed to wear stitched and tailored clothing, or any clothing at all. Although there are current claims for a modern human arrival in Australia as much as 65,000 years ago, they are not directly associated with human remains and would benefit from further verification.

The initial out-of-Africa tropical migrants traveled with stone tools similar to those of contemporary Neanderthal populations, with a Levallois prepared core technology in the west that was gradually replaced by a more informal "horsehoof-shaped core" technology in Southeast Asia and Australia. Along with the stone tools were taken bone and shell ornaments, the world's oldest rock art (currently in Sulawesi and Borneo), inhumation and cremation burial traditions, and an ability to cross ocean passages—in fact, many of the cultural and artistic markers of the Upper Paleolithic.[59] Absent in Late Pleistocene Southeast Asia and Sahul were the long blades, bifacial points, and backed tools that were to characterize a great deal of *sapiens* activity in western and northern Eurasia.

Temperate and cold latitudes still had some time to wait after the initial tropical migration of *Homo sapiens* before experiencing their own arrivals of what appear to have been the new elements of sapient behavior, except in southern Africa itself, where the emergence of backed tools and bifacial points occurred as early as 75,000 years ago, as I discussed above. This was about 25,000 years before anywhere else in the world. However, these tool types did not spread beyond Africa at that time, despite an earlier presence of cave burial and use of red ocher in the Levant. It was only after 50,000 years ago that the blade industries that eventually characterized much of the Eurasian Upper Paleolithic began to make an appearance outside Africa, assisting humans to penetrate into ever-colder latitudes. Eventually, at about 16,000 years ago, Upper Paleolithic humans reached the Americas.

Lingering Mysteries: A Personal Tale

I stated just before that bifacial points have never been found in Southeast Asia. This might not be entirely true.

It has long been known that stone flake industries of Middle Paleolithic and ultimately Oldowan inspiration were long retained within the *sapiens* prehistory of Southeast Asia, lasting until the beginning of the Neolithic, or into the mid-Holocene in the case of Australia. As we have seen, however, the edge-ground axes from Madjedbebe in northern Australia generate some uncertainty about this seemingly rather static situation, and they are not alone.

During the early 1980s, I was excavating a site at Tingkayu in southeastern Sabah (Malaysian North Borneo) with colleagues from the Sabah Museum in Kota Kinabalu. The site, which is located in figure 5.1, had been a workshop for the preparation of some remarkable bifacial points and knives made of chert, a flint-like rock (figure 5.2). Tingkayu produced no other artifacts (no blades, backed tools, or edge-ground axes, for instance), apart from the bifacial tools and the many flakes that were discarded during their production. The archaeological layers were in stiff clay close to the modern ground surface, with no survival of organic material such as charcoal or bone. Therefore, we could not radiocarbon-date the site to find out its age.[60]

We knew, however, that Tingkayu was located on a promontory that projected into what was once an extensive swampland or shallow lake formed when the Tingkayu River was blocked by a lava flow. Such a wetland (nowadays an oil palm plantation—the river cut through the blockage many millennia ago and the wetland drained away) would have been a promising place to hunt and gather. We managed to locate the blocking lava flow, still exposed in the side of the outlet gorge that today drains the former swampland, and to radiocarbon-date the layer of charcoal from the vegetation burnt beneath it. The result was a date of about 33,000 years ago for the lava blockage and the formation of the wetland. This, presumably, gives a maximum date for the stone tools, although they could be younger.

The Tingkayu tools are classically Upper Paleolithic bifacial points (some perhaps also used as knives) of a type that occurs widely in Japan and northeastern Asia, as I discuss in the next chapter. Yet in Sabah they exist in total isolation. A century of excavation in Southeast Asia and southern China, in dozens of archaeological sites, has never produced a similar stone tool industry. There are parallels in the stone points made in northern Australia ("Kimberley Points") during the Holocene, and especially in Upper Paleolithic contexts in Siberia and Japan (chapter 6). But these parallels are extremely distant, and even if they are relevant one must wonder why there are no intervening traces of human movement between these far-away places and Borneo.

It is unknown if the Tingkayu bifaces represent an isolated and independent burst of creativity, or if they reflect a remarkable instance of voyaging from (perhaps) Japan, somehow bypassing Taiwan and the Philippines and hitting the northern coastline of Borneo. If they do reflect some kind of maritime contact with Upper Paleolithic Japan, they will be useful hints to suggest that similar movements could have occurred from Japan toward the north. And guess what lay in that direction, awaiting discovery beyond the Beringian land bridge!

As with Madjedbebe, much is mysterious about Tingkayu. I am still unable to explain it. Archaeology can be fun, but often it is exasperating, and there is an awful lot that we do not know about the activities of our remote ancestors. It is time to head back into the cold.

6

Stretching the Boundaries

WE NOW EXAMINE the dramatic events that occurred in northeast Asia and the Americas during Act III of the Odyssey. I examine how modern human populations spread onward through steppes and mountains into the northern regions of Asia, including sea crossings to Japan, and eventually, by 16,000 years ago, through the Arctic Circle by boat or Beringian land bridge into the Americas (figure 6.1). These movements completed the *sapiens* settlement of the earth's continents, except for Antarctica, leaving only distant islands such as Madagascar and New Zealand, and many smaller islands, especially in the Pacific, as completely pristine landscapes for Holocene settlers to reach.

Braving More Cold: Northeast Asia and the Americas

During the Last Glacial Maximum, between 25,000 and 18,000 years ago, the massive ice sheet of moisture-rich northern Europe extended from the British Isles to Novaya Zemlya in the Barents Sea (see figure 5.1). Arctic Siberia was mostly too dry for major ice sheets to form on land. Instead, the ground was frozen all year round as permafrost, the melting of which under modern conditions of global warming releases large amounts of the greenhouse gas methane into the atmosphere from thawed and decaying vegetation.

During the Late Pleistocene, permafrost extended beneath the grass and shrub tundra landscapes of northern Russia, above 60° north, from the Ural Mountains in the west to the Bering Strait land bridge ("Beringia")

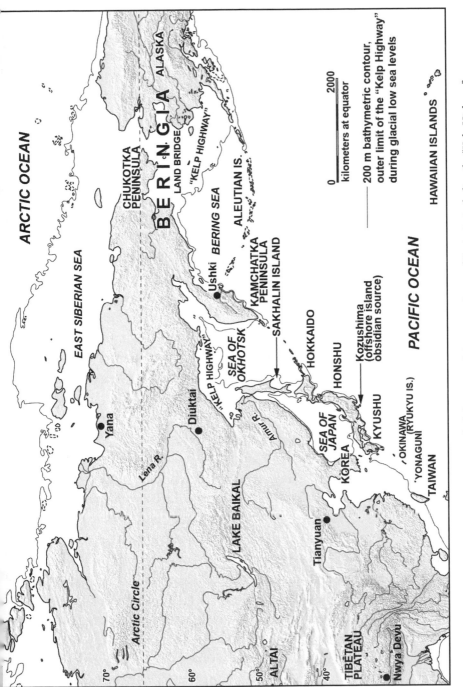

ARCTIC OCEAN

EAST SIBERIAN SEA

CHUKOTKA
PENINSULA

ALASKA

B E R I N G I A

LAND BRIDGE

"KELP HIGHWAY"

Ushki *BERING SEA*

ALEUTIAN IS.

KAMCHATKA
PENINSULA

SAKHALIN ISLAND

*SEA OF
OKHOTSK*

"KELP HIGHWAY"

Amur R.

Yana

Diuktai

Lena R.

Arctic Circle

LAKE BAIKAL

HOKKAIDO

HONSHU

Kozushima
(offshore island
obsidian source)

*SEA OF
JAPAN*

KOREA

KYUSHU

OKINAWA
(RYUKYU IS.)

YONAGUNI

TAIWAN

Tianyuan

TIBETAN
PLATEAU

Nwya Devu

ALTAI

70°

60°

50°

40°

PACIFIC OCEAN

HAWAIIAN ISLANDS

0   2000
kilometers at equator

—— 200 m bathymetric contour,
outer limit of the "Kelp Highway"
during glacial low sea levels

Figure 6.1. Map of northeastern Asia, to show Upper Paleolithic sites and the approach toward Beringia, including the "Kelp Highway."

in the east. This land bridge was a maximum of 1,800 kilometers wide from north to south during the Last Glacial Maximum, and it led onward into unglaciated Alaska. Humans were able to cross the open Siberian territory from the west and south when climatic conditions permitted, with no major glaciers to block their way until they met the massive Cordilleran and Laurentide ice sheets of North America just beyond Alaska (seen in figure 6.2). The weather was cold, but the hunting was good, including herds of mammoth, musk oxen, woolly rhino, reindeer, and saiga antelope.

By 40,000 years ago, hominins, presumably *Homo sapiens* (although their exact identity still remains uncertain—they might have been Neanderthals), had reached this Arctic Circle tundra landscape in the vicinity of the northern Ural Mountains, close to the Arctic coastline.[1] Farther east, people making Upper Paleolithic stone tools had arrived by 42,000 years ago in the Altai Mountains and northern China. *Homo sapiens* reached the Arctic Circle coastline of eastern Siberia at the site of Yana by at least 30,000 years ago.[2] Then a dramatic plunge in temperature and global sea level followed, a downward slide into the Last Glacial Maximum, toward the absolute nadir in temperature and sea level that occurred between 25,000 and 18,000 years ago.[3] Most of northern Asia above 55° latitude was abandoned during this time.

Who were the first modern human populations of northeastern Asia? The oldest *sapiens* individual found in this region, from Tianyuan Cave near Beijing, dates from about 40,000 years ago. This individual carried a mitochondrial DNA lineage that still exists today among many East Asians.[4] The Tianyuan genome also shared nuclear genes with some of the first Upper Paleolithic populations in Europe, with the First Americans, and with modern people in New Guinea, Australia, and the Andaman Islands.

Tianyuan thus carried some degree of ancestral status for both the cold-adapted and the tropical-adapted populations that had developed in East Asia by the end of the Pleistocene, as described in the previous chapter. Presumably, its ancestors migrated into what is now northern China from the south. During the Last Glacial Maximum, it is likely that this Tianyuan-related early *sapiens* population retreated into coastal

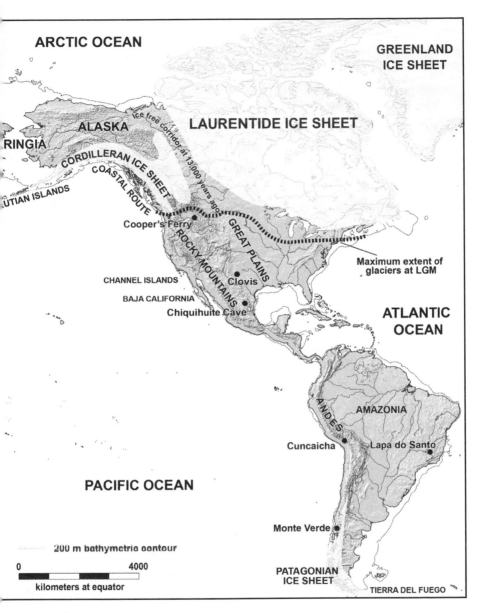

ARCTIC OCEAN

GREENLAND
ICE SHEET

LAURENTIDE ICE SHEET

ALASKA

RINGIA

CORDILLERAN ICE SHEET

COASTAL ROUTE

Ice free corridor at 13,000 years ago

UTIAN ISLANDS

Cooper's Ferry

ROCKY MOUNTAINS

GREAT PLAINS

Maximum extent of
glaciers at LGM

CHANNEL ISLANDS

Clovis

BAJA CALIFORNIA

Chiquihuite Cave

ATLANTIC
OCEAN

ANDES

AMAZONIA

Cuncaicha

Lapa do Santo

PACIFIC OCEAN

Monte Verde

200 m bathymetric contour

0   4000

kilometers at equator

PATAGONIAN
ICE SHEET

TIERRA DEL FUEGO

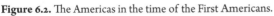

Figure 6.2. The Americas in the time of the First Americans.

refuge regions, including perhaps Japan. Recent research into ancient DNA from Heilongjiang Province in northeastern China has shown that a different population spread through the region as the climate warmed again after 19,000 years ago, from a likely origin in the general region of the Amur Basin in the Russian Far East.[5] A few millennia later, this Amur population contributed a degree of genetic input to the settlement of the Americas, and it still has many descendants in northeast Asia today, as I discuss in chapter 11.

At some point prior to 30,000 years ago, *sapiens* populations in East Asia also headed up to the Tibetan Plateau, emulating their Middle Pleistocene Denisovan predecessors. A recent report of Upper Paleolithic stone tools from an open site called Nwya Devu (located in figure 6.1), at 4,600 meters above sea level in the southern part of the Tibetan Plateau (thus only about 30° north, but at high altitude), makes it clear that *sapiens* had developed some resistance to low (hypoxic) oxygen levels by this time. Genetic research suggests a possibility that this ability to withstand hypoxia could have been acquired through hybridization with similarly cold-adapted Denisovans.[6] Presumably, like the Denisovans, early *sapiens* populations were also attracted to this high altitude by its hunting potential.

The oldest direct evidence for modern human settlement along the Arctic coastline of northeastern Siberia comes from the Yana Rhinoceros Horn Site at 71° north, near the delta of the Yana River, within the Arctic Circle and close to the shoreline of Russia's East Siberian Sea. Recent analysis of ancient DNA from the 31,000-year-old Yana burials reveals that these people became genetically separate from other *sapiens* populations in northern Eurasia at about 38,000 years ago, which fits well with other dates for *sapiens* expansion into northern Asia.[7]

Once people arrived in Yana, apparently from the west rather than the south in terms of their ancient DNA connections with other Paleolithic populations in western Eurasia,[8] they hunted Arctic tundra animals—mammoths, woolly rhinos, bison, reindeer, horses, and bears—and carved the ivory and bones of these animals, especially the mammoths, into spear points and ornaments, some carrying intriguing combinations of dots and incised lines, alas now incomprehensible to us. The importance of mammoths is evident, so much so that archaeologist Robin Dennell describes Asian Arctic Paleolithic sites as having a "mammoth economy."[9] Also intriguing were the many complete skeletons of Arctic hares discovered by the excavators, which were presumably killed for their skins rather than their meat. Tailored warm clothing stands out as an imperative in the Arctic Circle, where winter temperatures can drop far below freezing point.

After Yana, evidence of human activity in much of Siberia declined or vanished with the climatic stresses of the Last Glacial Maximum. It seems that humans waited out the extreme climatic conditions in sheltered refuge zones that would probably have included the Altai Mountains, Lake Baikal, the Amur Valley in eastern Russia, the Korean Peninsula, and especially the warm maritime islands of Japan (figure 6.1). Some geneticists and archaeologists also suggest that they survived in unglaciated southerly regions of the Beringian land bridge itself during the Last Glacial Maximum,[10] waiting in the immediate wings to enter North America as soon as the ice retreat would allow them to do so. However, there is as yet no compelling archaeological evidence in support of this view. So far, the oldest archaeological sites with stone tool industries in Beringia, including Alaska, date only to about 14,500 years ago.[11]

It is clear from genetic and archaeological research that the First Americans were a people of Asian Upper Paleolithic origin who moved through Beringia and then southward from Alaska, initially around the seaward edge of the retreating Cordilleran Ice Sheet of western Canada (figure 6.2), starting sometime between 17,000 and 15,000 years ago.[12] Where did they come from? Yana, occupied 31,000 years ago, lies a full 2500 km to the west of Alaska and its ancient DNA does not reflect a likely source for the First Americans. The most likely immediate source for them, especially in terms of access, was somewhere along the northeastern Pacific coastline of Asia—the Chukotka and Kamchatka Peninsulas, the Sea of Okhotsk, the lower Amur River basin, and Japan. When did *Homo sapiens* first penetrate these coastal regions? The events that are unfolding are summarized in figure 6.3.

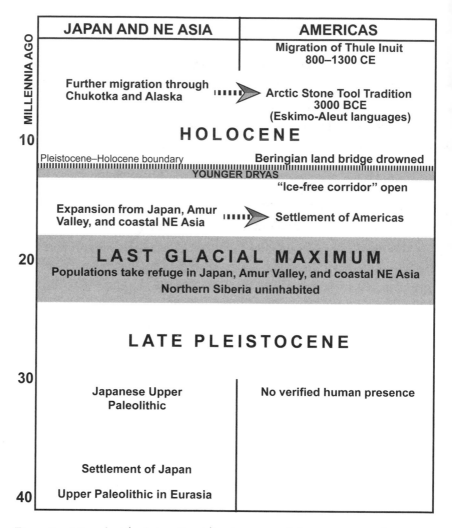

Figure 6.3. A time chart (scale in millennia) to show some of the main events in the preagricultural population history of northeast Asia, Japan, and the Americas.

Upper Paleolithic Japan

By 38,000 years ago, modern humans had reached the Japanese Islands. The sea level at this time would have been around 80 meters below that of the present, meaning that the northern island of Hokkaido was joined by a land bridge to Sakhalin Island, and from here directly to the Asian mainland near the mouth of the Amur River in eastern Russia. Most of

Japan—the major islands of Honshu, Shikoku, and Kyushu—formed a single island called Paleo-Honshu by researchers, separated by narrow sea passages from Hokkaido and the Korean Peninsula. The first human arrivals in Paleo-Honshu must have crossed one or both of these passages, most likely that from the south via the Tsushima Islands that lie between the southern tip of Korea and Japan.

Japan today also controls the Ryukyu Islands that extend south through the island of Okinawa toward Taiwan. These islands lie beyond the East Asian continental shelf, and sea crossings of up to 140 kilometers would have been required to reach them. Directly dated human remains from these islands indicate that such crossings were underway by at least 30,000 years ago—an extremely important observation that highlights some of the longest Paleolithic sea voyages on record, well out of sight of land. One intriguing and successful experimental exercise in 2019 involved paddling a five-person logboat (dugout canoe) for 200 kilometers across the northward-flowing Kuroshio Current from Taiwan (joined to mainland China during periods of glacial low sea level) to reach Yonaguni Island in the Ryukyus (figure 6.1), a forty-five-hour journey. This experiment does not prove that ancient people ever made such a voyage in this way, but it suggests the possibility of one, and it also suggests intentionality on the part of the ancient voyagers.[13]

The first people to arrive by sea on Paleo-Honshu were also crossing 40 kilometers of open sea to the small island of Kozushima, located off Honshu to the south of modern Tokyo, in order to collect obsidian for making their stone tools. No seacraft from the period of first settlement have yet been found in Japan, but the long Jomon period of Japanese archaeology that commenced with pottery manufacture around 16,000 years ago has produced a remarkable total of 160 dugout canoes from waterlogged deposits, together with many polished stone axes. The Jomon people certainly knew how to travel across water.

In terms of East Asian prehistory, Japan is a very important place. During the last glaciation it formed a temperate refuge from the interior Siberian cold, straddling latitude 35° north and washed by the warm Kuroshio and Tsushima ocean currents that flow northward from the western tropical Pacific.[14] For prehistoric people, Japan was clothed in deciduous and evergreen forests rich in resources for hunting and gathering. Where better to escape those Last Glacial Maximum Siberian winters, especially for those who knew how to hew out a boat with a Paleolithic edge-ground axe?

Japan is also important because we know a great deal about its prehistory. Modern Japan invests considerable resources in both rescue and research archaeology, meaning that these islands have a density of archaeological sites unparalleled elsewhere in the world. There are 500 or more sites already known from the oldest millennia of human settlement, and there are over 10,000 Japanese sites considered to be Upper Paleolithic, dating between 38,000 and 16,000 years ago, the termination marked by the appearance of the oldest Jomon pottery. This high density of Upper Paleolithic sites, with no convincing presence of any preceding Middle Paleolithic settlement, renders it likely that *Homo sapiens* was the first hominin to arrive, without archaic predecessors.

Unfortunately, the Japanese Islands have a problem as far as their record of modern human settlement is concerned. Hokkaido and Paleo-Honshu have thousands of archaeological sites, but they lack Upper Paleolithic skeletal and cranial remains, partly because there are few caves and human bones do not survive well in the neutral to acidic volcanic soils that occur in many Japanese open air sites. This means also that there is no Upper Paleolithic DNA from human remains in Japan that can be compared with ancient DNA from other regions, especially the Americas. By contrast, the Ryukyu Islands have many alkaline limestone fissures, with human remains dating back as far as 30,000 years ago on the islands of Okinawa, Miyako, and Ishigaki, but with a much less informative archaeological record and, so far, no reported Pleistocene whole-genome DNA.

The First Japanese, at 38,000 years ago, brought in a rather unique material culture to Paleo-Honshu.[15] As in northern Australia, there are edge-ground axes, although in the case of Japan there are so many of these in good dated contexts (over 900 in total) that no one disputes their antiquity. Perhaps this is an important point in favor of accepting claims for Late Pleistocene Australian edge-ground axes in Arnhem Land

and the Kimberley region, although the actual date for the oldest Australian specimens, whether 65,000 or 50,000 years ago, is a different issue.

Also occurring at the base of the Japanese archaeological sequence are so-called trapezoids: stone spearheads or arrowheads with transverse sharp edges. Japan had a native fauna that included deer and pigs when humans arrived. Interestingly, some of these animals appear to have been trapped in deliberately dug pits, as well as hunted on foot using these trapezoid-tipped projectiles.

And what of the Ryukyu Islands? Here, we have skeletons but few stone tools. One of the best preserved skeletons is that of an adult male from the Minatogawa limestone fissure on Okinawa, about 20,000 years old, with a mitochondrial DNA haplotype (but no preserved whole genome DNA) and cranial and facial features that relate it to other Late Pleistocene people of Southeast Asia, Australia, and New Guinea.[16] The Minatogawa individual was probably a hunter (dwarf species of deer and pig existed on Okinawa), and some of his contemporaries in Sakitari Cave on the same island fished using small circular shell fishhooks.

These fishhooks are similar in shape to Late Pleistocene shell fishhooks found in southeastern Indonesia (see figure 6.4), and also to fishhooks used by the First Americans when they reached small islands located off the Baja California coastline in northern Mexico.[17] This is perhaps not entirely coincidental in terms of the coastal locations of these populations. The use of fishhooks no doubt reflected a scarcity of terrestrial resources on small islands, and a need to focus on maritime substitutes.

A Japanese Origin for the First Americans?

All in all, the evidence from Upper Paleolithic and Jomon Japan, with the Ryukyu skeletons, renders it certain that people by 38,000 years ago had access to a maritime technology that could transport them across sea passages at least 40 kilometers wide. The earlier settlement of Australia carries similar implications, whichever route was used to get there. When it comes to asking that driving question "Where did the First Americans come from?," it must be admitted that of all the locations in

Figure 6.4. Artifacts from East Asia and the Americas. (*Top right*) A 2.9-centimeter-diameter shell bait hook from Tron Bon Lei Cave, Alor Island, eastern Indonesia. Similar Late Pleistocene/early Holocene hooks without line-attachment knobs or grooves are found in Timor, the Ryukyu Islands (southern Japan), on islands off the western coast of Baja California in Mexico, and in western South America. (*Main*) Closely matched stemmed projectile points from (*left*) Cooper's Ferry, Idaho (6.5 centimeters long), dated 16,500 to 15,000 years ago, and (*right*) the Upper Paleolithic Kamishirataki site on Hokkaido Island, Japan (4.8 centimeters long). Alor fishhook courtesy Sue O'Connor and Sofia Samper Carro. The bifacial points are reproduced courtesy of Loren Davis and *Science.*

northeastern Asia that were occupied by Paleolithic humans when the Americas were actually settled, approximately 16,000 years ago and after the Last Glacial Maximum, the Japanese Islands score on every point.

For instance, the Upper Paleolithic Japanese had a demonstrated maritime technology and a comparable stone tool technology to the First Americans, with bifaces and microblades (a specific variety of small blade, as discussed further below) in the main Japanese islands remarkably similar to those found in North America. The chronologies match well because both bifaces and microblades were popular in Japan when people first reached Alaska.[18] Japan also had an impressive density of human population, perhaps straining at the leash as postglacial climatic conditions improved. The major problem at the moment is the lack of ancient DNA from this population. However, a recent study of the mutation ages of mitochondrial DNA lineages among Minatogawa, Jomon, and living Japanese has indicated a significant increase in the human population size of this archipelago between 15,000 and 12,000 years ago.[19] Perhaps this is not a coincidence.

An intriguing Kelp Highway hypothesis, proposed by archaeologist Jon Erlandson and colleagues in 2007, also fits perfectly with a significant role for Japan in the settlement of the Americas.[20] After 18,000 years ago, postglacial sea levels rose and began to flood the continental shelf in the northern Pacific, from Japan around to California and down the coastline of South America. This zone was previously a large expanse of flat land that had been exposed above sea level during the Last Glacial Maximum. Inshore forests of kelp grew in the warming shallow waters, as they do today, harboring rich food resources of sea mammals, fish, and shellfish. Japan lay conveniently close to the southwestern limit of these kelp forests (figure 6.1), and at the time in question had human populations equipped with boats. Its shores were also washed by the warm Kuroshio Current from the south, as they are today. A great deal of food lay ahead for those able to penetrate the colder conditions.

Getting to America

Of course, it is unlikely that every migrant who crossed the Bering passageway from Asia into Alaska originated in Japan, and opinions about the homeland of the First Americans are constantly in flux. This is one of the most intensively researched fields in global prehistory, a topic of immense excitement for many people. It was also recent enough to be the first continental-scale migration in human prehistory that scholars think they can seriously understand. Australia and New Guinea were settled too long ago and the records are too faint for deep understanding. But the Americas are recent enough in time for scrutiny through finely differentiated Upper Paleolithic stone tool traditions, lots of ancient DNA, and, for the first time in our Odyssey, comparative linguistic evidence.

So let me move directly to the multidisciplinary evidence that is available on the First American migration, its timing, origin, and general direction. Before we start, all readers need to be aware of one important archaeological observation. It has recently been estimated that the First Americans had arrived between 15,500 and 15,000 years ago in Alaska.[21] Remarkably, they had already reached a location called Monte Verde in southern Chile, at more than 40° south latitude, by at least 14,000 years ago. This suggests that they moved quickly, possibly from north to south in as little as a few centuries.

A recent mitochondrial DNA observation that the First Americans increased their population numbers by a factor of sixty as they migrated south, within the first 3,000 years of settlement in the Americas, is highly illuminating as a relevant factor behind this success.[22] Another recent genomics paper describes this First American population boom as "one of the most substantial growth episodes in modern human population history."[23] New analyses of radiocarbon dates also identify significant population growth in North America between 15,000 and 13,000 years ago, and in South America between 13,000 and 9,000 years ago.[24] These dates overlap with those for population expansion in postglacial Japan, as I mentioned above. Settling the Americas was surely one of the greatest demographic success stories in the remote human past.

Evidence for the First Americans

Let me commence with summaries of the First American situation from archaeological, linguistic, and finally genetic sources. That order of presentation seems to work best in the current instance.

First, the archaeology, because this offers absolute dates for stone tool assemblages that can be unequivocally associated with human activity. The archaeological record as it is currently known commenced by 17,000 years ago on the Siberian side of Beringia, and perhaps later in the case of Alaska, after what appears to have been abandonment of these high latitudes through the Last Glacial Maximum after occupation ceased at Yana.[25] A large number of dated sites support this consensus.

However, as I was writing this chapter, a challenge appeared. Chiqui-huite Cave is located at almost 3,000 meters of altitude in central Mexico, thus a long way south of Alaska.[26] It contains a bifacial point industry indirectly dated to between 26,000 and 19,000 years ago, hence firmly within the Last Glacial Maximum if the dates are directly appli-cable to the stone tools. Chiquihuite would have been a cold place at that altitude and at that date, although the animal bones found in the cave suggest that hunting was undertaken and plant foods collected.

Are the dates (radiocarbon and luminescence) for Chiquihuite cor-rect? If yes, how did people circumnavigate the huge ice sheet that oc-cupied western Canada at the time in order to travel to Mexico? I am uncertain, although I will observe that Chiquihuite stands currently as an isolated claim. It is puzzling, and some archaeologists have already dismissed it.[27] Why are there no other definite sites of the same antiq-uity in the Americas, given the large number of other First American sites that have been thoroughly excavated, all much younger?

Currently, North American archaeologists have had little time to as-sess the Chiquihuite implications that the date for human settlement of the Americas needs to be pushed back beyond 26,000 years ago. In the rest of this chapter, I will follow the widely accepted date of settlement after 16,000 years ago, comfortably after the Last Glacial Maximum. I suspect that the First Americans traversed Beringia into North America during a fairly narrow window of time, as the ice retreat along the coast-line of British Columbia opened the route southward (figure 6.2) but before Beringia itself was completely drowned by the rising sea level, an event currently dated to about 12,000 years ago. Of course, if they used boats all the way to Alaska they might not have needed the land bridge, but we cannot be certain about this.

Regardless of the exact date of arrival, two different stone tool tech-nologies were involved in the process of settling the Americas, one fo-cusing upon the production of bifacial projectile points (like those claimed at Chiquihuite), and the other focusing on the production of microblades from a specific type of wedge-shaped (or "boat-shaped") core (these are absent at Chiquihuite). Both of these tool types occurred widely during and after the Last Glacial Maximum in Japan and afterward in northeastern Asia—for instance, at Diuktai Cave on the Lena River and in the Ushki Lake sites in Kamchatka (figure 6.1).[28]

In Alaska, it appears that the people who made the initial human passage favored the bifacial point technology because this was the one that was carried onward to create the oldest stone tools in the rest of the Americas, right down to Tierra del Fuego. Microblades never spread beyond northern and western North America, which implies that their arrival might have been secondary, eventually to become associated with the ancestral Eskimo-Aleut- and Na-Dené-speaking peoples who are discussed in the next section.[29]

Current archaeological observations beyond the ice suggest that the Americas were settled from Alaska by two major routes. The west coastal route, ice-free by 16,000 years ago, appears to have been used first, by people carrying a tool kit of stemmed and leaf-shaped bifacial points, and small backed tools called lunates. As a recent article points out, these tools are rather precisely paralleled in Japan at about the same time (figure 6.4). We also know that some were transported by boat within the Americas because they occur on the Channel Islands that lie off the coast of California.[30] These islands could only be reached by crossing sea passages, within the Kelp Highway.

By about 13,000 years ago, an inland ice-free corridor in Canada opened a second north-to-south route between the Laurentide and Cordilleran ice sheets (figure 6.2), and people possibly moved through it to develop what is now known as the Clovis culture (named after a site in New Mexico) in the central and eastern United States, with its magnificent basally fluted projectile points. Some archaeologists also think they might have gone the other way, given that similar but less refined bifacially flaked projectile points also occur in Alaska.

The most telling point, however, is that some sites in Texas have stemmed points that occur beneath the Clovis types, hence the stemmed points are older. This is important because Clovis was once thought to have been the oldest archaeological culture in the Americas. Now we know that it was not.

Languages and the First Americans

I now introduce some linguistics for the first time in this book, because we are entering a time span for which the comparative study of different languages across the world can reveal remarkable information about the human past.

In 1987, Stanford University linguist Joseph Greenberg published a book called *Language in the Americas*.[31] In it, he classified all of the indigenous languages of the Americas into three major divisions that he called Amerind, Na-Dené, and Eskimo-Aleut. Our immediate focus of interest here is on the Amerind language grouping, which includes all of the languages and language families of the Americas, apart from the Na-Dené and Eskimo-Aleut languages of the northern and western regions of North America. These last two were spread by Holocene migrations of relatively restricted extent, and they are discussed later. Interestingly, both of these later movements involved people who used microblades rather than bifacial points.

Greenberg's observations implied that the Amerind languages were the first to spread right through the Americas in ancestral form. Here, we have the makings of a correlation. According to the archaeological record in Alaska, bifacial points were brought from northeastern Asia prior to 14,500 years ago, and then were taken down the western coastline of Canada into the United States and the rest of the Americas. The ancestral Amerind languages are thus likely to have spread at the same time, given their distribution across both American continents.

Greenberg attracted a lot of criticism from other linguists in terms of his methodology of so-called mass comparison, using lists of words and their meanings shared between different languages. But in terms of historical and cultural reality, I have always suspected that he was right in his conclusions.[32] Unfortunately, because the First American migration was so long ago in language history terms, no linguist is able to reconstruct anything sensible about the nature or origin of early Amerind society. Linguistic reconstruction gets harder as time depth increases, to become virtually impossible by around 10,000 years ago, even if a few faint traces of common ancestry can still survive from beyond that time depth.

It is by no means certain, therefore, that all the Amerind languages descend from a single linguistic ancestor. But I see no reason to doubt the reality of a ~16,000-year-ago ancestral spread of people speaking Amerind founder languages, commencing from somewhere in northeastern Asia and moving through Alaska into the rest of the Americas.

Genetics and the First Americans

Let us now turn to the third topic of interest, ancient and modern DNA. This research field has undergone a revolution since 2014, when it was shown that the genome of a 12,500-year-old child of the Clovis culture from Anzick in Montana could be considered ancestral to the genomes of many Indigenous Central and South Americans today, thus supporting the correlation suggested above between the spreads of bifacial tools related to Clovis points and the ancestral Amerind languages.[33] Clovis was not the oldest archaeological culture in North America, but its genomic signature was obviously part of a major population movement through both American continents.

Since 2014, a large amount of ancient genomic literature has repeatedly driven home the conclusion that the majority of Native Americans, past and present across both continents, belong to a single First American population of northeastern Asian origin.[34] The Na-Dené and Eskimo-Aleut speakers, identified separately from Amerind speakers by Joseph Greenberg, descend from closely related but genomically distinguishable secondary migrations, in the Eskimo-Aleut case spreading along the newly ice-free northern coastline of Canada around 5,000 years ago.

According to the genetic evidence, where did the First American ancestral population come from, and by which route did it arrive in Alaska? One potential ancestral population, dating from both before and after the Last Glacial Maximum, has been identified from burials with ancient DNA near Lake Baikal in central Siberia, 4,000 kilometers away from Beringia.[35] However, Lake Baikal seems rather remote to be an immediate source, as does Yana, and neither are the DNA profiles of these two populations agreed to be the best fit with those of the First

Americans. The most recent genetic analysis suggests a First American homeland much closer to the eastern seaboard of Asia, centered around the Amur Valley of the Russian Far East, and not far inland from the land bridge that led during the Last Glacial Maximum from the lower Amur Valley through Sakhalin Island into Hokkaido in northern Japan.[36]

In this regard, several research teams have noted that the molecular clock genomic separation time between the ancestors of existing Siberian and Native American populations is older than the Last Glacial Maximum, dating to around 24,000 years ago, rather than to the ~16,000-year-ago date for the migration into North America itself.[37] This suggests that there was a long standstill of perhaps 8,000 years when the ancestral American population was waiting in the wings in relative isolation, hiding from the extreme cold. My preference for that waiting room is northern Japan and the adjacent northeast Asian coastline, although the current lack of Upper Paleolithic DNA from Japan that could support this suggestion remains a difficulty.

If the Amur Basin and the northern Japanese islands were part of the origin region for the First Americans, how did people reach Alaska from there? A coastal route around the shoreline of the Sea of Okhotsk from northern Japan and the Russian Far East using the resources of the Kelp Highway seems most likely, across the narrow neck of the Kamchatka Peninsula. The Kamchatka sites at Ushki, dating between 16,000 and 13,000 years ago, possibly marked the way.[38]

Population Y?

There is one puzzling aspect to the genetic ancestry of the First Americans. A few living South American populations share traces of genomic ancestry with living indigenous Australians and Papuans. And there is more to this puzzle than just the DNA of living people. Paleoanthropologist Noreen Cramon-Taubadel and her colleagues have also suggested that ancient Sahul (Australia and New Guinea) affinities can be identified in the morphologies of some early Holocene South American skulls.[39]

A series of 10,400-year-old skeletons excavated in Lapa do Santo Cave, in the Lagoa Santa region of east-central Brazil, carry such morphological affinities and have actually yielded ancient DNA with traces (less than 6 percent) of ancient Sahul ancestry, at least according to two research teams.[40] Two hundred individuals were buried here, in flexed, squatting, or contorted postures, some dismembered or with bones removed, one even decapitated. These practices are similar to those associated with late Paleolithic burials in southern China and Southeast Asia, and with Jomon burials dated to after 16,000 years ago in Japan.[41] In both regions, the burials were placed in the ground without grave goods.

Is this coincidence? The people who took this Sahul genetic signature into the Americas have been termed "Population Y" by geneticists, after an Amazonian Tupi language term *Ypykuéra*, meaning "ancestor."[42] However, there is no genetic trace of a Population Y signature in the ancient DNA of Pleistocene Siberian populations, or indeed in most ancient American populations, especially in North America or the Caribbean.[43] Furthermore, there is no evidence from archaeology or cranial morphology for the existence of a separate Population Y migration throughout the Americas. Paleolithic Southeast Asian and Sahul populations did not use the types of bifacial point or microblade that traveled to Alaska, unless the Tingkayu outlier described at the end of chapter 5 was more significant than is currently apparent.

What does the Population Y signature represent? In my view, it could represent a small survival in the Americas of a genetic signature derived from the initial modern human population that migrated out of Africa toward East Asia, Australia, and New Guinea. The 40,000-year-old Tianyuan individual found near Beijing and discussed previously was an early member of this population, with a genome related to those of early Australian and Papuan populations. The analysts of the ancient DNA of this long-dead person offer a most interesting statement about its significance:

> The Tianyuan individual, who lived in mainland Asia about 40,000 years ago, has affinities to some South American populations that is [*sic*] as strong as or stronger than that observed for the Papuan and Onge [Andaman Islands]. . . . [This] suggests that a population related to the Tianyuan individual, as well as to the present-day Papuan and

Onge, was once widespread in eastern Asia. This group or another Asian population related to this group persisted at least until the colonization of the Americas and contributed to the genomes of some Native American populations.[44]

A 14,000-year-old skeleton excavated in Red Deer Cave in Guangxi Province in southern China also shared some of its genes with Native Americans, more indeed than it did with Tianyuan, these genes also being absent in the Lake Baikal population that I mentioned above.[45] This implies that temperate East Asia still held populations with identifiable American genetic affinities at the time when the Americas were first settled, long after the lifetime of the Tianyuan individual. It does not prove that the First Americans came exclusively from temperate East Asia (including the Russian Far East, China, and Japan) rather than the Siberian interior, but it increases the likelihood that some people from that region played a role in the settlement of the Americas, maybe as the elusive Population Y.

South of the Ice

It is obvious that the two American continents beyond Alaska are unlikely to have been settled much before 16,000 years ago, even though greater ages for human settlement have had their champions, as discussed above for Chiquihuite Cave in Mexico. Why do I think this? Aside from the Glacial Maximum ice sheets blocking the way all around the western and northern coastlines of Canada, at least until they melted in warming postglacial conditions, the main reason is that the two American continents contain hundreds of caves with layer upon layer of archaeological record dating to within the past 16,000 years or less, never (except possibly for Chiquihuite) convincingly more. Unless all of these caves were only formed 16,000 years ago, which I gravely doubt, this situation leaves little possibility for archaic hominin or modern human arrivals long beforehand. It is hard to believe that pre–Glacial Maximum human arrivals in such food-rich locations as the American continents could simply have died out, especially since post–Glacial Maximum population trends went rapidly in the opposite direction.

Once they had arrived, the First Americans moved quickly. By 15,000 years ago they were already south of the ice, at a site called Cooper's Ferry in Idaho. Here, they made elongated bifacial projectile points with stems for hafting flaked around their bases. These were remarkably similar, as noted above, to the stemmed projectile points from Upper Paleolithic Hokkaido in Japan (figure 6.4).[46] As colonizing parties spread out through North America, hunting and trapping mammoths, mastodons, bison, horses, and other large mammals, the shapes of their projectile points began to vary. "Western stemmed" projectile points like those from Cooper's Ferry have been found down the western side of the continent, together with leaf-shaped points and small backed tools, and sometimes shell fishhooks. East of the Rocky Mountains and in South America, there were other developments.

After 13,000 years ago, people in the central and eastern United States specialized in the production of the remarkable Clovis points, with their fluted bases to aid fixing into a wooden spear shaft. Bifacial projectile points in various shapes also characterized the oldest archaeological sites in South America, one being the 14,600-year-old waterlogged site of Monte Verde, located at 42° south in southern Chile. While the large mammal species shared with Asia kept North American hunters busy, their South American counterparts were dealing with a different fauna of giant sloths, armadillos, camelids (related to llamas and alpacas), and rhino-like creatures.[47]

The First Americans did not lose the boating tradition that had assisted their migrations from Asia. By at least 12,000 years ago, they were paddling with their shell fishhooks across 10-kilometer-wide sea passages to reach small islands off the coast of California, to which they sometimes brought obsidian tools sourced 300 kilometers away in eastern California.[48] One mystery that puzzles me, however, is that there is no evidence for a human arrival in the Caribbean Islands until around 6,000 years ago. The shortest sea crossing from South America or Mexico to the Caribbean Island chain is about 150 kilometers, but all the way in warm tropical waters. People were crossing similar distances to reach the Ryukyu Islands by at least 30,000 years ago.

Why did it require 10,000 years for people to reach these islands after they first arrived in the Americas? Has important evidence for human settlement in the Caribbean simply not yet been found? I have no idea, but there is a major challenge here for future archaeological research.

Despite questions over their voyaging ability, First American human settlers did exceed 4,500 meters of altitude at Cuncaicha rock shelter in the High Andes by 13,000 years ago. Here, they hunted deer and vicuñas (wild ancestors of alpacas) with "fishtail points" of a type found widely in the southern part of South America.[49] This altitudinal achievement into oxygen-deficient circumstances paralleled the independent and much older altitudinal achievements on the Tibetan Plateau by Denisovans and Upper Paleolithic *Homo sapiens*.

Some of the First Americans also brought dogs from Siberia, where it is possible that wolves were first tamed and domesticated.[50] The source of all of our dogs today is the Eurasian wolf; the native American wolf species was never domesticated, according to genetic data. The dogs that originally reached the Americas gave rise to the dogs associated with the later American cultures, alas now all extinct and replaced by Old World dog breeds during European colonial times, apart from the sled dogs (huskies) used in the Arctic.

The Holocene Settlement of Arctic Canada: Paleo-Inuit and Thule Inuit

The Odyssey of Paleolithic hominins entering uninhabited landscapes is now approaching its end. The settlement of the Americas meant that all the continents in the world, apart from Antarctica, had been reached by hunter-gatherers. Yet not all parts of North America had been settled. The northern coastline of Canada remained under sea ice until the peak-warmth conditions of the mid-Holocene, as indicated in figure 6.2.

The first humans to settle the Canadian Arctic coastline and islands, including Greenland, were the ancestors of the modern Eskimo-Aleut-speaking peoples, including the modern Inuit. As the Pleistocene ice sheet retreated, this coastline, like the Arctic Sea coastline of Eurasia, began to offer a good living for those who could hunt off treeless tundra and sea-ice, targeting large mammal species such as seals, Arctic bowhead whales, walrus, polar bears, musk oxen, and caribou (reindeer). The key to survival was to withstand the winter, with its lean food resources, by means of efficient technology for clothing, hunting, food storage, and shelter.

The Arctic coastlines of Canada and Greenland opened for human settlement in the warmest part of the Holocene, around 5,000 years ago, when they were first settled by people who used what archaeologists term the "Arctic Small Tool Tradition," with its characteristic microblades. This originated in a homeland thought to be on the Siberian side of the Bering Strait and then spread rapidly east through Alaska and along the previously unoccupied northern coastline of Canada to become the first archaeological identity to reach Baffin Island and Greenland, by 2500 BCE.[51]

Genetically, the makers of the Arctic Small Tool Tradition were closely related to the First Americans as well as to the modern Inuit and other living populations such as the Yupik of western Alaska and the Aleutian Islanders.[52] Their geographical distribution makes it likely that they were early speakers of Eskimo-Aleut languages. They also traveled with dogs, and (as noted above) it has recently been demonstrated genetically that modern Greenland sled dogs are descended from wolves domesticated in Siberia over 9,500 years ago.[53]

The makers of the Arctic Small Tool Tradition appear eventually to have been forced into retreat by readvancing ice, and the high Arctic latitudes of Canada were almost devoid of human settlement during the first millennium CE. With the return of warmer conditions during the Medieval Warm Period (800–1300 CE), a second dramatic migration across northern Canada, that of the Thule Inuit, got underway.

By the thirteenth century, the Thule Inuit had separated from other Eskimo-Aleut speakers in Alaska and were migrating eastward, through territory crossed almost 4,000 years before by their Paleo-Inuit predecessors; eventually achieving a remarkably rapid 5,000-kilometer movement that again reached as far as Greenland. However, this time they found themselves not alone. The southern and western coastlines of this huge Arctic island were already inhabited when the Inuit arrived by Norse Vikings from western Europe, who had arrived from Iceland in 985 CE.

This was one of the first meetings to occur between Native Americans and Europeans (there had been an earlier one in Newfoundland, presumably involving Algonquians rather than Inuit), brought about by remarkable migratory achievements from both directions. However, interaction between the two populations appears to have been infrequent. It is worth remembering that only the Inuit survived continuously in Greenland to the present; the original Norse settlers having withdrawn during the fifteenth century.

We have now come to the end of the human story of primary settlement as far as the world's continental landmasses are concerned. The Eskimo-Aleut populations were by no means the last hunter-gatherers to migrate during human prehistory, but excluding faraway islands they were the last to settle continental terrain (excluding Antarctica) that had not previously witnessed the passage of human feet. What came next would be unprecedented within the five-million-year hominin career on earth.

7

How Food Production
Changed the World

THIS CHAPTER CONSIDERS the opening of Act IV of the Odyssey, the age of food production. Where and how did it start, and what were its impacts on human populations in terms of their numbers and densities? How did it spread, and why did it become so dominant in so many regions of the world during the past 11,700 years that have so far constituted the Holocene interglacial?

The rise of domesticated plant and animal food production during the Holocene, which commenced around 9700 BCE,[1] imposed major changes on humanity. Increasing supplies of food led to increasing numbers and densities of humans, eventually reaching the eight billion people that the world must support today. Food production formed the basis for the development of all state-level civilizations, both ancient and modern. The increasing transportability of major domesticated food species, both plant and animal, underpinned farmer migrations to new and suitable locations, including ones already occupied by hunters and gatherers. Wild hunted and gathered resources could not be controlled, increased, or transported to anything like the same extent as domesticated ones.

Perhaps the ancient Polynesians offer the most dramatic example of the transportability of a domesticated plant and animal economy. Between 2000 BCE and 1250 CE, their ancestors carried their domesticated crops and animals in boats across 16,000 kilometers of mostly tropical ocean, from homelands in Taiwan and the Philippines to as far as New Zealand and Easter Island, with some even reaching South America. Their transportable economy of pigs, dogs, chickens, and various nutritious fruit and tuber crops allowed them to increase their populations quickly and to settle on tiny and isolated islands that would never have been able to support purely hunting-fishing populations—the native resources were too few and too easily exhausted. The ancient Polynesians show us clearly that domesticated resources could be assembled, suitable for transfer by land or sea, and taken to places thousands of kilometers away from any agricultural homeland.

What Was Ancient Food Production?

In what follows, I use the general terms *food production, farming,* and *agriculture* interchangeably to refer to prehistoric subsistence systems that relied on domesticated plants and animals. Two more specific terms are *cultivation* and *domestication,* and these are central to the question of what happened to humanity between the cultural stages of hunting-gathering and food production.

Cultivation needs little explanation—farmers need to prepare fields for planting, protect growing crops from predators, organize harvesting, and construct storage facilities. The parallel concept for animals is husbandry. Domestication, however, is not merely synonymous with cultivation or animal husbandry. Instead, it involved morphological and genetic changes to the plants and animals that ancient humans chose to exploit for food and other purposes. Thus, while cultivation and husbandry were human actions applied during the management of plants and animals, domestication was a result of those actions. Some of the changes that marked ancient domestication were perhaps selected deliberately, especially when emergent farmers picked out favored plants or animals for planting or breeding. Plants were bred essentially for yields of food (including beverages) and fibers, and ease of harvesting and processing. Animals were bred for changes in size, color, degree of docility, strength for traction, and yields of milk and wool.

As far as the evolution of food production is concerned, it is obvious that the first developments would have been little removed from the hunter-gatherer background of exploiting the wild in terms of the levels of domestication that they expressed, given that all plant and animal food resources were wild in the first place. The road from exploiting the wild to exploiting the fully domesticated in different parts of the world required genetic changes that often took a millennium or more to come to fruition. Many prehistoric and ethnographic peoples straddled both hunter-gatherer and farming lifestyles, especially in environments marginal for full-scale food production.

Once it had developed, however, food production could be made more productive through "intensification," by investing more labor in the land to improve facilities such as fields and their yields. Rice shoots transplanted closely together into a sunken and plowed paddy field fed by an irrigation canal will produce much more food per hectare than a few grains dibbled into holes made with a digging stick on a dry hillside, waiting for monsoon rains that may or may not arrive. With the rise of complex cultures and civilizations during the Holocene, agricultural intensification increased as population sizes and densities grew.

The Advantages of Food Production

Subsistence farming feeds more people, who can live at much higher densities and under more sedentary circumstances, than does hunting and gathering. There is nothing new here. It is obvious from the ethnographic record, which indicates clearly that formerly mobile hunter-gatherers who have settled down and adopted agriculture have undergone rapid increases in birth rate, despite the increasing disease loads and infant death rates that come with sedentary and relatively unsanitary living.[2] The archaeological record supports this conclusion by revealing major increases in the numbers and areas of archaeological sites during the millennia of transition into full-scale agriculture, especially in regions with dense archaeological records such as Europe, the Middle East, East Asia, Mesoamerica, the central Andes, and southwestern Amazonia. Analyses of human birth rates from the ages at death recorded in human skeletons give the same picture.[3]

As for the background to population growth among farmers, agriculture provides soft cooked gruels that can be fed to infants, hence reducing the length of time needed for breastfeeding and thereby encouraging more frequent conception. Such weaning foods are more difficult for hunter-gatherers to find, especially if they do not store suitable cereal grains all year round. The mobile hunter-gatherer lifestyle meant that mothers needed to carry their infants until they could walk independently, and carrying two such dependent infants at once over large distances would not have been an attractive proposition.[4] Anthropologist Richard Lee once calculated that a !Kung (San Bushman) hunter-gatherer mother in southwestern Africa would have carried her child a total distance of 7,800 kilometers during its four-year period of dependency.[5]

Hunter-gatherer population numbers always remained low, constrained by the need for mobility and the finite supply of natural resources. Australian hunter-gatherer population densities during the early twentieth century ranged from one person per 2 square kilometers in fertile country down to one person per 80 to 200 square kilometers in deserts. Hunter-gatherers in particularly productive areas of California could apparently reach densities up to four persons per square kilometer.[6] Population densities before the Colonial Era population declines might have been higher, but this is not certain.

In contrast, traditional farmers cultivating fertile territory today can have population densities in the tens or hundreds per square kilometer of exploited land. The mathematics of compound interest, illustrated by Colonial Era and modern world situations, show what humans can be capable of. With a potential population growth rate of 2.4 percent per year—not uncommon among European colonial farmer-settlers in recent history—a founder population of fifty people will become 155 people in fifty years, a trebling. Fifty people increasing at a lesser rate of 0.8 percent per year will become 65,000 in 900 years.

Such figures make one think about consequences, even if growth rates on these scales were short-lived rather than constant, in response to opportunities provided by migration. We have many historical parallels in North America and Australia where European colonial populations who entered conquered landscapes with food production skills achieved family sizes that averaged between six and ten children, at least for a few decades before changing social conditions reduced the numbers. In European colonial circumstances, as presumably in any new and disease-free landscape, the vast majority of these children survived childhood. I discussed the case of Pitcairn Island in chapter 5, in which the Bounty population increased sevenfold in sixty years after arriving in 1790.

Colonial-Era immigrant populations often grew alarmingly from the perspective of dispossessed indigenous inhabitants, especially low-density hunter-gatherer populations, who also died from the lethal diseases such as smallpox, tuberculosis, and measles that the immigrants brought in, to which they had no resistance.[7] In the Americas, many agricultural populations suffered the same fate because they also lacked immunity to introduced diseases, likewise the Maori in New Zealand. The first migrating farmers might have had similar impacts on the hunter-gatherers they came into contact with thousands of years ago.

The Ancient Domesticated Species That Still Feed Us Today

What were the crops and animals that supported ancient food-producing cultures across the world? If you go to a supermarket you can easily observe the staple foods that we consume today, and have done for thousands of years. Virtually all meat in modern societies comes from chickens, cattle, pigs, and sheep. Other domesticated animals, such as dogs, goats, water buffalos, camels, llamas, and horses, pale into insignificance by comparison, although some of these were more important as food sources in the past than they are now. Let us also not forget fish: many ancient peoples raised them in artificial saltwater or freshwater ponds, and while they were not truly domesticated, they were certainly cultivated.

The most important staple food plants in the world today are protein-rich and easily storable cereals—wheat, barley, oats, rice, maize, and various millets. Legumes (podded plants that fix nitrogen in soil, also rich in proteins) include beans, lentils, peas, chickpeas, and peanuts. Important tubers include potatoes, sweet potatoes, yams, aroids (including taro), carrots, and manioc. Fruits and nuts include bananas, coconuts, breadfruit, citrus, apples, avocados, tomatoes, and many more, although, apart from bananas and breadfruit, few of these have ever served as staples, and many are difficult to store in a fresh state. Cucurbits include pumpkins and squashes, domesticated originally for their protein-rich seeds rather than their flesh. To these we can add many other categories of plant food, important in some ancient societies, including sago (starch extracted from the trunks of certain tropical palm species) and the grain-bearing leafy chenopods, of which the South American quinoa is perhaps the best known.

Of course, hundreds more plant foods, both domesticated and wild, are consumed by humans around the world today and have been throughout prehistory. But they do not equate in importance with those just named. Many of the plants listed above were also domesticated in different parts of the world from several different wild species—this is especially true for rice, millets, yams, cucurbits, and beans.

The Homelands of Food Production

If we now look at where these most ancient domesticated species originated, we quickly see that the vast majority came from several midlatitude and tropical regions that archaeologists have identified as having developed food production at early dates (table 7.1), especially the regions indicated in figure 7.1. It is no coincidence that four of these regions—the Fertile Crescent of the Middle East, East Asia, Mesoamerica, and the central Andes—were also the locations for the most ancient urban civilizations and empires in their regions of the world. Egypt was also an ancient civilization, but its farming was based on Fertile Crescent crops and animals.

I have selected several of these agricultural homelands for extended discussion in chapter 8 because of their high importance in two respects: for the origins of food production, and for the consequent human migrations that resulted in each case. I am not suggesting that these were the only places ever to be associated with the domestication of a plant or animal species, and a case can be made that the majority of the world's populations were shifting gradually toward food production as

TABLE 7.1. Seven significant homelands of food production and their essential characteristics.

	Latitude and climate	Altitude	Rainfall and growing season	Important native domesticated crops	Important native domesticated animals	Approximate era of agricultural transition[1]
Fertile Crescent	Temperate Mediterranean, winter rainfall	Intermediate	Winter	Wheats, barley, legumes	Cattle, sheep, goats, pigs	End Pleistocene/ Early Holocene
Yellow/Yangzi/ Liao River basins of China	Temperate monsoonal,[2] summer rainfall	Low	Summer	Japonica rice, foxtail millet, broomcorn millet, soya bean	Pigs (no cattle until the Late Neolithic)	Early to Mid-Holocene
Sahel and Sudan	Tropical monsoonal,[2] summer rainfall	Low	Summer	Sorghum, pearl millet, finger millet, African rice	None (sheep, goats and cattle from Fertile Crescent)	Middle to Late Holocene
Mesoamerica	Tropical, summer rainfall	Intermediate	Summer	Maize, beans, squash	Turkey	Mid-Holocene
Central Andes	Tropical, summer rainfall	High	Summer	Potato, quinoa, beans, squash	Llama, alpaca, guinea pig	Mid-Holocene
Southwestern Amazonia	Tropical, summer rainfall	Low	Summer	Manioc, squash, peanut	None	Mid-Holocene
New Guinea Highlands	Equatorial,[2] year-round rainfall	High	All year round	Banana, sugar cane, yam	None until after 3,000 years ago (Asian pigs and dogs)	Middle to Late Holocene

[1] For convenience, the Holocene is here divided into three phases: Early (9700–6000 BCE), Middle (6000–2000 BCE), and Late (after 2000 BCE).

[2] Monsoon climates exist in the warmer latitudes of Africa and Asia, and are affected by seasonal wind changes (the monsoons). Tropical latitudes lie between 5° and the Tropics of Cancer and Capricorn. Equatorial latitudes lie within 5° of the equator.

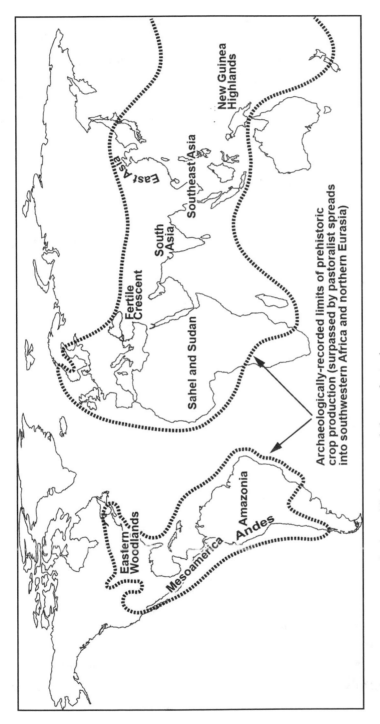

Figure 7.1. Major source regions for the world's domesticated plants and animals.

terrestrial resources expanded during the exit from the Last Glacial Maximum. But some populations shifted their behavior more conclusively than others, especially in environments with strong seasonality and large-seeded annual crop plants. In the long run, the descendants of those early opportunists had immense impacts upon the world.

Some of those early opportunist populations were backed immensely by their good luck in the lottery of wild plant and animal biogeography. The most productive and nourishing domesticated species, such as wheat, rice, maize, cattle, pigs, chickens and sheep, simply did not occur everywhere in the world in a wild ancestral state. Only a lucky few had access to them, although some species admittedly had more extensive wild distributions than others. Once they were domesticated, they spread, often rendering it unprofitable to domesticate similar wild species in other regions. This remains the case today for the species listed above—they have held their ground as core staples for many millennia and show absolutely no signs of being displaced any time soon.

Coincidence?

Did the agricultural homelands indicated in table 7.1 and figure 7.1 develop food production independently of each other? I believe that the answer is predominately (but perhaps not entirely) yes. One good reason is that the seven locations listed in table 7.1 were remarkably different from each other in terms of climate and environment, not to mention the crops and animals domesticated within them.

These differences ranged from equatorial to temperate in climate, highland to lowland in altitude, and winter rainfall to summer rainfall in seasonality, or all-year-round rainfall in the case of equatorial New Guinea. There were also major differences in the availability of important food categories such as cereals and large herbivores. These variations alone make it unthinkable that food production could have begun in just one place and then spread to everywhere else in the world.

Despite their initial independence, however, some of these seven homelands certainly developed contacts with other regions after food production had developed, such that important plant and animal species were quickly transferred between them. Two important examples dealt with later include the movement of Fertile Crescent domestic animals (sheep, goats, cattle) across the Sahara and Arabia into tropical Africa, and the movement of early forms of domesticated maize from Mexico into South America.

Dogs were special in this regard. As a major species domesticated from Eurasian wolves by Paleolithic hunters and gatherers, they spread virtually everywhere in the world, including to Australia (but not Tasmania) and right through the Americas.[8] No other domesticated species, plant or animal, was able to do this during prehistory. Dogs, potentially, would have been present in all regions where food production eventually developed, as examples of the benefits that might accrue from taming and domesticating an animal. Perhaps lessons learned from Paleolithic dogs were also applied during the Neolithic to other animal species.

What Did Humans Do to Plants and Animals in Order to Make Them Domesticated?

In the cases of the cereals and legumes, humans developed varieties that kept seeds on the stalks when they were ripe, rather than falling free ("shattering") and disseminating naturally. They also selected plants with large seeds and with loose seed cases that could easily be threshed and removed. Two other important characteristics of domestication were synchronous ripening of all seeds on a given plant, and the ability of those seeds to germinate during seasons that required dormancy in the wild state.

All in all, a fully domesticated cereal, legume, or tuber offered the potential for planting at any time of the year in any part of the world with a climate suitable for its propagation, to be harvested and processed easily without wastage, and to yield more protein and carbohydrate per hectare than any wild counterpart. We can also rephrase this sentence to fit animals, in terms of ease of breeding in captivity, manageable demeanor, productivity in terms of meat, milk, or wool, and even factors of shape and color that accorded with religious beliefs.

Did the First Farmers Promote Plant and Animal Domestication Deliberately?

Whether plant and animal domestication was deliberate is an important question, and I wish I knew the answer. I do think that some of our ancestors might deserve the benefit of the doubt. Presumably they would have had the intelligence to notice how a planted field of newly domesticated cereals behaved differently from a neighboring and unplanted stand of wild equivalents. Israeli researchers Shahal Abbo and Avi Gopher certainly agree when they observe that, in discussing early agriculture in the Fertile Crescent, "the domestication of the Levantine plant package was all about choosing the species for domestication and identifying and selecting the suitable mutants from the standing genetic variation, [and it was] thus knowledge-based, conscious, rapid."[9]

Nevertheless, many archaeologists and archaeobotanists currently believe that domestication was a gradual and unconscious process for most populations involved. For instance, some cereal harvests in the Fertile Crescent and China only consisted of fully domesticated grains after 1,000 or 2,000 years of relatively slow development from the wild state.[10] This slowness might have reflected the unconscious nature of the process, but it might also have reflected continued exploitation of wild stands alongside domesticated ones, with consequent mixing of grains during harvesting, processing, and replanting.

The debate over intentionality is likely to continue. In terms of archaeological reality, there are some sites that indeed reveal a gradual start to plant domestication. Yet there are others where it was undeniably rapid.[11] As so often in understanding human prehistory, some accommodation between the two opposed views might be needed.

Why Domestication?

The hunter-gatherer economy that created humankind no longer exists in a pristine state, but its crucial importance during more than 99.9 percent of the Odyssey is irreversibly stamped upon all of us. According to some commentators, farming just brought ill-health and hard labor to people who were previously carefree hunters and gatherers, affluent and free from oppression. If this were really true, why did ancient peoples bother to domesticate plants and animals? Can the ethnographic record provide any answers?

Unfortunately, the ethnographic record of the past century or so does not reveal any hunter-gatherer populations undergoing an internal shift from hunting and gathering into agriculture that was free from outside pressure, either from state authorities enforcing sedentism, or from farmers engaged in occupying their former hunting and gathering territories. There were hunter-gatherers at the time of European contact in California and Australia who were managing wild resources through regeneration of vegetation by burning, occasional replanting of wild tuber segments, and harvesting of wild cereal seeds in places where grasses were plentiful.[12] But these practices did not lead to domestication of the targeted species, and there was no selection through the conscious replanting of ripe seeds for desirable characteristics in future generations. The hunter-gatherers of California and Australia, despite their careful attention to the management of landscapes and resources, cannot illuminate for us the transitions into food production that occurred many millennia ago in other parts of the world.

Debates about why humans developed food production often turn to issues connected with the environment. A century ago, archaeologist Gordon Childe thought that farming developed when humans crowded into oasis situations during postglacial droughts and had to turn to farming to feed themselves. Nowadays, we know that postglacial climates across the world generally became wetter, not drier, therefore some archaeologists construct answers around improving postglacial environments that encouraged population growth, sedentary settlement, and consequent stress on wild food resources. Whatever the real answer, and I suspect that elements of both drought and deluge may have played their parts, it is a good bet that our modern human world could never have come into existence if we were still living under the climatic conditions of the Last Glacial Maximum.

At that time, between 25,000 and 18,000 years ago, ice sheets, cold tundras, and semideserts would have occupied many regions that have high agricultural productivity today. Human population numbers and densities during the Last Glacial Maximum were low, according to the archaeological record, as was overall global environmental productivity in terms of the resources available for hunting and gathering. Climates were also unstable, with dramatic changes in temperature and rainfall sometimes taking place over just a few decades.

The oldest transition to full cereal and legume domestication in human prehistory occurred in the Fertile Crescent, immediately after the Younger Dryas glacial readvance dated between 10,800 and 9700 BCE (see figure 1.2 and chapter 3). After the Younger Dryas, Holocene climates attained greater levels of warmth, rainfall, and general stability than previously. Unless this was all sheer coincidence (which I doubt), we must assume that the Last Glacial Maximum was lacking in essential elements as a driver of agricultural development, an observation made by others almost twenty years ago.[13] On the other hand, the Holocene, after the Younger Dryas, clearly had what it takes for independent agricultural developments to occur in several different parts of the world.[14]

One remarkable thing about the Younger Dryas is that, apart from bringing on a millennium of unexpected glacial weather in the midlatitudes, it ended incredibly quickly. Within forty years, around 9700 BCE, the world switched from cold and dry glacial conditions to warm and wet conditions almost like those of today.[15] I find this rapid ending to the Younger Dryas to be thought provoking. The vast bulk of human activity related to food production across the world has occurred within the past 11,700 years of warmth and environmental stability that followed this dramatic Younger Dryas downturn at the end of the Pleistocene. Was it entirely coincidental? In my opinion, it was not.[16]

The weather was not the only factor that changed with the beginning of the Holocene. Another major postglacial environmental change was the rise in global sea level, as glacial melt water flooded into the oceans. Last Glacial Maximum coastlines plunged deeply offshore beyond the edges of the continental shelves, with productive environments such as tropical coral reefs and temperate "kelp highways" confined to small refuge areas. Postglacial meltwater from ice sheets drowned vast expanses of those exposed continental shelves, producing millions of square kilometers of warm and shallow coastal ocean with greatly enhanced marine resources. I discussed this topic in chapter 6 in connection with the Pacific Ocean Kelp Highway as a factor in the human settlement of the Americas. With sea level rise, early Holocene coastal human populations could, and did, grow rapidly in numbers.

Were the enhanced environmental conditions of the early Holocene, after the end of the Younger Dryas, the basic background enablers of food production? Perhaps, in part, but there must be a proviso. If the warming climate and the increased climatic stability were to be the only causes for the development of food production, then one would expect the whole world to have gone through the transition at the same time, given that so many different regions went through the same climatic changes as those that ended the Younger Dryas and heralded the Holocene in northern Eurasia. Yet, agriculture did not begin everywhere at the same time. It is clear that independent developments occurred in different places, at different times, distributed through many millennia during the Holocene. There must have been other factors at play.

In my view, the resources available in specific locations, in terms of what crops and animals were eventually domesticated, were also crucial factors in agricultural origins. Not all potential resources were equal, as we saw above when considering just how few major crops and animals, domesticated in so few regions, feed such a vast proportion of the global population today. As Jared Diamond noted twenty years ago, only fourteen of the 148 species of mammalian herbivores in the world that weigh more than 45 kilograms were actually domesticated in prehistoric times.[17] Furthermore, the Fertile Crescent, the first region in the world to develop agriculture, had more of these species in the wild than anywhere else. It is not at all surprising that food production appeared first in those regions that were the homelands for the major species of plants and animals that still feed us today.

Here comes an essential question. Were seed selection, deliberate seed planting, and eventual domestication all reactions to a combination of both an increasing food supply and a periodically stressed one? Such situations could have developed as postglacial wild resources increased and became more heavily exploited by increasingly sedentary and "affluent" hunter-gatherers. However, if these people were increasing in number faster than their wild food resources, as we might expect in situations of relative affluence, they would soon have faced trouble, given that wild resources are difficult to multiply without switching to some form of cultivation and management.

Many archaeologists have certainly emphasized the importance of this kind of population pressure as a stimulus to agricultural development in the past.[18] In the Fertile Crescent, for instance, it appears that gazelles had been hunted almost to extinction at about the time that sheep and goats were first domesticated. In China, rice was domesticated close to the Yangzi River, right at the northern edge of its range as a wild plant, where small annual changes in the duration of the growing season and the monsoon rains would have had large impacts on wild rice distribution. Such situations reflect both opportunity, and stress.

I am tempted by this combined response-to-affluence and response-to-stress explanation. From this perspective, the ultimate cause of food production was, on the one hand, the postglacial warming that led to denser and more sedentary human populations. On the other hand, these populations were afflicted from time to time by stresses in food supply and always on the lookout for food resources that could be managed and increased in quantity. By developing food production, they could take advantage of both types of situations.

And life was not only about food. Other persuasions toward domestication of plants and animals would have come from the needs for fiber and textile materials, as societies became more complex in terms of the social roles that needed to be announced through costume and body ornamentation.[19]

In the final resort, food production developed independently in several different parts of the world, initially in those regions that had the largest repertoires of productive resources that were amenable to domestication. As the process developed in favored homeland regions, the populations indigenous to those regions and their languages, along with the crops and animals they domesticated, expanded farther and farther afield. These trends overwhelmed, in demographic terms, the many independent trends toward agriculture in intermediate and environmentally less favored regions. The following chapters are essentially about these expansions and the human populations they produced.

With food production, the world of humanity has never really looked back. Our global task today is to build upon our unity as a species in order to share our resources across the whole of the human population, and also to protect those resources from the malignant repercussions of uncontrolled exploitation.

8

Homelands of Plant and Animal Domestication

IT IS NOW NECESSARY to examine what actually happened with the rise of food production in different parts of the world. I do this in two stages. This chapter discusses the Holocene unfolding of food production in the major homeland regions of agriculture and their hinterlands, drawing mainly on archaeological records. Thus, we examine in turn the Fertile Crescent, East Asia, the African Sahel and Sudan, the New Guinea Highlands, the Andes, Amazonia, Mesoamerica, and the Eastern Woodlands of the United States. For each, we follow the course from the beginnings of domestication to the development of large and mostly sedentary agricultural societies.

The following chapters will then examine the crucial question of what happened once people had developed efficient food production in these homeland regions with growing and land-hungry populations. I use data drawn from multiple sources because we must trace not only the archaeological records but also the movements of the people themselves and their languages.

We begin this chapter with one of the most important homelands of food production, not just in terms of its status in the history of intellectual enquiry, but also in terms of its impact upon the world.

The Fertile Crescent

The Fertile Crescent of the Middle East was the homeland for one of the world's most important agricultural repertoires. Agriculture with domesticated crops and animals developed there by 8500 BCE, with staple food species that included cattle domesticated from the Eurasian aurochs, sheep, goats, pigs, einkorn and emmer wheat (two different species), barley, lentils, peas, chickpeas, and broad beans.[1] Fruits such as figs, grapes, olives, and dates followed between 6500 and 3500 BCE.[2]

It is perhaps no exaggeration to state that the societies that pulled this repertoire together had a longer-lasting impact on the population history of western Eurasia than Alexander the Great, the Roman Empire, the Ottoman Empire, and the British Empire all rolled into one. This powerhouse of food energy evolved in a Mediterranean climate with winter rainfall and summer drought, meaning that the food plants had annual (as opposed to perennial) growth habits, with seed dormancy during the hot dry summer. In practical terms, ancient Fertile Crescent farmers harvested the grains of these plants in the spring, selected some to store through the summer, processed the rest as food, and then planted the stored grains when the rains returned in the autumn.

The human population that grew out of this agricultural system was enormous and highly migratory. One could stretch world history to its limits and suggest that the outpouring of populations from Europe during the post-1492 CE Colonial Era was a replay of what happened during the Neolithic, between 6,000 and 9,000 years before, because it was based essentially on the same Fertile Crescent food species. By 4000 BCE, Neolithic farmers from the Fertile Crescent had migrated through Europe to as far as Ireland and Scandinavia, into Africa to as far south as Sudan, through the Middle East toward the Indus Valley, and around the Black Sea toward the steppes of central Asia.

These migrations out of the Fertile Crescent are well attested in the archaeological record, in ancient DNA from buried skeletons, and in the histories of two of the world's most important language families—Indo-European, spoken from the British Isles to Bangladesh (discussed in chapter 10), and Afro-Asiatic, spoken throughout most of the Middle

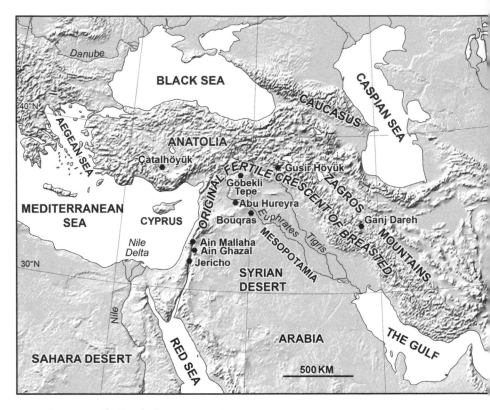

Figure 8.1. The Fertile Crescent.

East and North Africa (discussed in chapter 12). I return to these languages families later because their internal genealogies provide invaluable frameworks for understanding the course of human prehistory.

The original concept of the Fertile Crescent came from the writings of James Henry Breasted, an early twentieth-century Egyptologist. He defined it in 1916 as the crescent of fertile land between the desert and the mountains that curved from Palestine in the west, across the upper courses of the Euphrates and Tigris rivers in northern Syria and Iraq, and then down the western foothills of the Zagros Mountains of Iran toward the head of the Persian/Arabian Gulf (figure 8.1).

By the mid-twentieth century, archaeologists were beginning to investigate how agriculture had developed in this region. An early idea favored in the 1920s by W. J. Perry and V. Gordon Childe offered Egypt as a likely source, even though Egypt was not strictly a part of the original Fertile Crescent as defined by Breasted. They chose the Nile because it flooded reliably and regularly in Egypt every late summer and autumn (mid-August through September), fed by summer monsoon rainwater from its central African source region close to the equator. As Perry commented, in what was to become a rather famous statement,

> When the flood of the Nile came on at the end of the summer any grains of barley and millet that had escaped the attentions of the birds would be embedded in the mud, and, when the flood subsided in the autumn, would sprout and grow rapidly in the genial warmth of the Egyptian winter. . . . Thus year after year the gentle Nile would, by means of its perfect irrigation cycle, be growing millet and barley for the Egyptians. All that would be necessary, therefore, would be for some genius to think of the simple expedient of making channels to enable the water to flow over a wider area, and thus to cultivate more food.[3]

Would that it were all so easy. I am not sure if Perry realized that barley was not indigenous to the Nile Valley, and that although sorghum millet was indeed domesticated in the monsoonal (summer rainfall) upper Nile Basin in Sudan, there is no evidence that the Egyptians living far downstream in the winter rain belt were also responsible for its domestication. Egypt is no longer regarded as a source of early food production by archaeologists and natural scientists, who now agree that all of its major domesticated plants and animals were introduced from the Fertile Crescent, with the singular exception of the donkey, domesticated from the North African wild ass.

What is more, the Nile Delta and the valley-bottom alluvium that nurtured one of the world's greatest ancient civilizations hardly existed in 8500 BCE, when agriculture was already beginning in the more promising rainfall conditions of the Fertile Crescent. During the Last Glacial Maximum, the lower course of the Nile was incised to meet a Mediterranean sea level more than 100 meters below that of the present; even at 8500 BCE that sea level was still down between 40 and 50 meters. The lower Nile Valley in Egypt did not build up enough alluvium to support intensive irrigation-based food production until possibly 6000 BCE,[4] when Fertile Crescent farmers were perhaps already starting to move in.

What about Mesopotamia, close to the opposite end of the Fertile Crescent, and another center of ancient civilization? The Mesopotamian lowlands along the Tigris and Euphrates rivers in Iraq gave the world the Sumerians, Elamites (in Iran), Akkadians, and Babylonians. But, as with the Nile, the Mesopotamian delta lands at the head of the Gulf also did not exist in 8500 BCE, when farmers were emerging in the Fertile Crescent to the north and west. As in Egypt, the climate in lower Mesopotamia was too dry to support farming by rainfall alone; and because the Tigris and Euphrates (unlike the Nile) rose in the northern hemisphere, in what is now Turkey, they flooded inconveniently from winter rains and snowmelt in the spring, after the winter growing season was over. Agriculture in the Mesopotamian lowlands required canals to be dug through river levee banks in order to tap the low autumn water levels for irrigation when planting took place. The first farmers did not find these lowlands attractive, and we have no evidence for agricultural activity here until the ancestral townships of the Sumerians were founded about 6000 BCE, by which time irrigation techniques were well under development.

The correct answer to the question of where agriculture began started to appear in the 1950s. American archaeologists Robert Braidwood and Bruce Howe confronted the archaeological community in 1960 with one of its most significant questions: "How are we to understand those great changes in mankind's way of life which attended the first appearance of the settled village-farming community?"[5] Braidwood specified what he termed the "hilly flanks" of Breasted's Fertile Crescent as the key location for early agriculture and village life, rather than Egypt or lowland Mesopotamia. The hilly flanks had sufficient winter rainfall for cultivation to occur without irrigation. Most importantly, they were the homeland region for virtually all of the major crops and animals that formed the Fertile Crescent domesticated repertoire. Braidwood began to realize this through his excavations in Iraqi Kurdistan during the 1950s and 1960s, and two generations of archaeologists and other scientists since his time have given us what today is a fairly clear picture.

The Natufian

The road toward food production and domestication in the Fertile Crescent was the first in the world to be traversed successfully. A key archaeological culture in the western part of the region, termed the Natufian by British archaeologist Dorothy Garrod, is believed to contain the oldest evidence for increasing population density and settlement sedentism, dating between 12,000 and 10,000 BCE.

Natufian settlements covered up to 3,000 square meters. One large example at Ain Mallaha (also known as Eynan) in northern Israel contained an estimated fifty oval or circular houses with stone wall foundations, although it is perhaps unlikely that all were actively inhabited at any one time. They were arranged in a circle around a central open area that contained pits used for storage and human burial. Ain Mallaha had all the hallmarks of a sedentary settlement that was lived in continuously, perhaps through several generations.

The Natufians were not yet fully-fledged farmers with domesticated resources, although they did have domestic dogs at Ain Mallaha (one was buried alongside a woman). They hunted onagers (the Asian wild ass), gazelle, deer, and wild boar. Sharp stone blades and microliths were used for harvesting wild cereals and legumes in the spring, a practice recognizable through a shiny area of silica gloss left behind from the plant stalks on the tool edges. Hollowed stone mortars were used to grind cereal flour, and there has recently been a remarkable discovery of a carbonized piece of Natufian flat unleavened bread dating from 12,000 BCE in Jordan.[6]

The Natufians represent a fundamental foundation in the emergence of agriculture in the Fertile Crescent. It is easy to forget that they lived during the Pleistocene, not the Holocene, and that in western Eurasian archaeological terms their stone tools would put them in an "Epipaleolithic" (or Mesolithic) pigeonhole. However, a recent analysis of the Natufian archaeological record indicates that there was perhaps a tenfold increase in the size of the human population in the northern Fertile Crescent between 11,400 and 9700 BCE, the latter date marking the end of the Younger Dryas glacial advance.[7]

This comes as something of a surprise, because the Younger Dryas, as a millennium of renewed cold, has long been regarded as a negative period in human demographic history. As I noted at the end of chapter 7, it is more likely to have been the warm, wet, and stable conditions that commenced rapidly when the Younger Dryas ended that provided the main stimulus for plant and animal domestication to develop.

However, I also suggested in chapter 7 that periodic stresses might have persuaded people to turn to planting and cultivation to maintain food supplies. Is it possible, therefore, that Natufian hunters and gatherers increased their population density in the warm and favorable conditions that preceded the Younger Dryas, then reacted with more efficient methods of wild resource cultivation as the renewed cold made its mark, before finally emerging to commence plant and animal domestication in earnest at the beginning of the Holocene?[8]

The Fertile Crescent Neolithic

After the Natufian and the Younger Dryas, climatic conditions in the Fertile Crescent improved markedly. For the next few millennia, summer rains from the African and Indian Ocean monsoons spread northward, bringing summer moisture to the Saharan and Arabian deserts and overlapping with the existing belt of winter rainfall in the Middle East. The Natufians and their contemporaries occupying Braidwood's hilly flanks were ready to go.

The Fertile Crescent Neolithic, which emerged from the Natufian and related contemporary cultures, lasted between 10,000 and 5500 BCE, after which time increasing numbers of copper implements in archaeological sites announced the coming of the Chalcolithic and Bronze Ages. The Neolithic spanned two main phases: Pre-Pottery Neolithic and Pottery Neolithic, with pottery coming into common use after 7000 BCE.

The transition from wild plant and animal exploitation into a dependence on domesticated resources took place almost entirely during the Pre-Pottery Neolithic. Domesticated (nonshattering) cereals began to appear around 8500 BCE and accounted for about 50 percent of all plant food intake in the Fertile Crescent by 7500 BCE. By 6500 BCE that percentage had become almost 100 percent.[9]

The Pre-Pottery Neolithic in the Fertile Crescent began with a bang in terms of community construction projects, especially stone buildings that were unparalleled elsewhere in the world at that time. The first major surprise came from excavations during the 1950s by British archaeologist Kathleen Kenyon, who excavated into the side of the 14-meter-high *tell* (Arabic and Hebrew for a stratified settlement mound, also *tepe* in Persian and Kurdish, and *höyük* in Turkish) at Jericho in the Jordan Valley.[10] Near the base of this large mound, Kenyon exposed part of a 2.5-hectare town that contained circular houses built of sun-dried mud bricks, flanked by a rock-cut ditch and a stone defensive wall at least 4 meters high (it is not clear if the wall extended all around the settlement). The wall adjoined, on its inside, a monumental stone tower 10 meters in diameter and 8 meters in surviving height, with a staircase of twenty-eight steps inside leading to its top (figure 8.2C). This Jericho Pre-Pottery Neolithic settlement was rebuilt several times through many centuries, and we now know that it was initially constructed, with the tower, at a rather astounding date of 9000 BCE.

Surprisingly, perhaps, the initial builders of the Jericho settlement were still at an economic stage prior to any observed domestication of animals and plants. They were certainly cultivators, however, and carbonized remains of wild cereals and legumes have been found in the site. But the people of Pre-Pottery Neolithic Jericho, even with their wall and tower, were still only at an early stage in the transition from a hunter-gatherer to a food-producing lifestyle.

Contemporary with the Jericho tower, but located over 700 kilometers to the north near Sanliurfa in southeastern Turkey, lies Göbekli Tepe, close to the fertile agricultural plains of the Balikh River, a tributary of the Euphrates. This remarkable site contains what must be the world's oldest multibuilding religious sanctuary built of stone. Göbekli Tepe was also founded around 9000 BCE, as a group of circular (and later rectangular) stone structures with enclosing walls, circumferential stone benches, and T-shaped stone pillars (figure 8.2A and B).[11] There are some smaller circular buildings that perhaps served as houses, but it is the ritualized sanctuary aspect of the site that is the most striking. Originally, the circular sanctuaries were constructed on the rock surface of the hilltop, with their stone pillars set in slots. Later, as the quantity of soil and debris deposited on the site increased, they were dug down into the accumulating deposit and lined by enclosing walls.

The T-shaped pillars were set radially inside each circular enclosing wall, with two standing separately in the center. These central pillars are up to 5 meters high and can weigh up to 8 tons. They are amazing in their workmanship, given that each was shaped from a single rectangular block of stone using hard stone pounders and abraders. It is thought from the use of ground penetrating radar that there are between 100 and 200 of them in the whole site, from different construction periods. Many are decorated in low relief with birds and animals, often poisonous or dangerous ones like scorpions and lions, and some have representations of human arms and hands.

The original excavator of the site, German archaeologist Klaus Schmidt, believed that the enclosures were used for ancestral and funerary rituals, with the corpses perhaps exposed for defleshing by crows and vultures. Bones of these species were found in the filling deposits,

Figure 8.2. Pre-Pottery Neolithic architecture in the Fertile Crescent. This figure illustrates the Pre-Pottery Neolithic shift from circular to rectangular architecture. (A) A decorated pillar at Göbekli Tepe, carved with enigmatic rectangular items (the three "handbags"), raptorial birds, a threatening scorpion, and a headless ithyphallic human. Note the absence of a "mother goddess": she came later, with full agriculture. (B) The main excavated area at Göbekli Tepe with its circular buildings and T-shaped pillars. The cultivated Balikh Valley lies in the background. (C) The tower at Jericho, showing the location of the inner staircase, entrance at bottom, and the exit under the metal grille on top. (D) Circular Pre-Pottery Neolithic house foundations at Khirokitia, Cyprus. (E) The Skull House foundations at Çayönü, eastern Turkey. At top is an older circular building, overlain by a younger rectangular structure with cubicles for human bones from 400 individuals and a cluster of seventy skulls (removed by excavators). (F) The sun-dried mud brick walls of a multiroomed rectangular Pre-Pottery Neolithic house at Bouqras, Euphrates Valley, with low arched crawl-ways (blocked by the excavators) linking the rooms, and the round opening of a bell-shaped underfloor storage pit. Göbekli Tepe carved pillar reproduced courtesy of Verlag C. H. Beck, photograph by Klaus Schmidt. Göbekli Tepe panorama reproduced courtesy of Deutsches Archäologisches Institut Berlin (negative D-DAI-IST-GT-2010-NB-5845, photo by Nico Becker), Çayönü Skull House reproduced courtesy Mehmet Ozdogan. All other photos by the author.

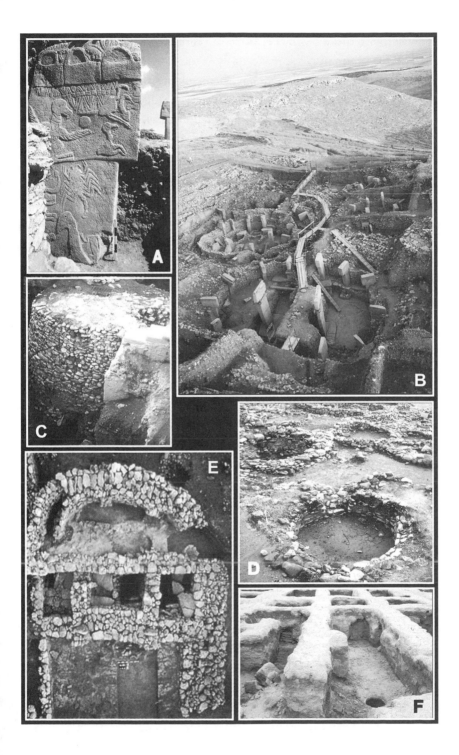

together with those of gazelle, aurochs (cattle), and onager, all wild animals.[12] Few human remains actually remained inside the enclosures, suggesting that the bones might have been taken elsewhere for veneration, but it remains likely that the Göbekli carved monoliths represented important human ancestors, perhaps imbued with supernatural powers.

Göbekli Tepe is not entirely unique, and other similar but less grandiose circular and rectangular constructions with carved pillars exist elsewhere in the northern Fertile Crescent. Furthermore, the site was not a temple complex in the middle of a township, as we find so commonly among the later civilizations of the Middle East. There is no sign of a central supervising authority. It was instead a prominent hilltop sanctuary or ritual site, where populations in the transition between hunting-gathering and early agriculture must have assembled for communal ceremonies.

Like the builders of the Jericho wall and tower, the people of Göbekli Tepe were capable of bringing together a large labor force under some form of leadership, which raises the question of how they were all fed. Perhaps they engaged in intensive cultivation of wild cereals, because charred remains of them have been found in the site and large numbers of hollowed-out stone troughs appear to have been used for grinding them to make porridge.[13] Was Göbekli Tepe a sanctuary complex for a community that, like Jericho, would fit the scenario that I presented at the end of chapter 7? In other words, was it a community in which a growing and fairly sedentary human population was exploiting a finite supply of wild food, thus creating a situation, fueled by population growth, sufficient to encourage it to invest more time in agricultural activity?

Jericho and Göbekli Tepe underline the power of the subsistence economy in the Fertile Crescent at about 9000 BCE, at a time when hunters and gatherers were settling into large sedentary communities but before the actual domestication of plants and animals had been able to progress far. The major Neolithic migrations out of the Fertile Crescent into Europe were still 2,000 years in the future, apparently because they required transportable and fully domesticated plant and animal resources. But does this mean that no migration occurred before this time?

Cyprus

In recent years, some dramatic and unexpected finds have been made on the Mediterranean island of Cyprus. Around 9000 BCE, hence long before full domestication had developed in the Fertile Crescent yet contemporary with the oldest constructions at Jericho and Göbekli Tepe, settlers crossed a roughly 50-kilometer sea gap from the southern coastline of Turkey (sea levels were then lower than now) to reach this island, together with their crops and animals. Their Anatolian homeland is fairly clear because they brought into Cyprus thousands of stone tools of shiny black obsidian (volcanic glass), quarried 250 kilometers inland from the Turkish coast in the Cappadocia region of central Anatolia.[14] At this early date, we have no other evidence for significant migration out of the Fertile Crescent, although Cyprus had been reached slightly earlier by a Mesolithic population that played a role in the extinction of an endemic Cypriot fauna of pygmy hippos and elephants.

By 8500 BCE, the Pre-Pottery Neolithic settlers of Cyprus had introduced many species of the northern Levant repertoire of plants and animals to the island, some already domesticated, and some still in the process of becoming so.[15] One can imagine skin boats or dug-out canoes being paddled across the sea carrying seeds for planting, trussed cattle (perhaps calves rather than heavy adult animals), goats, and pigs, together with dogs and cats. Mice and foxes hitched a ride as well, and the cats would have been useful to catch the mice, which would have eaten stored grain.[16] Sheep arrived a little later, as did fallow deer, the latter presumably translocated for hunting given that deer provided much of the Cypriot meat supply during the Neolithic. Cyprus had none of the wild ancestors of the Fertile Crescent plant and animal package, except possibly for wild barley. If the native pygmy hippos and elephants had already been removed by earlier Mesolithic settlers, there would have been little else to hunt, at least until the arrival of the deer.

In sum, Cyprus offers one of the clearest bodies of evidence about the commencement and transport of food production in the Fertile Crescent, both because of its relative isolation, and because most or all of its wild, semiwild, and domesticated plant and animal resources had to be transported from the Asian mainland. There is no confusion here between indigenous and introduced resources, even if some of those resources were still in the process of becoming domesticated when the Neolithic settlers arrived.

The puzzle, of course, is why Cyprus was reached so early, given that no Neolithic migrations from the Fertile Crescent entered Europe, including Greece and its islands, until after 7000 BCE, two millennia after early Pre-Pottery Neolithic settlers reached Cyprus. In theory, a large and relatively uninhabited island like Cyprus should have offered a bonanza to farmer-settlers, especially because it has been observed that einkorn wheat and barley developed domesticated characteristics more quickly on this island than in the Fertile Crescent proper, apparently because there were no wild wheat populations there to create wild/domesticated admixture situations.

Nevertheless, the Neolithic Cypriots appear to have been slow to enter into a full reliance on domesticated cereals, which might explain why they did not immediately migrate further. A recent report even suggests that Neolithic agriculture remained generally less successful on the island than in contemporary Fertile Crescent sites.[17] The cattle were also not very successful in terms of numbers. There are potential contradictions here that still defy full understanding.

A Land of Demographic and Cultural Growth

Between 12,000 and 7000 BCE, the Fertile Crescent witnessed social and cultural developments that the world had never seen before, starting with the Natufian and the Pre-Pottery Neolithic. We are now examining the roots of one of the oldest and most extensive landscapes of linked ancient cultures and food-producing economies anywhere in the world.

Apart from the fundamental development of food production itself, the Fertile Crescent at this time saw a major increase in the overall size of its human population, and in the average sizes of individual settlements.[18] The largest settlements increased dramatically in size during the later Pre-Pottery and Pottery Neolithic periods, perhaps reflecting an increasing need for safety in numbers in situations of increasing social instability. One major development at this time was a shift from the older circular dwellings that were suitable for single family units toward the large multiroomed and presumably extended-family rectangular structures that characterized Middle Eastern architecture thereafter (figure 8.2).

By 7000 BCE, townships of such close-set multiroomed rectangular dwellings, up to 12 hectares in extent, were relatively common in the Fertile Crescent, each housing several thousand people. One of the most remarkable examples is Çatalhöyük, near Konya in central Turkey, with its almost solid network of conjoined single-story rectangular rooms entered from above via ladders, probably with open activity areas on the flat earth-covered rooftops.[19] An example of this kind of architecture from Bouqras on the Syrian Euphrates can be seen in figure 8.2F. The larger rooms at Çatalhöyük were used for sleeping, burial (often in large numbers, in crouched postures under the sleeping platforms), food preparation, storage, and activities related to superstition, visible to us today in the form of bull horn cores set into low walls, "mother goddesses" in baked clay, and mysterious wall paintings. Open spaces left by abandoned rooms were used for rubbish disposal and as latrines.

Other settlements of a similar size include Abu Hureyra in northern Syria (circa 7200 BCE), and Ain Ghazal in Jordan (7000–6500 BCE), the last also noted for its twenty or so almost one-meter-tall human mannequins of lime plaster, built around frames of reeds and sticks. Some of these had bitumen eye pupils, and, as presumed representations of important ancestors, they appear to match the human skulls from Jericho and other sites that were plastered with the facial features of their original owners. Like the pillars of Göbekli Tepe, perhaps these plastered skulls and mannequins were a way of representing settlement founders and key ancestors in the genealogies of high-ranking families.

The Transformation of the Fertile Crescent Neolithic

By 6500–6000 BCE, during the late Pre-Pottery Neolithic and continuing into the Pottery Neolithic, there is evidence for a considerable decline in the number of occupied settlements in the Fertile Crescent.[20]

Populations appear to have been congregating, perhaps for protection, into ever-larger townships, some of which in turn then became abandoned. This decline may have been caused in part by an episode of climatic instability, known from paleoclimatic records in different parts of the world to have been occurring around 6200 BCE. It could also have been promoted by human impact on the environment, such as through deforestation for agriculture and for the firewood needed to burn limestone or chalk to produce lime mortar, used from the Natufian onward for house floors and to cover burials.[21]

It is probably no coincidence that the centuries around 6500 BCE also saw the commencement of Neolithic population movements out of the Fertile Crescent into Europe, North Africa, and Central Asia. We examine these movements in chapter 10, but I wonder if an overexploitation of resources in the Fertile Crescent led some populations to move out in search of new land. Such boom and bust situations were not uncommon during the later Neolithic, especially in the Middle East and Europe, although it is often frustratingly difficult to determine whether complex human activities such as migration resulted purely from environmental changes or from the compounding and uncontrolled results of human impact. Almost certainly, both sets of potential causes were in play in many circumstances.

These outward migrations would have been assisted greatly by the increasing birth rates that would have occurred as the transportable economy of domesticated resources was taken into new and fertile lands, just as such increases occurred in the early years of European Colonial-Era settlement in Australia and North America. The population boom emerging from the Fertile Crescent would have been assisted by an eventual reliance on pottery, useful both to prepare weaning foods for infants and to process dairy products for adults, who at that time lacked the genetic ability to digest the lactose in fresh ruminant milk. In order to be easily digestible for Neolithic populations, this milk needed to be boiled or fermented in pots into cheese or yogurt to break down the lactose.

Earthenware pottery often seems like a humdrum material to archaeologists because it is virtually indestructible, like stone, and thus common in sherd form in most archaeological sites in which it was used. However, it is striking that the only other regions of western Eurasia to have a definite Pre-Pottery Neolithic, beyond the Fertile Crescent itself, were Cyprus and the South Asian borderlands in Baluchistan (western Pakistan). Everywhere else appears to have received the cultural skills of pottery manufacture with the first Neolithic arrivals. It is almost as if humble pottery vessels were a crucial factor behind Neolithic dispersal out of the Fertile Crescent.

By 6500 BCE, the Fertile Crescent was ready to open like a flower, or perhaps explode like a wild legume pod. Its peoples and their transportable economies were ready for launching. We will follow their activities in chapter 10 because at this point another homeland of food production awaits our attention.

Early Farmers in East Asia

The second major Eurasian homeland of food production was focused on the Yellow, Yangzi, and Liao drainage basins in what is now China. In combination, these formed the homeland by 6500 BCE for pigs (a separate domestication here from that in the Middle East), foxtail and broomcorn millet, and short-grained rice of the *Oryza sativa japonica* subspecies. Soybeans and silkworms came later. This repertoire evolved in a temperate monsoonal climate characterized by summer rainfall, the opposite of the Mediterranean pattern of winter rainfall that characterized the Fertile Crescent. Again, the major crops were annuals, although *japonica* rice appears to have been domesticated initially as a perennial wetland species.

The human population that grew out of this food production system was also enormous, like that propagated by the Fertile Crescent, and highly migratory, even though the number of staple plants and animals domesticated in East Asia was smaller than in West Asia. Nevertheless, by 2000 BCE, farming populations from the East Asian homeland region had migrated throughout western and southern China, including the Tibetan Plateau, together with the mainland of Southeast Asia and northern South Asia, below the Himalayas. They

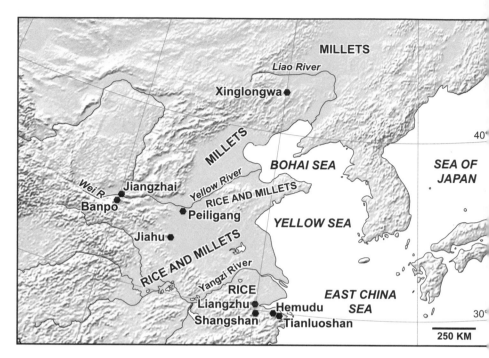

Figure 8.3. Agricultural homelands in East Asia.

also spread northward and eastward into the Russian Far East, Korea, and Japan, across ocean passages through Taiwan and the Philippines into Indonesia, and eventually into the empty islands of Oceania beyond New Guinea and the Solomons. Some of them ultimately became the Polynesians (including the New Zealand Maori) and the Malagasy of Madagascar (chapter 11).

The East Asian homeland as a whole straddles a similar average latitude to the Fertile Crescent, between 30° and 45° north (figure 8.3).[22] Rice and millets matured here during the wet summer monsoon season, and the high rainfall and more extensive forests in this homeland region meant that Neolithic people built predominantly in wood and other organic materials, rather than in stone or sun-dried mud brick. Hence, Chinese archaeologists excavating early Neolithic settlements trace postholes rather than freestanding stone foundations, except in lower Yangzi wetland sites where wood preservation can sometimes be remarkable (figure 8.4E).

The East Asian Neolithic also had pottery from the beginning, unlike the Fertile Crescent, reflecting no doubt a preference for the preparation of rice and millets by whole-grain boiling rather than by grinding into flour. All in all, the Fertile Crescent and East Asian routes into food production were culturally very different, as we might expect, given the 7,500 sometimes inhospitable kilometers of the Asian continent that lay between them.

The East Asian homeland stretched 2,000 kilometers from north to south, mainly around the inland edge of the vast North China alluvial plain. This agricultural powerhouse of the modern Chinese nation is drained by the lower courses of the Yellow and Yangzi rivers, and much of it today, toward the sea, is covered by recent alluvial sediments. The major Neolithic sites occur on slightly raised areas of older alluvium around the inland edges of this plain and in small tributary valleys. Indeed, when agriculture was just beginning, much of the plain was under a shallow sea owing to the postglacial sea level rise.

Millets and Rice

The East Asian homeland comprised three overlapping sub-homelands of food production, with millet agriculture in the Liao and Yellow River valleys in the north, and *japonica* (mostly short-grained) rice agriculture in the south, in and around the low-lying plains of the middle and lower Yangzi. These three homeland regions were clearly interconnected; both rice and millets were often grown together, although the Liao River valley was climatically beyond the early range of rice. Each region had pottery and stone tools that overlapped in style and shape all the way from modern Shanghai to Beijing, and beyond.

The inhabitants of the Liao and Yellow River basins started to domesticate broomcorn and foxtail millet by 7000 BCE, and domesticated (nonshattering) millet grains appear in archaeological sites by 6500 BCE. According to Israeli archaeologist Gideon Shelach-Lavi, the domestication process here began when warm postglacial climatic conditions encouraged a shift into sedentary hunter-gatherer settlements, similar to but a little later in time than the situation with the Natufian in the Fertile Crescent.[23]

Matters were more complicated for rice. The wild perennial ancestor of *japonica* rice was a native of warm-climate permanent swamplands in southern China. Its distribution today only reaches as far north as Jiangxi Province, well south of the Yangzi. However, the climate was a few degrees warmer than now during the early and middle Holocene, and perennial wild rice was able to spread north for a relatively short time. It was eventually domesticated close to its temporary northern limit in the Yangzi and nearby Huai Basins, where humans began to grow it in monsoon season swamps that encouraged an annual habit and larger grains. The oldest rice remains in Neolithic village situations in China occur just south of the lower Yangzi Basin at about 7000 BCE, although the grains were still predominantly wild in morphology at that time.[24]

Major Trends in the East Asian Neolithic

Unfortunately, China has yet to produce excavated hunter-gatherer settlements with surviving house plans equivalent to those in the Fertile Crescent Natufian. The settlement record in northern China begins with early Neolithic agricultural villages with pottery (figure 8.4). The oldest of these in the north, dating to about 6000 BCE, belong to the Xinglongwa culture of the middle Liao River valley and the Peiligang culture of the middle Yellow River valley.

The largest settlements of these cultures covered more than a hectare and contained close-set, round, or roughly square single-roomed wooden houses with sunken earth floors, usually grouped around an open central space. Such villages were sometimes enclosed within an outer defensive ditch. These settlements are comparable in size to many earlier Pre-Pottery Neolithic settlements in the Fertile Crescent, for instance, that at Jericho with the stone wall and tower.

By 5000 BCE, the Yellow River villages were growing rapidly in size. Two major sites in the Wei Valley, a western tributary of the Yellow River, are Banpo and Jiangzhai, both belonging to the Yangshao culture. Both have outer defensive ditches, and Banpo (figure 8.4A) had about twenty-five single-roomed square and circular houses at any one time,

Figure 8.4. The Chinese Neolithic. (*A*) Model of a Yellow River Neolithic village. Note the defensive ditch, large central house, and the pottery kilns at rear left outside the ditch. Fifth millennium BCE. Banpo Site Museum, Shaanxi Province. (*B*) Stone reaping knife with a serrated edge, of a type used for both millet and rice harvesting in the Yellow River Neolithic. Sixth millennium BCE. Jiahu Site Museum, Henan Province, central China. (*C*) Earthenware pottery dish with rice chaff temper in its clay from Shangshan. Seventh millennium BCE. Shangshan Site Museum, Zhejiang Province. (*D*) A reconstructed pile dwelling of the Hemudu culture. Fifth millennium BCE. Hemudu Site Museum, Zhejiang Province. The staircase is not part of the reconstruction—a notched pole would have been more likely in 5000 BCE. (*E*) Tian-luoshan, Zhejiang Province: exposed house piles and a wooden walkway, and (*foreground*) an earth oven left unexcavated with baked clay heating balls. Tianluoshan Site Museum, Zhejiang Province. Fifth millennium BCE. Photos by the author.

within a 3-hectare area. Jiangzhai was larger, with about fifty houses in several separate groups, each associated with a large square community house. At Banpo, a separate cemetery for adults and a group of domed pottery-firing kilns were located outside the defensive ditch. Deceased children were buried in pots inside the village.[25]

The oldest village sites in the lower Yangzi region, associated mainly with still not fully domesticated rice, belong to the Shangshan culture and date to about 7000 BCE. As well as eating the rice, these people also used the husks for tempering the clay they used to make their pottery. Larger villages were established around 6000 BCE on the riverine lowlands and deltas that were coming into existence, in both the middle and lower Yangzi, as the postglacial sea level stabilized. By 5000 BCE, rice was being grown in small enclosed paddy fields that could retain rainwater, remote ancestors to the much larger rain-fed or canal-irrigated paddy fields that one can see widely in eastern Asia today.[26]

The villages associated with these rice fields in the Yangzi region contained long timber houses raised above the ground on posts, as revealed most clearly by the fifth millennium BCE remains at Hemudu and Tianluoshan, both located in waterlogged alluvial lowland settings south of Hangzhou Bay. Hemudu was an eye-opener for the world when it was first excavated in 1973, being a village of rectangular raised-floor timber houses, up to 7 meters wide by over 23 meters long, constructed using skillful carpentry techniques with dowels, mortises, and tenons (figure 8.4D).

In one area of the Hemudu excavation, a solid mass of rice husks, grains, straw, and leaves formed a layer, perhaps once a threshing floor, with an average thickness of 40 to 50 centimeters. This rice, still not completely domesticated, was consumed alongside a range of nondomesticated plants, including foxnuts, water chestnuts, and huge quantities of acorns stored in large pits. It is possible these nuts were from planted trees, and some were perhaps fed to domesticated pigs. Indeed, while on the topic of the Chinese Neolithic diet, I should add that the site of Jiahu in the Huai Valley (between the Yellow and the Yangzi rivers; see figure 8.3) has evidence for the raising of carp in ponds, use of millet for pig fodder, and consumption of fermented beverages ("beer") made from rice, honey, and fruit.[27] Millet beer was also widely brewed in the Yellow River valley.[28]

The East Asian Neolithic Population Machine

Between 6000 and 3000 BCE, Neolithic villages in China developed into some of the largest urban agglomerations in the contemporary world. One striking example is the 3-square-kilometer (1.9 by 1.7 kilometers) city located at Liangzhu (see figure 8.3), to the south of Shanghai, dating from 3000 BCE and defended by an earthen city wall on a stone foundation. Liangzhu has a central complex of massive earthen mounds and also elite cemeteries with jade jewelry. It is thought to have contained between 15,000 and 30,000 people, fed by rice grown in paddy fields using irrigation water from adjacent dammed valleys. The dams, some of which still stand, contained almost 3 million cubic meters of earth stacked in woven grass containers like modern sand bags.[29]

It is not my purpose here to discuss the rise of Chinese civilization in full, but I would like to stress that the Liao, Yellow, and Yangzi River basins, where millet and rice agriculture originated, might together have been one of the world's most densely populated regions by 3000 BCE, just as they are today. As a contemporary of Liangzhu, the city of Uruk in lowland Mesopotamia (Iraq) in 2900 BCE had walls enclosing a slightly greater 5 square kilometers of urban space; Uruk was the largest city in Mesopotamia at that time. However, the extent of fertile farmland in the agricultural lowlands of China in 3000 BCE was arguably greater than in Mesopotamia, and one presumes that the total population might also have been greater.

As in the Fertile Crescent, the increases in population number that accompanied the development of food production in China can be calculated, on a relative basis, from settlement numbers and their areas plotted through time. For the period between 6000 and 2000 BCE, a large number of archaeological surveys, backed by the detailed records of archaeological site discoveries kept by Chinese provincial authorities, suggest population increases between ten and fifty times in the Liao, Yellow, and Yangzi River valleys. As in the Fertile Crescent, the North

China plain did not take long to become very crowded indeed, and it is not surprising that so many sites had enclosing defenses.

As a rather thought-provoking example, one study of the Wei Valley, where Banpo and Jiangzhai are located, has estimated an increase in the number of archaeological sites from 24 to over 3,000 between 6000 and 2000 BCE, and a corresponding increase in the population from 4,000 to 1,550,000 individuals. Also during this time period, the percentage of the valley floor used by humans for food production is estimated to have increased from 0.2 percent to 12 percent, and the average population per archaeological site from 160 to 481 persons.[30] Such figures rank the East Asian homeland of agriculture as one of the most potent arenas of human population growth at that time, anywhere in the world.

The African Sahel and Sudan

The indigenous development of food production in Sub-Saharan Africa took place around 3000 to 2000 BCE in the Sahel and Sudan vegetation zones, sparsely wooded grassland and open woodland, respectively, that extend across the continent in its northern tropical latitudes between the southern edge of the Sahara and the northern edge of the rainforest (figure 8.5). The main domesticated cereals were millets, especially sorghum and pearl millet. The western part of this region also gave rise to African rice, Guinea yam, oil palm, and cowpea (a legume). Domesticated cattle, sheep, and goats were brought in from the Fertile Crescent. Native North African cattle were not domesticated independently according to their DNA, but they were hunted in the wild in northeastern Africa during the early Holocene.[31]

The development of Sahel agriculture is agreed by many archaeologists to have been independent of that in the Fertile Crescent, even though the arrival of domestic animals from the Levant must suggest some potential contact with that quarter. To understand African plant domestication in perspective it is necessary to consider, first of all, two preceding events: the early Holocene occupation of the "Green Sahara"

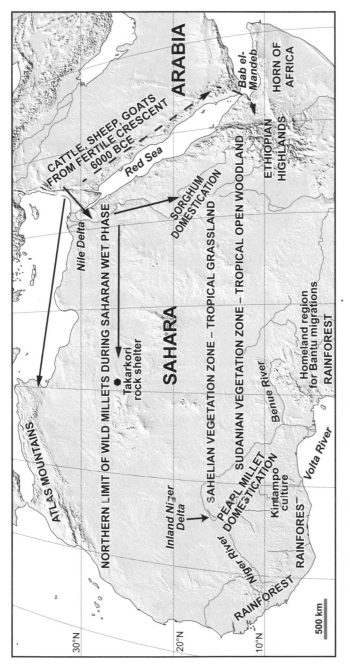

Figure 8.5. Early Sahelian and Sudanian agriculture in Africa.

and the spread of Fertile Crescent farmers and herders through Sinai into Egypt.

The Saharan Humid Phase

During the Last Glacial Maximum, around 20,000 years ago, the Sahara Desert was not an attractive region for human settlement, given that it extended from the Mediterranean Sea to at least 500 kilometers south of its present southern boundary. Even the Nile Valley appears to have been inhospitable owing to the lower rainfall, encroachment by sand dunes, and the absence of a delta.

As in the Fertile Crescent, conditions improved markedly with increasing humidity after about 14,500 years ago, and especially after the end of the Younger Dryas at about 11,700 years ago. Summer monsoon rains spread into the Saharan and Arabian deserts from the south and southeast, creating grasslands with wild sorghum and pearl millet up into the central Sahara. These grasslands attracted wild animals such as antelopes, gazelles, cattle (mainly in the north and east), and hippopotamuses, together with aquatic creatures such as fish and crocodiles.[32] The highland massifs that exist in many parts of the central Sahara attracted the largest human populations, perhaps because of their higher rainfall. These people moved in from the south and left behind hunting camps with animal bones and decorated pottery, as well as cave walls decorated with justly celebrated paintings and engravings of animals and humans.

By 3000 BCE, the good times were shrinking, and the Saharan humid phase was soon over. The hunters moved either into the Nile Valley, to amalgamate with the emerging Egyptians whose farming ancestors had recently moved in from the Fertile Crescent, or they moved southward into the Sahel, where they became part of the ancestry of the modern speakers of Niger-Congo and Nilo-Saharan languages (chapter 12). Those hunters who moved south must have played a role in the commencement of Sahelian and Sudanian agriculture. I return to them later, after first considering the role played by the Nile Valley in the beginnings of African agriculture.

Farmers and Herders from the Fertile Crescent

During the sixth millennium BCE, if not before, migrants from the Fertile Crescent
reached the Nile Valley and established
Neolithic cultures on the newly forming
Nile delta and in the Faiyum depression.[33]
The exact date of their arrival is a little obscure. This is because the Nile Valley has
only a Pottery Neolithic in terms of its excavated settlement record, but discoveries of
spear and arrow points of Fertile Crescent
Pre-Pottery Neolithic types have also been
made in Egypt, especially of "Helwan
points," which occur in Göbekli Tepe and
contemporary sites in the Levant and Arabia
(figure 8.6). Possibly, there was some Pre-

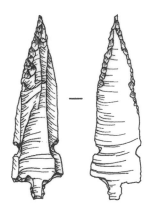

Figure 8.6. A Helwan Point
(7 centimeters long) from Tell
Halula, Syria. Drawn by Mandy
Mottram, Australian National
University.

Pottery Neolithic occupation from the Levant that is now buried out
of reach under deep Nile alluvium. Genetic evidence also suggests
that Natufian populations once moved along the northern coastline
of Africa as far as Morocco, possibly taking advantage of improving
postglacial climatic conditions.[34]

Whatever the date of Neolithic arrival into Egypt, the domesticated
crops that they brought from the Levant did not spread into Sub-
Saharan Africa—Fertile Crescent cereals and legumes did not thrive
under African summer monsoon rainfall conditions. Crops such as
wheat and barley were taken into the highlands of Ethiopia in later prehistoric times, but they were not able to spread farther south except
when taken to temperate South Africa in colonial times.

The situation was different with the domesticated animals, which
were rather less tied to specifics of rainfall seasonality. Cattle, sheep, and
goats were taken through the still-humid Sahara by pastoralist populations,
and herders had reached the central Saharan massifs by 5000 BCE; milk
residues on potsherds found in the Takarkori rock shelter in the Tadrart

Acacus region of southwestern Libya (see figure 8.5) indicate that they were milking cattle by this date, as well as keeping sheep and goats and harvesting wild sorghum.[35]

Domesticated cattle, sheep, and goats were taken farther toward the Niger River in the western Sahel by 2500 BCE, and they reached the African Rift Valley in Kenya at about the same time. Many of these pastoralists were probably speakers of Afro-Asiatic languages, entering Africa from the Fertile Crescent, as I will describe in chapter 12. However, there was also a significant, and indigenous, Sub-Saharan development of plant-based food production in West Africa and Sudan that was underway by 2500 BCE. We turn to it now.

Savanna and Parkland

Apart from the Fertile Crescent cattle, sheep, and goats, African food production in the Sahelian and Sudanian vegetation zones was based on the domestication of indigenous monsoon-climate cereals, legumes, and tubers. The overall homeland region lay between 10° and 20° north latitude, hence within the tropics. According to present evidence, there were three focal regions here for early agriculture: the Niger and Volta basins in West Africa, the upper Nile Valley and its tributaries in Sudan, and the Ethiopian Highlands.[36]

It is possible that plant domestication in these three focal regions occurred independently, but the archaeological record in tropical Africa is simply not detailed enough to allow a decision to be made. Domesticated pearl millet was present by 2000 BCE in the Niger Basin, and domesticated sorghum by 3000 BCE in the Atbara tributary of the upper Nile in Sudan, according to a new analysis of domesticated sorghum chaff used as temper in pottery clay.[37] Unfortunately, the chronology is not yet so clear for early agriculture in Ethiopia. Pearl millet and sorghum were transferred to the Indus Valley (Harappan) Civilization in Pakistan by 1800 BCE, suggesting that these crops were already domesticated in Africa by this time and had spread as far as its eastern seaboard.

The Domesticated Economy behind the Bantu Migrations

The archaeological record in Sub-Saharan Africa lacks detail when compared with those of the Fertile Crescent and China, and so far it does not contain detailed information about demographic changes in the human population through time. Nevertheless, the last three millennia of Sub-Saharan prehistory in Africa tell us, in no uncertain terms, that indigenous crop domestication there was a driver of human migration on a massive scale.

The genesis of the Bantu migrations occurred in West Africa from an immediate origin about 4,000 years ago in Cameroon, but with a background in the agriculture that developed beforehand in the Sahel region. Village life with millet farming began in the Niger and Volta basins prior to 2000 BCE. As an example, the Kintampo culture of the Volta Basin in Ghana was associated with wattle and daub (saplings and mud) house construction in village-sized settlements, with decorated pottery and ground stone axes and ornaments. The Kintampo people kept sheep, goats, and cattle, and they grew domesticated pearl millet and the leguminous cowpea. Millet chaff was used for tempering their pottery clay.

A recent analysis of almost 2,000 archaeological sites in West Africa dating from 2000 BCE until recent times reveals that there were two foci of early village life here at the beginning of this time span, one being the Kintampo culture just described and the other located in the middle Niger region of Mali, close to the large inland delta complex of that river (figure 8.5).[38] By 1000 BCE, domesticated crops and animals were widespread in West Africa, including the Bantu linguistic homeland in eastern Nigeria and Cameroon.

The major Bantu migrations that eventually covered most of Sub-Saharan Africa took place mainly between 1000 BCE and 500 CE, and were dependent on pearl and sorghum millet, legumes, and sheep, goats, and cattle. During the first millennium CE, the indigenous African domesticated crop repertoire was boosted by an arrival of Southeast Asian tropical crops, such as bananas, taro, the greater yam, and sugar cane.[39] The Bantu migrations were also greatly assisted after 500 BCE by a knowledge of ironworking, especially in the longest migratory leg southward from Lake Victoria into eastern South Africa. I return to the Bantu migrations in more detail in chapter 12.

Highland New Guinea

More than fifty years ago, I traveled from the United Kingdom to my first lecturing position at the University of Auckland in New Zealand. One lunchtime I was talking in the Anthropology Department tea room with linguist colleague Andrew Pawley. I do not now remember our exact words, but we were discussing New Guinea and the peopling of the Pacific.

Most people who inhabit the vast region of islands between Madagascar and Sumatra in the west to the far limits of Polynesia in the east speak a set of closely related Austronesian languages. In the middle of the distribution lies the large island of New Guinea. A few coastal pockets of New Guineans speak Austronesian languages, but all interior languages, spoken by the majority of the island's population, belong to other diversified language families known collectively to linguists as "Papuan."

Andrew remarked to me how interesting it was that the Papuan languages should exist through most of New Guinea and in a few adjacent islands, amid the huge swath of Austronesian speech that extended more than halfway around the world. Austronesians had settled every other region of Island Southeast Asia and the Pacific, with the exception of Australia, but they never penetrated significantly into the interior of New Guinea.

We both knew that the Austronesians had arrived in their islands as farmers (chapter 11), and we suspected that the New Guinea Papuan speakers had maintained control of most of their island because they were also food producers when Austronesians arrived. It has since turned out that we were essentially correct. Demographically, New Guinea populations were large enough to be able to retain most of their territory, allowing Austronesian settlement on the island in only a few coastal pockets.

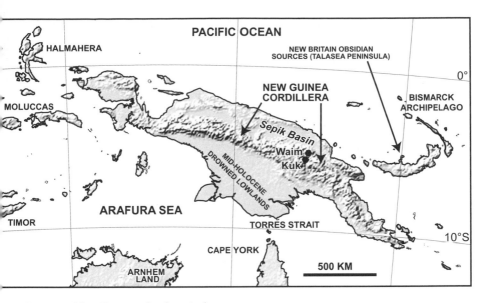

Figure 8.7. New Guinea and early agriculture.

As Andrew and I were discussing the Papuan languages in that Auckland tea room, my future Australian National University (ANU) colleague (I moved there in 1973) Jack Golson was planning his research at the Kuk Tea Station in the Wahgi Valley, located in the Western Highlands of what was then the Australian Territory of Papua and New Guinea (now Papua New Guinea) (figure 8.7). Jack and his colleagues, working at Kuk between 1966 and 1977, revealed that New Guinea Highlanders had been controlling water levels in swamps with drainage ditches by at least 2000 BCE, and constructing mounds for cultivation purposes as early as 5000 BCE.[40] As confirmed through later research at Kuk by another ANU colleague, Tim Denham, these highlands were the scene of an indigenous agricultural system, based on bananas, yams, aroids (especially taro, *Colocasia esculenta*), and sugar cane.[41] Although the taro appears to have been an introduction from Southeast Asia,[42] this complex of domesticated crops predated the arrival of Austronesian-speaking communities around the New Guinea coastline.

As a result of this research, the world acquired another agricultural homeland. Some archaeologists were rather skeptical to start with, partly because the evidence consisted mainly of ditches dug in swamps, without the benefit of a rich archaeological record similar to that of the Neolithic cultures of Africa and Eurasia. New Guinea, like Amazonia, has a hot and wet tropical climate that breaks down organic remains quickly.

Furthermore, the fruits and tubers that were domesticated in New Guinea tend to leave few identifiable traces in archaeological sites, especially because their reproduction was by vegetative planting of stems and tuber segments rather than seeds. Evidence for their existence comes from phytoliths and starch (microscopic particles that exist in plant tissue), and there were no native grain-bearing cereals or domesticated animals on the island, although native legumes were consumed. Pigs, dogs, and chickens were only introduced to the New Guinea coastline from Austronesian sources in the Philippines and Indonesia around 3,000 years ago.

Recent excavations at a site called Waim in the Jimi Valley, about 50 kilometers north of Kuk, have added some interesting settlement data to the Kuk evidence. Waim is an open site on top of a ridge, with a major archaeological layer dating to around 3000 BCE. The artifacts include stone pestles, a stone carving of a human figure, a fragment of a polished stone axe, and starch from yams, sugar cane, and banana. There was no pottery, but there are postholes suggestive of wooden domestic buildings.[43]

When all the evidence is brought together, it becomes clear that New Guinea was an indigenous homeland of food production. Virtually all of its inhabitants were farmers at European contact, except for a few lowland groups who subsisted on starch extracted from the trunks of wild sago palms, and even these generally had access to minor agricultural produce and domestic pigs.[44] Papuan-speaking New Guineans certainly have not migrated into New Guinea from elsewhere, at least not for a very long time. They are an indigenous western Pacific population who, in terms of their archaeology, biology, and genomics, have descended continuously from the Pleistocene settlers of the island.[45] Papuan languages are related to no others in the world, not even to Australian languages, despite the deep biological connections between New Guineans and Australians. New Guinea agriculture was undoubtedly indigenous in origin.

An Equatorial Homeland of Agriculture

Let us have a closer look at what is special about New Guinea. This island and Australia may well occupy the same continental shelf, but they could hardly be more different geologically. Australia is an ancient, weathered, and mostly rather dry continent. New Guinea is a tectonically active cordillera of folded mountains that has been uplifted through tectonic subduction above the northern edge of the Sahul continental plate within the past five million years. As a true cordillera, it resembles the Himalayas, the Alps, and the Andes. Nothing like the New Guinea cordillera exists elsewhere in the islands of Southeast Asia. The volcanoes of Indonesia form individual mountains separated by lowlands rather than a cordillera, and the mountains of central Borneo lie at a much lower altitude, without broad fertile valleys like those in the New Guinea Highlands.

New Guinea is an equatorial island that lies between the equator and 10° south. Its highlands have no major dry season and are well watered throughout the year, although average temperatures drop by about 5°C with every thousand meters of altitude, meaning that inhabited valleys above 2,000 meters can have frost at night. The highlands are also rather sharply differentiated from the lowlands in terms of altitude, and they are arduous to access on foot from the coast, as Japanese and Australian soldiers discovered along the Kokoda Trail in the Owen Stanley Range (southeastern Papua New Guinea) during the Second World War. Some rivers in northeastern Papua New Guinea provide relatively easy routes, and a few potsherds here and there indicate fleeting contacts with the coast after 3,000 years ago, but pottery was otherwise absent throughout the prehistoric New Guinea Highlands.[46] In general, the New Guinea Highlands had a remarkably self-contained prehistory.

The clues to the success of New Guinea agriculture are three in number: fertile highland valleys above the altitudinal reach of malaria, year-round rainfall, and geographical remoteness from food-producing outsiders such as Austronesians. The broad fertile valleys of the New Guinea Highlands lie between 1,500 and 2,500 meters above sea level and were only reached by rather amazed European explorers during the

1930s, who found that they contained dense human populations, far denser than in most malarial lowland situations. It is from these valleys, so far mainly the Wahgi and Jimi valleys of Papua New Guinea, that the archaeological evidence for a New Guinea agricultural homeland has been recovered.

Whether similar developments were also occurring in the lowlands of New Guinea is still uncertain. Some tropical plant species restricted to the lowlands are likely to have been domesticated in Island Southeast Asia or western Melanesia, including coconuts, sago, and breadfruit, but there is currently little evidence for domestication of these before the arrival of Austronesian-speaking populations with Lapita pottery (chapter 11) about 3,000 years ago.

One final observation can be made about the date of commencement of food production in New Guinea. Had it developed in lowland New Guinea before 9,000 years ago, when a dry land connection across the Sahul Shelf still existed with the Cape York Peninsula of Australia, then we might expect to find some traces of agriculture in Australia, as indeed some archaeologists have claimed.[47] However, the current dates for the appearance of food production at Kuk and Waim postdate the postglacial sea level rise, and this might help to explain why agriculture never became permanently established in northern Australia. The drowning of Torres Strait severed the land connection, and indigenous New Guineans and Australians went their separate ways.

The American Homelands of Agriculture

When Europeans began to enter the New World, starting in 1492 CE, the American continents supported food-producing populations over a north to south expanse of about 6,000 kilometers, from southeastern Canada to Argentina and Chile. At that time, most of Canada, the Great Plains, the Rocky Mountains, and the west coast of North America were still occupied by hunter-gatherer populations, as was the southern part of South America (figure 7.1).

Over much of the agricultural area the main food staple was maize, and when the Spanish conquistadors arrived this crop was fundamental in feeding the ill-fated Aztec and Inca civilizations of Mexico and Peru. Indeed, it had served this role in Mexico and Peru since at least 2000 BCE, long before the Aztec and Inca states existed. As the most widely grown crop in the world today, for human and animal food as well as nowadays for fuel ethanol, maize was grown in the prehistoric Americas between 47° north and 43° south, and up to 4,000 meters in altitude. Together with Amazonian manioc (cassava) it fed millions, and I know from my own experience in many regions of Southeast Asia that both maize and manioc, as well as sweet potato, are extremely important crops of American origin today in dry or infertile regions where rice cannot easily be grown.

On the level of the individual crops, the homelands of domestication in the Americas were remarkably scattered (figure 8.8). Many plants, such as squashes and beans, were domesticated more than once. Whether all of these individual crop domestications represented independent homelands of agriculture or additional domestication initiatives by people who were already farmers is not always clear. Archaeologists are agreed, however, that agriculture developed relatively independently from a hunter-gatherer background in at least four areas: the central Andes and adjacent parts of Amazonia, northwestern South America, Mesoamerica, and the Eastern Woodlands of the United States.

I state "relatively independently" because there are some major hints in the archaeological record that these regions were occasionally in contact during prehistory, just as were the various regional contributors to the domestication process in the Fertile Crescent, Africa, and East Asia. I give two examples. First, maize originated in Mexico and was taken before 5500 BCE, presumably by humans, into Central and South America, possibly via a west-coastal seaborne route that appears to have been in use by at least the Formative (early village farming) Era.[48] Second, American archaeologist James Ford published in 1969 a massive work illustrating how closely related in archaeological detail were the Formative cultures of the southeastern United States, Mesoamerica, the Andes, and other regions of the Americas.[49] I have always found Ford's observations most intriguing—these cultures did not evolve in total isolation from each other.[50]

Figure 8.8. Regions of ancient plant domestication in the Americas, and the locations of sites discussed in the text or shown in figure 8.9.

I see no reason, therefore, why the inhabitants of the various regions of the Americas where plants were domesticated should not have been in contact, even if intermittently. My main interest is not to determine who behaved independently of whom because I do not think we can ever know about such a diffuse concept as "cultural influence" so far back in the past in any fine detail, at least not without a written record.

It is sufficient to know that domestication happened in a given region so that we can turn directly to the repercussions in terms of human population prehistory.

In terms of those repercussions, we know that food production in Mesoamerica and the central Andes led to some of the most striking civilizations ever to develop in human history, created without a use of metals for industrial or military purposes. The Maya, Aztecs, Incas, and other celebrated American civilizations were Neolithic in technological orientation, which makes their architectural achievements (see figure 8.9) even more remarkable.

In this regard, heating and hammering of native copper spearheads occurred as early as 7500 BCE in the vicinity of the North American Great Lakes, but the tradition disappeared by 3500 BCE, perhaps as easily available native copper sources became harder to find.[51] Metallurgy that involved the hammering and smelting of copper, gold, and silver was present in the Andes by at least 3,500 years ago, but the results were used purely for body ornamentation and ritual purposes. Indigenous American metallurgy did not protect its makers from invasion by Spanish conquistadors; indeed it had the opposite effect, by arousing their interest in bullion.

America's First Farmers

The development in the Americas from careful management of wild plants by hunter-gatherers to a stage of agricultural dependence with fully domesticated crops took a long time. The clearest evidence exists for maize. Genetic comparisons suggest that this plant began to be selected from its ancestor, a western Mexico cereal grass called *teosinte*, at around 7000 BCE. Initially, it is possible that maize was used for its stalk juices that were fermented into beverages, including for the weaning of infants.[52] By 3300 BCE, maize cobs found in dry caves in central Mexico were in the process of losing their encasing sheaths, making the grains easier to extract, although they still had a shattering habit when ripe and they were very small.[53] Maize cob domestication was underway, if slowly.

By 2000 BCE, maize cobs were eventually showing signs of becoming a major food source, with cobs up to 6 centimeters long being grown in central Mexico and Honduras.[54] It was also around this time that maize was introduced into the US Southwest, and it began to appear in increasing quantities even earlier in Peru. By soon after 2000 BCE, maize had become a staple food for sedentary village-based cultures with high population densities in many regions of the Americas.[55]

The maize chronology gives about 5,500 years for the progression from wild to fully domesticated status, far longer than the 2,000 years or so required for the development from wild to domesticated cereals

---

Figure 8.9. Parallel universes: some archaeological achievements based on food production in the Americas. The sites are located in figure 8.8; I celebrate them here because they illustrate parallel achievements by Old and New World populations who experienced minimal contact during the 16,000 years since the Americas were settled, except for activities around the Bering Strait involving Eskimo-Aleut and Athabaskan-speaking peoples, and minor contacts involving Polynesians. During those 16,000 years and prior to 1492, no domesticated crops (except for the sweet potato in Polynesia) or animals, major language families, or large human populations crossed between the two regions. For me, this figure illustrates the essential unity of *Homo sapiens*, as expressed through the material side of our behavior. (*A*) A reconstructed Late Preceramic stepped platform with multiple rooms at El Paraiso, near Lima, Peru, circa 2000 BCE. (*B*) A 110-meter-diameter and 25-meter-high circular stepped platform ("pyramid") at Cuicuilco, Valley of Mexico, first millennium BCE. (*C*) A basalt head, La Venta, Tabasco, Mexico, first millennium BCE (Olmec culture). (*D*) A Maya ruler carved in stone at Copan, Honduras, eighth century CE. (*E*) The Avenue of the Dead, stepped platform bases and the Pyramid of the Sun (*left background*) at Teotihuacan, Valley of Mexico, early first millennium AD. (*F*) Toltec "Atlantids" on the summit of Edificio B at Tula, Hidalgo, Mexico, Postclassic circa 1000 CE. Edificio B at Tula is closely paralleled in shape and plan by the Temple of the Warriors at Chichen Itza; see photo J. (*G*) The remarkable ceremonial arch at Labna, Yucatan, in the Maya Puuc style, ninth century CE. (*H*) The Castillo at Chavin de Huantar, Peruvian Highlands, first millennium BCE (Chavin culture). This building contains an internal labyrinth of passages with remarkable human/feline carvings. (*I*) The Mayan "Palace" at Palenque, Chiapas, with its unique four-story tower. Seventh/eighth centuries CE. (*J*) The Temple of the Warriors at Chichen Itza, Yucatan, Postclassic Maya (900–1200 CE), but constructed with strong Toltec influence from the Valley of Mexico, more than 1,500 kilometers away. (*K*) The Inca town of Machu Picchu above the Urubamba River in the southern Peruvian Highlands, fifteenth century CE. (*L*) Fifteenth-century CE Inca stonework at Sacsayhuaman, Cuzco, Peru. The blocks are multiangular and fit closely without mortar. Similar stonework at Vinapu on Easter Island supports contact with South America late in prehistory (see the discussion in chapter 11). All photos by the author.

and legumes in the Fertile Crescent and East Asia. If stalk juices were the initial goal for harvesters (and this still remains a hypothesis rather than a certainty), then eventual selection for a large cob would have required a change in behavior. During the early stages of domestication, the teosinte ears needed to be removed when still unripe if stalk sugar was the desired product—this removal would have concentrated the sugar in the plant, but it would not have assisted production of a cob. To do this, the ears needed to be left to ripen on the plant so that their domesticated characteristics could be gradually selected. Perhaps maize was unusual in being domesticated for two different purposes, which could explain why it all took so long.

There may be other reasons for the relatively slow progression into full agriculture in the Americas. Many American crops were condiments rather than staples (chili peppers and tomatoes, for instance), and ancient American economies were never much stimulated by animal domestication, which in the Old World required extra production of crops to be used as winter or dry season fodder. Neither was there any dairy production. Llamas and alpacas were used for meat, cartage, hides, and wool, but were never taken far from their original regions of domestication in the Andes. This meant that meat protein was mostly obtained from wild animals, such as the deer and rabbits hunted by the inhabitants of the Mesoamerican civilization of Teotihuacan (300–600 CE). Perhaps the rabbits were kept and bred in captivity, just as were guinea pigs in South America.

Although there is some rather scattered and sometimes ambiguous evidence, especially from microscopic silica phytoliths and starch grains extracted from plant remains, that attempts toward domestication were occurring as early as the beginning of the Holocene in Mesoamerica and northern South America, food production in the Americas did not become prominent until after 2500 BCE, and in many cases long after that. It was mainly after 2500 BCE that the first platform and pyramid complexes with large villages made their appearances in the Andes, and somewhat later in Mesoamerica, heralding the rise of the Formative and Classic civilizations.

South America: The Andes and Amazonia

The current archaeological record in the Americas suggests that sedentary village life first appeared in coastal and lower riverine situations in Ecuador and northern Peru. In the Zaña Valley of northern Peru, there is evidence from phytoliths and starch grains dated to about 3000 BCE for a number of crops, including beans, squash, manioc, peanuts, and other tubers (but not yet maize cobs), stated to be associated with fields and canals for irrigation as well as small villages.[56] By 2500 BCE, a large village covering 12 hectares with perhaps 1,800 inhabitants was in existence at Real Alto in coastal Ecuador, here with pottery and now a presence of maize. The Real Alto settlement was arranged around three sides of a rectangular open court about 300 meters long, with oval settings of post holes marking house locations. On either side of the court were two large earthen mounds that were used as a cemetery and for communal activities.[57]

One of the difficulties with this early evidence of food production in South America is that most of the data come from phytoliths and starch, rather than larger-sized remains of actual seeds and cobs. It is difficult to be certain from such microscopic evidence that all species were fully domesticated or to determine how important they were in diets. The village-like contexts from which much of this evidence has been derived suggest that these plant species were being deliberately cultivated. However, the villages themselves are not defined in layout and extent with the clarity that is often available in the Fertile Crescent or East Asia. Nevertheless, a presence of maize anywhere in South America certainly suggests intentional human distribution, and phytoliths of maize recovered in sites in Panama suggest that it had left its Mexican homeland by 6000 BCE.[58] It had already arrived in northern Peru by 4500 BCE.[59]

The Amazonian lowlands that lie immediately to the east of the Andes, especially in the upper Madeira drainage system of eastern Bolivia and Rondônia province in Brazil, also have considerable phytolith and starch evidence for early cultivation and associated soil mounding. This evidence goes back almost to the beginning of the Holocene, and is associated with microscopic remains of plants such as manioc, squash, beans, and even cacao (chocolate), but initially without maize.[60] Here also, as in Ecuador and northern Peru, large settlements and irrigation systems appear to be much younger in time than the first evidence for domestication. In all these regions, we have a situation in which potential evidence for domestication based on microscopic plant remains in soils and on human teeth exists for perhaps six millennia before there is any substantial evidence for population growth.

When did significant population growth commence in the archaeological record in South America? The answer is probably during what archaeologists term the Late (or Cotton) Preceramic phase in Peru, between 2500 and 1800 BCE. Llamas, alpacas, white potatoes, and quinoa were becoming domesticated by this time in the Andes, and productive varieties of maize had arrived. As a result, some large Late Preceramic ceremonial centers based on irrigation agriculture, such as the 66-hectare site of Caral in the Supe Valley with its stepped platforms and pyramids, were established in the desert-flanked valleys that drain the Andes in northern Peru (see the stone platform reconstruction at El Paraiso near Lima in figure 8.9A).[61]

Peruvian sites became larger and more numerous during the Initial Period (1800–900 BCE), now with pottery and with increasing contact with the neighboring Amazonian lowlands. One estimate suggests population grew in coastal Peru by fifteen to twenty times during the Late Preceramic and Initial Periods (2500–900 BCE), fueled by a combination of agriculture and the rich maritime resources of the cold Humboldt Current.[62] Other archaeological estimates are in accord.[63]

Mesoamerica

Surprisingly, given the eventual importance of maize, the establishment of food production in Mesoamerica on a scale sufficient to support large villages and ceremonial centers did not occur, apparently, until after 2000 BCE. Indeed, the first large monuments equivalent to those of the Late Preceramic and Initial Periods in Peru were only constructed in the Gulf Lowlands of Mexico after 1500 BCE, and in the Maya homeland region of Guatemala by 1000 BCE. The reasons for this apparent lateness compared with the Andes are not clear, but one answer may be that the most productive varieties of maize were first developed in South America and then taken back to Mesoamerica at a slightly later date.[64]

Examples of early farming communities in Mesoamerica are widespread, with numerous villages excavated in the fertile valleys of Mexico, Oaxaca, and Guatemala. In the Valley of Oaxaca, settlements between 1 and 3 hectares in size, with rectangular post and thatch houses, bell-shaped storage pits, pottery vessels, and pottery human figurines made their appearance by 1700 BCE. The Valley of Oaxaca population trebled between 1200 and 900 BCE, increasing perhaps tenfold during the whole of the second millennium BCE. Similar estimates of a population doubling every 250 years have been made for the Valley of Guatemala.[65]

All of this implies a solid development of agricultural life with similar pottery styles in most parts of Mesoamerica during the second millennium BCE, associated with population growth[66] and an increasing interest in the construction of ceremonial centers, with their implications for the rise of power, authority, and increasing warfare. One result of these widening horizons was the remarkable Olmec Horizon (circa 1200–500 BCE), famous for the colossal stone heads uncovered at the ceremonial centers of San Lorenzo and La Venta, near the Gulf Coast of Mexico (figure 8.9C). Equivalent developments during the first millennium BCE occurred in Oaxaca, the Valley of Mexico (figure 8.9B), and in the Maya region of eastern Mesoamerica, as well as with the contemporary Chavin horizon in the central Andes (figure 8.9H).

As a consequence of these Formative developments, both Mesoamerica and the central Andes witnessed, during the first and second millennia CE, the rise of a series of remarkable cultures that were to succeed each other, through many wars and punishing droughts, until the fateful arrival of the Colonial Era. In Mesoamerica, these Classic and Postclassic cultures included Teotihuacan (Valley of Mexico, figure 8.9E), the Zapotec remains at Monte Alban (Oaxaca), the Toltec remains at Tula (Hidalgo, figure 8.9F), Aztec Tenochtitlan (Mexico City), and the many highly decorated constructions, statues, and inscriptions of the literate and erudite Maya (figure 8.9D, G, I, and J).

In South America, parallel cultures included the archaeological sites and remarkable tombs of the Moche, Wari, and Chimu cultures of northern and central Peru, together with the amazing stone constructions of the celebrated highland Incas (figure 8.9K and L), contemporaries of the Aztecs, who also experienced the fateful arrival of the Spanish conquistadors in the early sixteenth century. I return to some of these peoples and their languages in chapter 12, noting here that the American Classic and Postclassic civilizations always left me with a sense of puzzlement on those occasions when I was able to travel among them. I was both astonished by their creativity but also puzzled by the extent of religiously inspired human sacrifice that was apparently required to fuel their continued existences. What would they have evolved into had the conquistadors never arrived?

The Eastern Woodlands of the United States

Before we leave the Americas, it should be noted that there was also a development of plant domestication in the middle Mississippi, Missouri, and Ohio basins within the well-watered Eastern Woodlands of the North American continent. Many of the crops domesticated here have not survived as common ingredients of modern cuisine; these included cucurbits, goosefoot (a chenopod), and other seed plants. Some, such as sunflower seeds, are still grown today. American archaeologist Bruce Smith has shown that seed sizes within these species became larger and seed coats became thinner, starting about 3000 BCE, the implication being that humans were cultivating them and selecting their seeds for planting.[67]

This Eastern Woodlands focus on domestication may have been behind a record of earthen mound and village construction associated with "Woodland" cultures dating between 1000 BCE and CE 1000, especially those termed Adena in the Ohio Valley and Hopewell in the Greater Mississippi Basin by archaeologists. However, the major developments in population density and massive mound construction in the Eastern Woodlands came later, after 800 CE, with the rise of the maize-based Mississippian civilization.

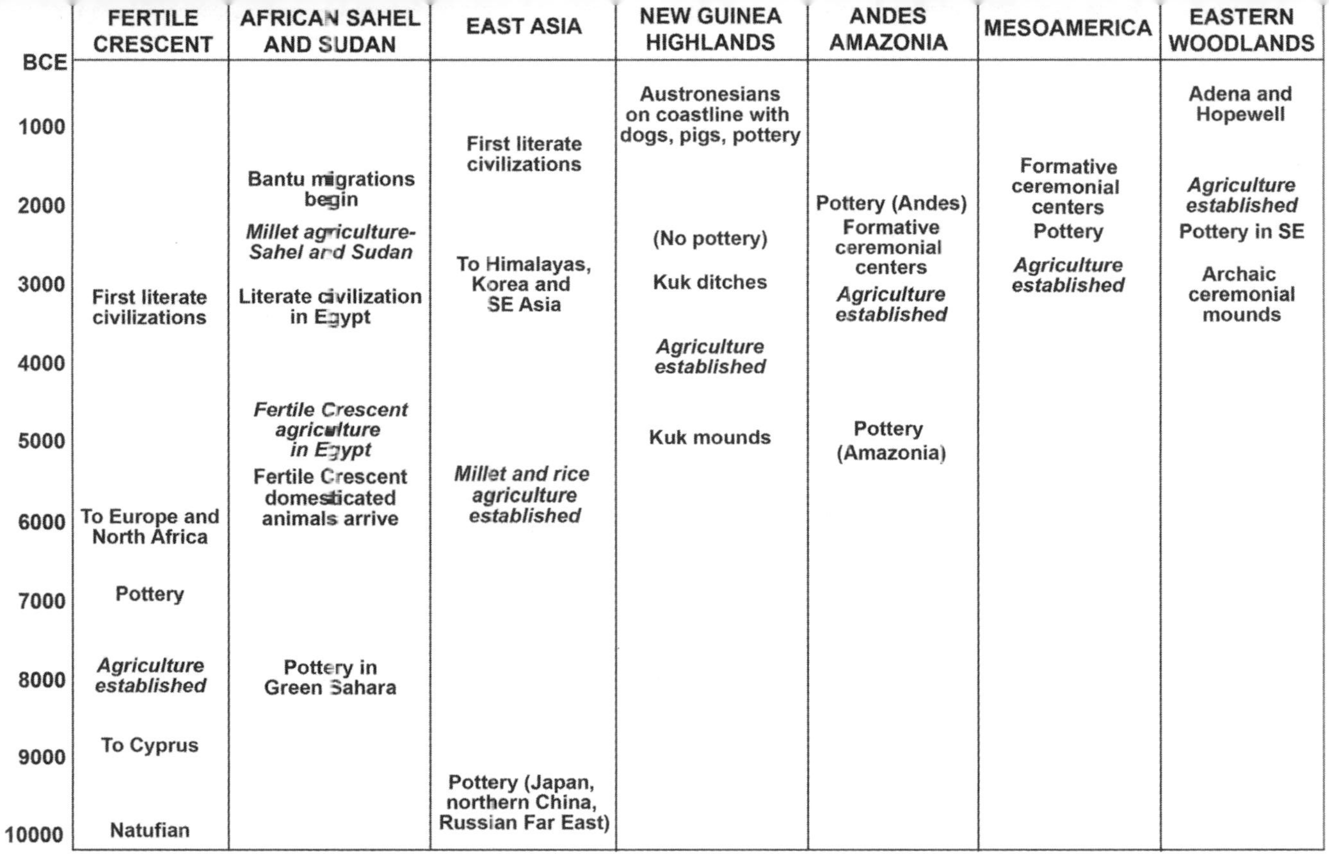

BCE	FERTILE CRESCENT	AFRICAN SAHEL AND SUDAN	EAST ASIA	NEW GUINEA HIGHLANDS	ANDES AMAZONIA	MESOAMERICA	EASTERN WOODLANDS
1000			First literate civilizations	Austronesians on coastline with dogs, pigs, pottery			Adena and Hopewell
2000		Bantu migrations begin; *Millet agriculture- Sahel and Sudan*	To Himalayas, Korea and SE Asia	(No pottery); Kuk ditches	Pottery (Andes); Formative ceremonial centers; *Agriculture established*	Formative ceremonial centers; Pottery; *Agriculture established*	*Agriculture established*; Pottery in SE; Archaic ceremonial mounds
3000	First literate civilizations	Literate civilization in Egypt			Pottery (Amazonia)		
4000				*Agriculture established*			
5000		*Fertile Crescent agriculture in Egypt*; Fertile Crescent domesticated animals arrive		Kuk mounds			
6000	To Europe and North Africa		*Millet and rice agriculture established*				
7000	Pottery						
8000	*Agriculture established*	Pottery in Green Sahara					
9000	To Cyprus		Pottery (Japan, northern China, Russian Far East)				
10000	Natufian						

Figure 8.10. A simplified chart of the main developments within seven selected agricultural homelands across the world. The developments relate to agricultural establishment with domesticated resources, pottery manufacture, the rise of social complexity, and subsequent population expansions.

As a final comment to close this section, the real demographic significance of the American homelands of agriculture is unfortunately hidden from us, at least in part, because of the catastrophic decline in the American population after 1492 CE due to the introduction of Old World diseases. Despite this tragedy, the American food-producer population expansions were on scales similar to many Neolithic expansions in the Old World, as we will see in chapter 12.

The Story So Far

In this chapter, my focus has been mainly on the archaeological record—the crops and animals domesticated, the associated human settlements, and the increases and sometimes decreases in the sizes of associated human populations. We have followed several courses of development from small and mobile hunter-gatherer communities to large agricultural societies with sedentary settlements. These developments are summarized in figure 8.10.

The next stage, which I take up in the following chapters, will be to examine the repercussions of these developments on human population and language family distributions through the succeeding millennia. Once food production had been established with dependence on a transportable repertoire of domesticated plants and animals, human populations began to search for new territory, a process often fueled by rapid increases in population size. The results were migrations of peoples, with their languages and their lifestyles, often on continental scales that were far greater than any recorded in later historical accounts of conquest and empire, at least prior to the Colonial Era.

9

Voices from the Deep Past

THE RISE OF FOOD production during the Holocene allowed increasing populations of farmers to spread into territories formerly occupied by hunters and gatherers, as well as to many previously uninhabited islands. How can prehistorians best trace those spreads? Archaeology, genetics, and biological anthropology each have perspectives on human migration. However, within the past 10,000 years another field of research comes to the fore, this being the study of the major language families of the world. This chapter focuses especially on the issue of how language families have originated and spread with food-producing populations, in some cases on transcontinental and transoceanic scales.

The Early Farming Dispersal Hypothesis

I begin this chapter with a hypothesis that has informed my own research for most of my career as an archaeologist. More than thirty years ago, a few archaeologists, linguists, and biologists began to realize that the origins of many of the world's most widespread language families, genetic ancestries, and Holocene archaeological cultures were linked to a growing reliance on food production. Major spreads of farming, genes, and associated languages had clearly occurred in such circumstances, deep in the realms of prehistoric time and long before the rise of any literate civilizations and historical empires.

The hypothesis first grew in my mind during the 1980s. Cambridge archaeologist Colin Renfrew and I were working independently on early versions of it during this time, he on the spread of the Indo-European language family from Anatolia into Europe, and I with linguist colleagues on the spread of the Austronesian language family from Taiwan to Polynesia.[1] Renfrew's greatest contribution, to my mind, was to point out that the beginning of the Neolithic involved a major shift in the archaeological record across the whole of the European continent, except beyond the climatic range of agriculture in the far north. The beginning of the Neolithic was the most fundamental change in the whole of European prehistory, not remotely equaled by movements in the Bronze or Iron Ages, or even early history. On such a scale, the Roman Empire was almost a local phenomenon.

This observation was, and still is, of immense significance, especially for the origins of the Indo-European language family that dominates most of Europe, much of the Middle East, and most of northern South Asia today. There can be absolutely no doubt that the spread of the Neolithic through Europe involved a massive replacement of population, although this is far more obvious now, thanks to ancient genomics, than it was in the 1980s. Such a replacement can only have gone hand in hand with a spread of related languages. We might never know precisely *which* languages because we are looking at the realm of prehistory, but the only language family in Europe today with an extent similar to that of the Neolithic spread is the Indo-European family. It has no competitor.

During the 1990s, Renfrew and I became aware of the convergence of our views and began to apply them to a worldwide canvas. In 2001, Colin initiated a major conference on the theme of what he then called the "farming/language dispersal hypothesis," which was held in the McDonald Institute for Archaeological Research in Cambridge.[2] My next contribution to this debate, prior to the appearance of my *First Farmers* in 2005, was an article on language family homelands and population spreads coauthored with *Guns, Germs, and Steel* author Jared Diamond, published in the journal *Science* in 2003.[3]

In 2005, I elaborated upon what I henceforth termed the "early farming dispersal hypothesis" in my book *First Farmers*.[4] A major underpinning for this hypothesis was the observation that the homelands of the largest language families associated with food-producing populations overlapped with the main homelands of agriculture. The situation for the Old World is demonstrated in figure 9.1 (and see figure 12.4 for the Americas). I still regard these observations as being of immense significance.

The early farming dispersal hypothesis today has both supporters and detractors. Both sides work in a world of fast-flowing and often revolutionary scientific results from archaeology, linguistics, and evolutionary biology/genomics. In particular, the analysis of ancient DNA has recently transformed our understanding of early farmer population demography and the consequent importance of migration.

My view of the course of Holocene prehistory through the eyes of this hypothesis leads me to suggest that the spreads of the major language families identified in figure 9.1 occurred hand in hand with the migrations of populations of their native speakers. Any claims for language family spreads on the scales, for instance, of the Indo-European, Bantu, or Austronesian language families that do not involve substantial quantities of human migration will need to propose other convincing mechanisms. Do any other convincing mechanisms for such impressive spreads exist? I return to this question later.

Understanding the Human Past through Language

We are now looking ahead toward our recent past, toward the creation of the global tapestry of humanity as we know it today. As we come forward in time within the Holocene toward the present, linguistic reconstructions of prehistory begin to play an ever more important role. With linguistics now added to the mix it is necessary to juggle the conclusions of four, rather than three, independent sources of information (namely, archaeology, linguistics, biological anthropology, and genomics) in order to understand the migrations that created the human past.

Therefore, we might usefully ask what is the best way to document the last 10,000 years or so of that past, between the various inceptions of food production and the cultures and civilizations of recent history. One way might be to list the findings of each field of study separately, as I have tended to do in several of my own previous books.

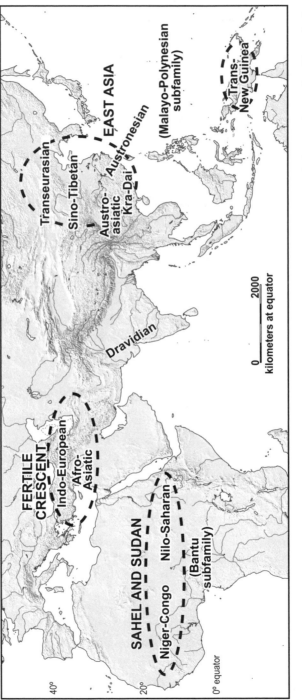

Figure 9.1. Map showing how widespread Old World agriculturalist language families might overlap in origin with the major agricultural homelands. The Bantu subfamily is the most expansive segment of Niger-Congo, and the Malayo-Polynesian subfamily is the most expansive segment of Austronesian.

For instance, as an archaeologist, shall I go through the relevant archaeological cultures of the world, one by one, listing their achievements and failures? Perhaps, but archaeological cultures lack precision about ancient lifestyles owing to loss of information with the passage of time, and they vary in quality of information in accordance with the distribution of wealth and scientific investment across the nations of the world. I need not give examples, but it is obvious that some relatively wealthy countries have dense archaeological coverage, whereas other politically unstable countries have almost none. Archaeology by itself does not give a full story of the human past.

What about those contentious entities that scholars and scientists for many years have referred to as "races." In scientific parlance, a race is a subspecies, in our case a subspecies of *Homo sapiens*. But the term carries immense biological and social baggage. The Caucasoids, Negroids, Mongoloids, and Australoids that graced the pages of anthropology text books in my youth have long since passed into limbo. Nevertheless, there are undoubted physical and genetic differences between Indigenous Africans, Europeans, Asians, Americans, and Australians today, as one would expect given their long histories of independent evolution and continuous exposure to different sources of natural selection. One can acknowledge these population differences without having to use the term "race," which carries false implications that sharply bounded morphological categories exist within humankind.

Can comparisons of living populations in terms of their physical characteristics and DNA therefore provide all the answers that we need? Again, I think not, in company with geneticist David Reich:

> The present day structure of human populations cannot recover the fine details of ancient events. The problem is not just that people have mixed with their neighbors, blurring the genetic signatures of past events . . . we now know, from ancient DNA, that the people who live in a particular place today almost never exclusively descend from the people who lived in the same place far in the past.[5]

Can ancient DNA solve these problems? Research in this field dominates all else in terms of scientific publication at the moment, but it still presents the same underlying problems as archaeology. Some parts of the world have yielded enormous quantities of ancient DNA, and others have yielded none, sometimes because of a lack of sample collection and sometimes because of the sensitivities of indigenous peoples toward analysis of ancient human bones. Ancient DNA also survives better in cold climates than in tropical ones, although new recovery techniques are closing the gap. Even so, we do not have a complete time and space ancient DNA map of the world. The further back we go in time, the smaller the samples become, and the more gaps there are between them, exactly as with the archaeological record.

What about language? Again, we have a problem with loss of information as we go backward in time. The families of related languages that exist in the world today only offer coherent information about human prehistory during the past 10,000 years, or less in many cases. This is because languages change continuously as people innovate and abandon habitual expressions, and eventually all traces of any shared ancestry between different languages may be lost. And ancient words from long lost and unrecorded languages without scripts cannot survive in the way that ancient genes can survive in bones, or ancient artifacts in the ruins of ancient settlements. Unwritten words are sadly ephemeral in any direct sense (ignoring comparative reconstructions from present-day attestations), just as are the unwritten languages to which they once belonged.

However, languages do retain traces of connectedness within the past 10,000 years, and it is precisely this span of time that is of immediate interest in the following chapters. I believe that we can usefully tie a reconstruction of Holocene human migration to a global prehistory of whole language families. I have followed this line of thought for most of my career as an archaeologist, and in this chapter I will explain why.

Why Are Language Families Important for Reconstructing Prehistory?

Languages have one useful advantage over archaeology and ancient DNA. They come packaged within language families, and these carry enormous amounts of information about the past migrations of their human speakers. Language families are well defined because their living speakers use complete languages, many spoken by thousands of people, that can be recorded in full rather than as disconnected fragments. They also can be compared in order to reconstruct their history. The major language families of both the Old and New Worlds, with their approximate distributions prior to the Colonial Era, are shown in figures 9.2 and 9.3, and listed in table 9.1.

Furthermore, all language families have one other important characteristic, that the languages classified within them are usually transparent in their family membership, by virtue of their co-inherited grammatical and lexical features. Even nonlinguists will be aware that the English language shares a recent origin with Dutch and German, but not with Tibetan or Navajo. Linguists can demonstrate this clearly by identifying the large number of commonly inherited and *cognate* linguistic items, in phonology, vocabulary, and grammar, shared between English, Dutch, and German (and, for that matter, all other languages in the Indo-European family), but not with languages in unrelated families.

Let me explain this a little further. In linguistic classification, English is a member of the Germanic subgroup of Indo-European languages in terms of its basic vocabulary and grammar. Its linguistic roots were planted in England by Germanic-speaking migrants from the far shores of the North Sea as the Roman Empire waned; these migrants were both mercenary soldiers in the late Roman army and free settlers who wanted new farmlands after the Roman legions withdrew. After these migrations, the English language developed on the English side of the North Sea, whereas the Dutch, German, and Danish languages developed on the continental side.

However, English vocabulary today also contains many borrowed words of medieval French origin, as a result of the Norman conquest of England in 1066 CE. French is a Romance language derived from Latin; English is a Germanic language. The borrowed words are obvious to knowledgeable linguists, and indeed to anyone who wishes to consult a dictionary (or who understands French).

English is therefore classified as a Germanic language in terms of its mainstream descent from ancestral Germanic languages spoken in

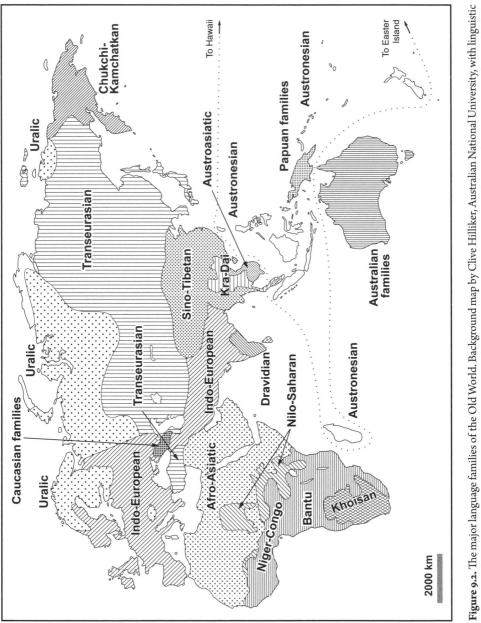

Figure 9.2. The major language families of the Old World. Background map by Clive Hilliker, Australian National University, with linguistic

Figure 9.3. The major language families of the New World. Background map by Clive Hilliker, Australian National University, with linguistic boundaries redrawn from Michael Coe et al., *Atlas of Ancient America* (Facts on File, 1986).

TABLE 9.1. The world's most extensive agriculturalist language families, ordered by approximate territorial extent at 1500 CE.

Language family	Number of languages spoken today (*Ethnologue*)[1]	Longitudinal extent[2]	Latitudinal extent[2]	Proto-language homeland
Austronesian	1,258	210°	65°	Taiwan
Indo-European (table 9.2)	445	110°	55°	Anatolia, Pontic Steppes (homeland debated)
Transeurasian	79	125°	45°	Northeast China (Liao River basin)
Niger-Congo	1,542	60°	45°	West African Sahel (the Bantu homeland was in Nigeria/Cameroon)
Afro-Asiatic	377	75°	35°	Levant, Northeast Africa (homeland debated)
Sino-Tibetan	457	60°	30°	Middle and lower Yellow River valley
Tupian	76	35°	30°	Southwest Amazonia
Uto-Aztecan	61	30°	35°	Central or western Mexico
Nilo-Saharan	207	30°	25°	East African Sahel/Sudan
Arawakan	55	20°	30°	Western Amazonia
Trans-New Guinea	481	43°	12°	New Guinea Highlands
Austroasiatic	167	25°	20°	Southern China or northern Southeast Asia (homeland debated)
Kra-Dai (Tai-Kadai)	91	15°	20°	Southeast China
Dravidian	86	10°	10°	Pakistan or Deccan Peninsula (homeland debated)

[1] *Ethnologue: Languages of the World*, http://www.ethnologue.com.

[2] In degrees, and approximately.

northern Europe at the end of the Roman Empire. Unlike French, it is not a member of the Romance subgroup of Indo-European languages derived from Latin, but it has borrowed vocabulary from that source as well as from many other sources, such as the related Germanic languages spoken by the Scandinavian Vikings. If we are to use a human metaphor for the English language, we could say that it is genealogically

Germanic in terms of its core, but partly French (and Latin) in terms of its history of contact with other peoples.

The example of English should reinforce for us that languages and language families

- have origins within episodes of human expansion,
- borrow from and influence each other,
- sometimes die and are replaced, and
- sometimes expand across vast regions, replacing other languages.

The latter was, of course, the case for Colonial-Era English, 1,500 years after its Anglo-Saxon foundation in England.

Because languages are mainly transmitted within human populations from one generation to the next, the families that they belong to have a coherence that can be of enormous assistance in understanding the recent prehistory of humanity. The internal evolutionary histories of language families can be unraveled through comparison and reconstruction. Because of these advantages, language families can be regarded as useful identifiers of the human populations who have spoken and migrated with their component languages through time, even if those human populations were sometimes fairly diverse in their genetic and archaeological attributes. Language families act as a skeleton, a scaffold, upon which human prehistory and history can be displayed.

Do Language Families Equate with "People"?

The answer to this question is that languages can equal people—or ethnic groups, in common parlance—at least in part. It is obvious from recent history that languages and the named populations who speak them have often spread together, as was the case with the English, Spanish, Dutch, and French during the Colonial Era. However, I should make it clear that I am not suggesting that all the people who speak a given set of related languages will carry an identical genotype or have a single unified cultural history. Past and present, humans can be presumed to have mixed whenever circumstances allowed or encouraged them to do so. Sometimes, they must have switched to new languages and abandoned their former ones, even if this did not happen often on a large scale. I give some examples of this kind of switching later.

Nevertheless, all significant migrations recorded in history, from Julius Caesar's account of the attempts of the Celtic Helvetii to migrate from Lake Geneva into Gaul in 58 BCE[6] through into the Colonial Era and the modern diasporas from war and famine, have been strongly constituted around core ethnic groups with shared languages and stated regions of origin. Languages matter today and surely mattered in the past as badges of community cohesion and membership. They have rigorous tales to tell about human prehistory and history, as we are about to see.

The Origins of Language Families

Within the past 10,000 years of expansion of food production and farmers, we come face to face for the first time in human prehistory with the significance of actual languages within language families, and their reconstructed linguistic ancestors. These ancestors are termed "proto-languages" by linguists, and they consist of sounds, words, and word meanings reconstructed through comparison of living languages within families, or ancient languages for which there are literary sources. The meanings attached to words related to material culture and the economy in proto-languages are especially significant for archaeologists. Many major language families, for instance, have multiple terms related to food production in their proto-languages, implying that their one-time speakers also knew food production.

I find the concept of the proto-language to be an exciting one. It implies expansion from a relatively localized geographical source region, thus shining a spotlight on the role of human population expansion as the main vector of language spread. An opposing hypothesis, that languages within large families have all somehow converged together in place from unrelated linguistic ancestors, simply does not work. Charles Darwin realized this 150 years ago when he commented on the importance of divergence rather than convergence with respect to language families: "the same language never has two birth places ... if two languages are found to resemble each other in a multitude of words and points of construction, they would be universally recognised as having sprung from a common source."[7]

The concept of an ancestral proto-language implies that there was once a single language that existed as the root of any given family or internal linguistic subgroup. The concept need not imply that a single household or settlement was the total source for a whole modern language family, but there would have been limits to the number of related linguistic populations who could have contributed to the formation of a proto-language. If they spoke more than one language, then it is likely those languages were related and intercomprehensible to some extent.

Linguists have little information on exactly how much geographical space a proto-language might have occupied because all are reconstructions from far back in time, beyond the existence of historical records. However, one well-known example from recent European history is the Latin ancestor of the Romance subgroup of the Indo-European languages, which includes modern Italian, Spanish, Portuguese, French, and Romanian, all derived from the Latin of the Roman Empire.

Despite common assumptions, Latin is by no means a dead language, except through its medieval use in a fossilized form for ecclesiastical, medical, and legal purposes. The immediate ancestors of the living Romance languages were the Latin dialects spoken in different parts of the Roman Empire. The foundation language was the Latin spoken in central Italy when Roman civilization was forming during the first millennium BCE.

In its source area, this original Latin was, of course, quite localized compared with its eventual extent under the Roman Empire. Precisely how many people spoke it when Romulus and Remus founded Rome circa 750 BCE I have no idea, but I would guess that they numbered in the thousands rather than the millions. Presumably, at the time in question, ancestral Romance (early Latin) was a single language, probably with intercomprehensible dialects across much of central Italy. Then it spread, replacing the neighboring language of the Etruscan civilization and several others in the process. The Romans thenceforth lost little time in spreading their language onward through a remarkable expanse of territory, as did English-speaking settlers two millennia later in North

America and Australia. The major Romance subgroup of the Indo-European languages was born, and history tells us a great deal about it.

The Spreads of Language Families: A Comparative Perspective from Recent History

How did languages and their speakers spread on the scales necessary to create the major language families? The Romance languages originated during the Roman Empire, when Latin was established as an ongoing vernacular through the settlement of army veterans in conquered lands, albeit in less than half of the extent of the empire. Most populations in that empire simply kept the languages they had before the Romans conquered them (Greek, Egyptian/Coptic, Aramaic, Berber, Gaelic, and so on), becoming bilingual in their natal language and in some version of imperial Latin.

Given this observation, we can turn to a far more significant question. How did the whole Indo-European family (its members are located prior to 1492 CE in figure 9.4 and listed in table 9.2) spread from its long-ago homeland? This is a far more complex issue. By 1492, the whole Indo-European family had spread to encompass a vast extent of territory, from Iceland to Bangladesh, and from the European Arctic to Sri Lanka—far more territory than could conceivably ever have belonged to some hypothetical Neolithic or Bronze Age superempire. The Romans hardly competed on this scale at all, and written history tells us absolutely nothing about the main elements of the Indo-European spread.

Many language families that exist today, including Indo-European, cover such enormous geographical extents, as listed in table 9.1. Their expansions represent a force in prehistory for which our only historical parallels lie in the Colonial Era of recent historical times. That era saw massive population movement, especially from the British Isles in connection with the spread of English into North America and Australasia, and with the earlier spreads of Spanish and Portuguese (both with smaller population movements) into eastern Asia and the Americas. Many of the largest language families must reflect the operation of similar processes far back in time, long before writing was invented.

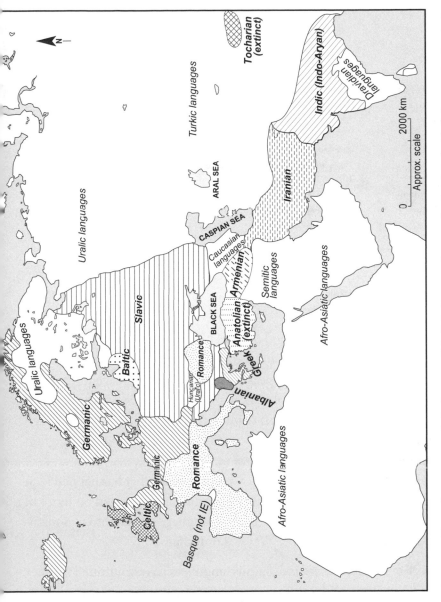

Figure 9.4. The twelve major recorded subgroups of the Indo-European language family, plotted with their approximate extents prior to the Colonial Era. Background map by CartoGIS Services, Australian National University, with linguistic boundaries redrawn from M. Ruhlen, *A Guide to the World's Languages* (Stanford University Press, 1987).

TABLE 9.2. The language subgroups and major languages within the Indo-European language family.

Anatolian[1]	Hittite, Luwian (perhaps the language of Troy), Lycian, Lydian, of what is today Turkey. Hittite names are recorded on clay tablets from ca. 2300–2000 BCE.
Tocharian[1]	Two languages used in late first millennium CE Buddhist and commercial texts, Tarim Basin, Xinjiang, China.
Greek	Modern and Ancient Greek, plus Mycenaean Greek from Linear B clay tablets of the later second millennium BCE.
Armenian	Today a single-state language.
Albanian	Today a single-state language.
Indic	Sanskrit, and the Prakrit ancestors of Hindi, Urdu, Bengali, Punjabi, Marathi, and others.
Iranian	Avestan,[1] Persian, Kurdish, Urdu, Pashto.
Italic	Latin and the Romance languages, Oscan,[1] Umbrian.[1]
Celtic	Irish, Scottish Gaelic, Welsh, Cornish, Breton.
Germanic	German, English, Dutch, Scandinavian (including Icelandic), Gothic.[1]
Baltic	Lithuanian, Latvian, Old Prussian.[1]
Slavic	Russian, Bulgarian, Polish, Czech, and many Balkan languages spoken in former Yugoslav nations.

Data from Thomas Olander, "The Indo-European homeland," in Birgit Olsen et al., eds., *Tracing the Indo-Europeans: New Evidence from Archaeology and Historical Linguistics* (Oxbow Books, 2019), 7–34.

[1] No longer exists in spoken form.

The propellant for this kind of whole-population language transmission at a vernacular level was the settlement of new territory by a migrating population, initially the speakers of the proto-language from which the future family was to evolve. Languages could never have spread by themselves, like brands of soft drink or models of car, solely through adoption by unrelated linguistic populations, at least not over significant distances and for no good reason.

How do we know this? Let us ask, first of all, what has happened in history when the speakers of one language have impinged on those of another language. In cases of recorded language spread into regions already settled, there have been several outcomes:

- The immigrant and indigenous languages have continued to be spoken side by side, as in much of the Roman Empire.
- The immigrant language has replaced that of the indigenous population through dominance of speaker numbers, as eventually

occurred with Latin in some parts of the Roman Empire, and in many parts of the Americas and Australia during the Colonial Era.

- The language of a more numerous indigenous population has replaced/absorbed that of a small but powerful immigrant elite, as occurred with virtually all premodern conquest empires in history that did not engage in the importation of large numbers of farmer-settlers (including those of Alexander the Great, Genghis Khan and the Mongols, the Ottomans, the Normans in England, and so forth).
- Indigenous and forcibly transported populations have created pidgins under recent Colonial-Era circumstances, using their own languages and those imposed by distant authorities. A good example is Tok Pisin of Papua New Guinea, with its indigenous grammatical structure and English vocabulary. Pidgin languages and creoles, however, are not known to have been significant before the Colonial Era.

Perceptive readers might notice that there is one kind of situation not listed here, despite its long-lived popularity among some prehistorians. It arises, in theory, when the language of a small but high-status immigrant elite *replaces* the language of a much larger indigenous majority, the reverse of the third situation described above. Archaeologists term such hypothetical situations "elite dominance."

I hold deep reservations about the linguistic relevance of elite dominance in prestate circumstances, without the pressures of literacy and central government propagation of a national lingua franca. I hold these reservations because I am unable to find any large-scale examples of such replacement within global human history, at least not on the geographical scale of the whole language families that I will be discussing in the following chapters. Why would any functioning and healthy society wish to give up its language, except under the most severe conditions of indigenous population decline and oppression?

For many people in the world, a language is a proud possession and a badge of identity. As linguist Marianne Mithun has pointed out, "When a language disappears, the most intimate aspects of culture can disappear as well: fundamental ways of organising experience into concepts, of relating ideas to each other, of interacting with other people. . . . The loss of a language represents a definitive separation of a people from its heritage."[8] Many situations involving linguistic resistance to "elite" foreign domination can be read from historical circumstances across the world during the Colonial Era. One eloquent historical statement about such matters is the account of *Ecological Imperialism* by historian Alfred Crosby.[9]

Crosby's main contribution was to divide the world, as conquered or colonized by European nations after 1492, into two parts. First came the "Neo-Europes," regions of mainly temperate climate with low indigenous population densities, often hunter-gatherers or kinship-based agriculturalists. The classic examples are North America, Australia, and New Zealand. Here, European settlers (not always free) arrived in their thousands aboard ships, especially between 1820 and 1930, taking up land, bringing in diseases, pushing back indigenous peoples, and replicating the lifestyles and food production systems that they were accustomed to at home.

The British were the most successful at this due to their massive population growth, a spin-off from the industrial and agricultural revolutions that encompassed the British Isles in the eighteenth and nineteenth centuries—a population growth so potent that even the convicts had to be exported. The Spanish had similar success in the southern parts of South America, especially in Argentina and Uruguay, where again the indigenous populations were few in number. The Dutch settled South Africa in much the same way, by focusing on an area of Mediterranean winter-rainfall climate with an indigenous hunter-gatherer and pastoralist population that lay beyond the range of the more numerous and resistant Bantu-speaking farmers.

At the other end of the scale from Crosby's Neo-Europes were the densely settled tropical and warm temperate regions that had their own home-grown and often literate civilizations, highly developed agriculture, and diseases such as malaria that were equal in their malignance to any that the European colonists brought themselves. Europeans were not able to settle in any numbers in most of tropical Africa or Asia— European empires and trading companies existed here to generate wealth for investors rather than to expropriate farmland for settlers. As Crosby described European Crusaders in the Middle East, "The conquerors, taken collectively, were like a lump of sugar presiding in a hot cup of tea."

Travelers nowadays will find few people speaking Dutch in Indonesia, French in Vietnam, or Spanish in the Philippines. Elite dominance certainly did not work there. They might find more people speaking English in India, the Philippines, Malaysia, and some parts of Africa, but this is not because of elite dominance. The British went home, or died, long ago. English is popular today in these countries, but it has never become the first language of the total population in any of them because its arrival was not linked to a wave of human settlement from the British Isles, or the United States in the case of the post-Spanish Philippines. It is used nowadays as a widespread lingua franca in many ex-colonial modern nations with high levels of indigenous linguistic diversity—for instance, India and the Philippines—but this is because it no longer carries adverse colonial connotations and because it is already so widely spoken.

There is an intermediate example in all of this. The Spanish and Portuguese who conquered the tropical Americas in the sixteenth and seventeenth centuries, from Mexico to the Caribbean Islands and down to Bolivia and northern Chile, met dense populations who were often organized into large empires, such as those of the Aztecs in Mexico and the Incas in the Andes. Guns, germs, horses, and military alliances got the Spaniards through, as detailed by many authors from Bernal Diaz to Jared Diamond.[10]

However, rather than arriving as a wave of colonist families like the later British in North America and Australia, the conquistadors and their followers formed a small and mostly male landowning elite that rapidly intermarried with the indigenous population. By the standards of the Old World tropics, the Spanish and Portuguese languages were lucky to have emerged so strongly into the present. How did they do so?

As Jared Diamond explains in his *Guns, Germs, and Steel*, the Americas presented fewer hazards to colonizing Europeans than tropical Asia or Africa because of their healthy isolation from the Old World with its disease-generating primates and domesticated animals. The New World was indeed quite healthy for European conquerors. But, in the reverse direction, the epidemic diseases brought into the Americas by Europeans during the sixteenth century, such as smallpox, tuberculosis, and measles, ensured that up to a staggering 90 percent of indigenous peoples died in some regions, together with their languages. They lacked the necessary resistance to the Old World pathogens.

This tragic genocide, that began in the Caribbean Islands with Columbus in 1492, paved the way for an eventual domination by the Spanish and Portuguese languages among the admixed European, African, and indigenous populations of Latin America during the Colonial Era.[11] Where Native American communities were able to survive some of the impact of the introduced diseases, as in the Arctic and in remote parts of Mesoamerica, Amazonia, and the Andes, their languages survived, and they still do so today.

Indeed, even with indigenous depopulation due to disease, the spread of Spanish in Latin America was not quick and easy. Early Spanish administrators in South America were reluctant to allow native peoples to learn their Spanish language owing to its importance as a vehicle for their rule—that is, native ignorance of Spanish aided that rule. In turn, indigenous people were resentful about learning Spanish because for them it was a symbol of oppression.[12]

The result was that the Spanish language was not adopted widely by indigenous South American communities for two centuries or more after the sixteenth-century arrival of the conquistadors. Indigenous languages, especially Quechua (the major language of the Inca state) and Aymara, were instead used as lingua francas and for Bible translation, and they still exist as important vernaculars among indigenous populations today.

Did Elite Dominance Spread Languages?

Given my observations about the inability of elite dominance to spread the languages of conquering minorities among indigenous majorities in the Colonial Era, it comes as no surprise to find that all of the ancient empires of the Old World, with the partial exception of the Romans in those areas that eventually adopted Romance languages through soldier settlement, suffered much the same linguistic fate as Crosby's crusaders. The example of Alexander the Great always strikes me as one of the most poignant.

Alexander and his armies conquered a 5,000-kilometer expanse of territory stretching from the Balkans and the Nile to the Indus River in Pakistan. Cities were founded in his name and occupied by Greek soldier settlers, often taking indigenous wives. The Greek language and script survived in the conquered regions of central and South Asia into the first few centuries CE, but then disappeared. The Indo-Greek kingdoms that succeeded Alexander had major cultural and artistic impacts in South Asia during the foundation centuries of Buddhist art and civilization (300 BCE to 100 CE), but their rulers were increasingly of Turkic or Scythian (Iranian) origin, not Greek. The Buddhist and Hindu religions that have dominated South Asia for the past 2,500 years existed before Alexander arrived, and they borrowed almost nothing from the Greeks while they were there.

As far as languages are concerned, the indigenous Persian, Kurdish, and Prakrit languages (the spoken contemporaries of Classical Sanskrit), all members the Indo-Iranian linguistic subgroup of the Indo-European language family, won the language battle in Alexander's former empire. So also did Aramaic, a Semitic language used in the lands of the former Persian (Achaemenid) Empire. These indigenous languages simply had far too many speakers unwilling to abandon their linguistic identities in order to adopt Greek on a permanent basis. The military prowess of Alexander was irrelevant in the linguistic context.

Alexander certainly had Greek colonization in mind when he founded all those nascent Alexandrias, but, paraphrasing Plutarch of Chaeronea, there were simply never enough Greeks to overcome the ultimate strength of indigenous revival, and many went home in despair. "They were not happy, so far from the Mediterranean, and on at least two occasions—both after a report of Alexander's death—the homesick veterans decided to go home."[13] Alexander lacked the population numbers required for successful linguistic replacement in his conquered territories, with the partial exception of Egypt, which was located much closer to Greece than was central Asia. Greek did indeed replace Egyptian in the final resort (except for Coptic in the south, a lineal descendant of Ancient Egyptian), until Arabic in turn replaced Greek.

The Ottomans, the Mongols and the Mughals fared no better than Alexander in this regard. To drive this point home, I turn to a book entitled *Empires of the Word*, by linguist Nicholas Ostler, who writes globally about the spreads of languages recorded in written history, from Egyptian, Sumerian, and Akkadian of 2500 BCE onward.[14] Ostler's observations are striking. I paraphrase some examples here:

- The many cases where serious language change failed to follow on from conquest expose the hollowness of much military glory. The conquests in Western Europe by Franks, Vandals, and Visigoths, and even the conquest in Britain by Romans and Normans, indicate this to be true.
- The simplest, biological, criterion for success in a language community is the number of users a language has.
- The most eminent judgment to emerge is that migrations of peoples, the first force in history to spread languages, dominates to this day.
- One factor that is often credited for language spread, trade, has had little demonstrated long-term effect.

And to trade I will add religion. The world's major religions spread virtually without attached spoken vernacular languages, as one can understand if one considers which languages are spoken by Muslims, Christians, Buddhists, and Hindus today.

The vast majority of the world's Muslims do not speak Arabic as a modern vernacular language, and modern Christians do not all speak the Hebrew, Greek, or Aramaic in which their biblical books were written. Neither do the world's Buddhists all speak languages derived from Pali, nor all Hindus from the Prakrits, at least not outside northern India. Religion, writing, trade, military conquests, and powerful dynasties do not explain much about language family distributions across the world. Neither does elite dominance.

Onward toward a Global Prehistory
of Human Populations

We have now reached an important point in the human Odyssey. Food production has developed in several parts of the world, as I explained in chapter 8. Human populations, their ways of life, and the languages they speak are standing on the cusp of some major expansions, ultimately covering thousands of kilometers of the earth's surface. These expansions become understandable to us through the histories of the world's major language families, using the principles that I have outlined in this chapter.

In the following chapters, I discuss the most significant of these expansions in terms of their impacts on the basic geographical structure of humanity, a structure that continued with few major changes until the year 1492. Of course, ancient civilizations and medieval conquests sometimes redistributed populations, but only on relatively small scales. Excellent examples are provided by the Romans and Han Dynasty Chinese, together with the medieval migrations of Seljuk Turks into Turkey and Magyars into Hungary. However, these phenomena were relatively local events compared with the total spreads of the major language families to which they belonged.

10

The Fertile Crescent and Western Eurasia

AFTER 6500 BCE, migrating farmers out of the Fertile Crescent spread through much of Europe, large areas of western Asia, and the northern half of Africa (Africa is discussed in chapter 12). The records of these expansions, which created the roots of so many existing populations in the western half of the Old World, are revealed to us through detailed records in archaeology, linguistics, and genetics. There are major controversies, especially surrounding the origins of the Indo-European-speaking peoples, and I attempt to give a balanced view of the arguments from different sides. This chapter also deals with the prehistory of South Asia (the Indian Subcontinent), with its Indo-European- and Dravidian-speaking populations.

By 6500 BCE, where we left the story in chapter 8, the Fertile Crescent Neolithic subsistence economy had reached a high level of dependence on fully domesticated crop and animal species that could be transferred easily into receptive new environments. Between 6500 and 4000 BCE, there was an unprecedented burst of migration that streamed out of the Fertile Crescent in several directions. Neolithic farmers from Anatolia, Greece, and the Balkans spread through all agricultural regions of Europe, both north and south of the Alps. Others from Iran and the southern Caucasus spread northward and eastward into the Eurasian steppes and toward the Indus Valley. Still others spread from the southern Levant into North Africa. Apart from the initial spread of *Homo sapiens* out of Africa, these movements formed the greatest combined episode of documented population dispersal in the *sapiens* prehistory of western Eurasia and northern Africa. Nothing since has come anywhere close.

Early Fertile Crescent Villagers

What kind of societies were undertaking these spreads? By 6500 BCE, Fertile Crescent Neolithic farmers were increasingly using earthenware pottery, which had several important roles. As I noted in chapter 8, earthenware pots and strainers could be used to process milk from cows, sheep, and goats into yogurt, ghee (clarified butter), or cheese. This would have made milk products more easily digestible for populations who lacked the genetically based ability (known as lactase persistence) to synthesize lactose sugar into glucose for energy beyond childhood. This ability was apparently first developed in pastoralist populations about 3000 BCE, meaning that lactase persistence was not a driver of the main Neolithic migrations out of the Fertile Crescent.[1] Pots were also less labor-intensive than stone vessels to manufacture, and this would have improved the availability of porridges and gruels for mothers wishing to wean infants. Pottery was thus a major invention for population health and fecundity.

Circumstantial evidence suggests that Fertile Crescent farmers by this time would also have been using domesticated and perhaps castrated oxen for pulling simple soil-furrowing implements known as ards (rather than the sod-turning plows with moldboards of more recent times). Whether or not they knew the use of the wheel in 6500 BCE is unknown; solid wooden wheels and carts appeared in the archaeological record of western Eurasia around 3500 BCE, usually interred with elite Copper and Bronze Age burials in well-protected shafts dug deep beneath large burial mounds. Occasionally, a few such earthen mounds, even some of Neolithic date in northern Europe, also fortuitously buried pre-existing ruts that suggest a possession of wheeled vehicles of some kind.[2] In terms of rotary motion, Fertile Crescent Neolithic farmers would certainly have known the principles behind the use of the drill and the lathe for working wood.

Also known to Fertile Crescent Neolithic villagers was the loom, used to weave textiles from spun flax fiber (linen) and eventually sheep's wool. Exactly when sheep were first bred to produce wool of a sufficient quantity and quality for weaving is uncertain, due to the rarity of textile preservation. The discovery of how to process, hammer, melt, and cast metals, especially copper and gold, also occurred gradually during the Fertile Crescent Neolithic, although metallurgy did not become a major factor in human societies until after 4500 BCE.

All in all, the Fertile Crescent economy by 6500 BCE was a highly productive one. Yet, and perhaps paradoxically, after almost 3,000 years of Pre-Pottery Neolithic growth, the Fertile Crescent by this time is reputed to have been undergoing environmental and climatic decline and settlement collapse. Many large towns such as Çatalhöyük and Ain Ghazal were abandoned in favor of smaller settlements, and many populations were turning away from the cultivation of arable land toward pastoralism.

Why did this happen? I am not totally convinced by frequent arguments that climate change was the major cause. Climatic and sea level fluctuations at this time were generally rather trivial compared with the Younger Dryas return to glacial conditions that occurred just before the start of the Holocene. Yet the Younger Dryas, despite its apparent malignance, seems to have imposed no brakes on the development of Natufian society in the Fertile Crescent.[3] It is hard to imagine that the less severe climatic changes that occurred during the Holocene would have caused major damage to complex Neolithic societies, unless they were living at the edge of agricultural viability.

Was human action itself the source of the problem? Did humans simply overdo it in the fragile and water-starved landscapes that existed in many parts of the Neolithic Fertile Crescent? Resource collapse has been suggested as a cause of decline at Ain Ghazal in Jordan, which reached an area of 13 hectares with a population of several thousand inhabitants and then rapidly shrank in size before being abandoned about 6200 BCE.[4]

What about pandemics? When large towns such as Çatalhöyük and Ain Ghazal were occupied by populations in their thousands during the peak centuries of the Pre-Pottery Neolithic, certain malevolent trends can be expected.[5] Lack of sanitation would have led to increasingly dangerous episodes of poor health, exacerbated by intimate contact with domesticated and commensal animals such as rats.[6] Viral pandemics were no doubt a possibility for both humans and their domesticated resources, and the plague bacterium *Yersinia pestis* was certainly at large during the Neolithic in parts of Europe and Asia according to analysis of surviving pathogen DNA in human bones.[7] However, there is so far no direct indication of any plague-related pandemic disaster in the Fertile Crescent at the time in question.

What about social changes? With increasing populations living cheek by jowl in villages and towns, social stresses would have increased if there was no regulatory authority to control disputes beyond the level of the kinship group. At Çatalhöyük, for instance, the Pre-Pottery Neolithic complexes of wall-to-wall rooms with large numbers of underfloor burials and mural art were replaced during the Pottery Neolithic, close to the end of site occupation, by multiroomed free-standing houses around courtyards, with burials in separate locations.[8] This change need not indicate social disruption, but it surely does indicate social change of some kind, especially toward an increasing independence of housing units from each other, and perhaps an increasing freedom for people to move away from the settlement, if desired. What is more, this social change was occurring at precisely the time of initial Neolithic migration through western Turkey into Europe, after 6500 BCE.

To be honest, however, I am unable to answer the question of exactly why settlement sizes declined in the Fertile Crescent during the seventh millennium BCE. I prefer to focus on the outcomes, in which groups of people with highly productive agricultural economies, perhaps unable through population growth to inherit land at home, saw reason to look elsewhere for better prospects. Let us not forget that this was a time of expansion, of powerful population growth in new environments. These migrants were not the plague-afflicted or starving remnants of collapsed societies. They were highly fecund, with a transportable economy of food production to match. Let us begin their story in Europe.

Neolithic Migration across Europe, 7000 to 4000 BCE: The Archaeology

The Fertile Crescent origins of agriculture were concentrated in the Levant, extending into Anatolia and the northern Zagros foothills of Iran. Pre-Pottery Neolithic village life was well established here and in Cyprus by 8000 BCE. On current evidence, the spreads of farmers beyond the Fertile Crescent into Europe, the Caucasus, and Armenia began around 7000 BCE, just when pottery was making its first appearance.[9] The approximate details can be seen in figure 10.1.

By 6500 BCE, pottery was in common use in the Fertile Crescent, and Neolithic villages were well established in western Anatolia and along the Aegean coastline, also around the Bosporus.[10] The movement toward the Balkans and the lower Danube Valley was underway in earnest. By 5400 BCE, farmers had reached the fertile loess (glacial-era wind-blown dust) soils of the North European Plain in Germany, negotiating the Carpathian Mountains along the way and settling the fertile Great Hungarian Plain that straddles the Danube River. At the same time, they spread east toward the Pontic Steppes beyond the northern coastline of the Black Sea, recently expanded from a large freshwater lake by the rising postglacial sea level that flooded into it through the Bosporus.

By 6000 BCE, farmers were also paddling westward in dugout boats along the northern coastline of the Mediterranean to reach Dalmatia, Italy, the western Mediterranean islands, and Iberia. One remarkable specimen of an early Neolithic log boat, more than 10 meters long, has been recovered from the waterlogged site of La Marmotta in Lake Bracciano near Rome. Presumably, movements were occurring at about the same time toward the North African coastlines of Tunisia, Algeria, and Morocco.

The British Isles and southern Scandinavia were finally reached by Neolithic farmers around 4000 BCE. The first settlers to reach Britain and Ireland appear to have arrived from the Atlantic coastline of Europe, bringing in a tradition of erecting communal burial chambers of wood or stone (megaliths) under earthen mounds. Another major element within the British Neolithic was apparently derived from the North

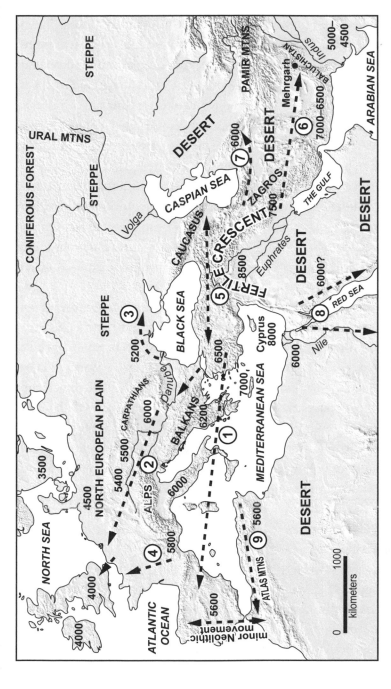

Figure 10.1. Population expansions from the Fertile Crescent Neolithic. Dates are all BCE. Arrows 1 to 4 are schematic representations of Neolithic movements into and through Europe. Arrow 5 represents communication across the northern Fertile Crescent, and arrows 6 and 7 represent movements into Turkmenistan and Baluchistan. Arrows 8 and 9 represent Neolithic movements from the southern Levant into North Africa (chapter 12).

European Plain, involving the construction of circles of upright stone or wooden pillars within enclosing ditches.

Such circles are somewhat reminiscent of the much older pillared enclosures at Göbekli Tepe, and British archaeologists term these monuments "henges." The most famous example, reconstructed several times from the Neolithic into the Bronze Age, is known to the world as Stonehenge, located on Salisbury Plain in western England. This cultural amalgamation in Britain suggests that the separate Neolithic movements south and north of the Alps might have met and mixed in what is now France before crossing the sea to the British Isles.

Even reduced to its essentials, all of this still seems rather overwhelming. But we must not forget that the whole process took at least 2,500 years to progress, segment by segment, from the Anatolian coastline and Greece to the British Isles and southern Scandinavia. As these Neolithic migrations progressed, so the environmental and cultural scenery changed.

In the wetter conditions of temperate Europe beyond Greece and the Balkans, the sun-dried mud brick architecture of the Fertile Crescent gave way to buildings of earth, wood, wattle, and daub. Building skills were adapted to new conditions—the inhabitants of La Marmotta, for instance, raised their wooden houses on stilts along the Lake Bracciano shoreline. Most houses in southeastern Europe were smaller than in the Fertile Crescent, with fewer conjoined rooms, but on the fertile loess soils of the North European Plain there was a striking development of wooden longhouses. In function, these structures may well have housed multiple related families, and because they were built with wood rather than brick, they would have been easy to dismantle and remove to a new site should circumstances require. After all, farmers were frequently on the move at this time.

In summary, between 6500 and 4000 BCE, Neolithic settlers in two major movements spread 3,500 kilometers westward, both north and south of the Alps, from western Anatolia and the Levant to Ireland and Portugal, exposing the fertile agricultural soils of Europe to their permanent field systems and their plowing, manuring, and foddering practices. Cattle, sheep, pigs, and domesticated crops provided secure food supplies that could ensure dramatic population increases in the early centuries. Estimates for initial Neolithic birth rates in the Balkans and on the North European Plain range in some regions up to 2.5 percent per annum, giving more than a trebling of population numbers within two generations. One recent archaeological investigation in the Neolithic Balkans indicates an average of up to eight children per woman.[11] Descendants of Mesolithic hunters and gatherers no doubt retained footholds for a time in the less fertile areas, but eventually they must have blended with the burgeoning farming populations.

All of this Neolithic migration and fecundity was no doubt rather wonderful, at least for a time. But eventually, as occurred in the Fertile Crescent, circumstances changed for the worse, especially on the North European Plain. Again, archaeologists are not sure why, and explanations connected to climate change, loss of soil fertility, and epidemic diseases are frequently put forward. By soon after 5000 BCE, defensive ditches, massacres, and even hints of cannibalism began to throw a pall over Neolithic village life, at a time when the population on the loess soils might have reached between 1 and 2.5 million.[12] Apparent population decline, abandoned settlements, and increasing investment in pastoralism characterized the later Neolithic in much of northern Europe, especially after 3500 BCE.

It seems, therefore, that there were decisive limits to growth under the systems of land management and social control that were available in many regions of Neolithic Europe at the time, just as there had been beforehand in the Fertile Crescent.[13] These changes were not lost on some of the contemporary pastoralist populations who lived on the steppe lands that extended north and east of the Black Sea. Like some of their descendants at the end of the Roman Empire, they began to look west after 3000 BCE in search of new opportunities.

Before we come to these Copper and Bronze Age upheavals, however, we must first ask the geneticists what they can tell us about Neolithic migration into Europe from the Fertile Crescent. Do the genes match the archaeology?

Neolithic Migration across Europe: The Genetics

I had little doubt when I was writing my *First Farmers*, published in 2005, that migration was extremely important in explaining the European Neolithic. It was a massive cultural change from the preceding Mesolithic, and I could not conceive how agriculture could have spread through a continent full of hunter-gatherers if there had been no farmers bringing in the crops and knowledge in the first place. I still hold this view today.

However, DNA results available at that time, based mainly on mitochondrial DNA comparisons of living people across Europe, steadfastly refused to reveal anything coherent about the issue. As many geneticists have since pointed out, it is difficult to understand events that occurred 8,000 years ago purely from the DNA of living people, more so if that DNA is only a small portion of the whole genome, derived only from female-inherited mitochondria. Not only did people move around frequently in prehistory, erasing local patterns in language and culture, but mitochondrial lineages are suspected by some geneticists to be subject to frequency variations in time and space due to genetic drift and various selective factors. Even two decades ago, it was clear that the future for genetic prehistory lay in the discovery of techniques to scan nucleotide sequences across whole human genomes, and to extract DNA samples from the directly carbon-dated bones of the long dead.

The floodgates opened in 2015 with the publication of a paper in *Nature* on ancient DNA across Europe, taken from skeletons dating from the Mesolithic through to the end of the Bronze Age.[14] The implications were crystal clear. Early Neolithic farmers had spread through the agricultural areas of Europe, both north and south of the Alps, replacing Mesolithic hunter-gatherers with an almost clinical efficiency. As stated recently for the Danube Valley Neolithic, "it is hard in the light of current bioarchaeological discoveries to give any explanation for the process of initial Neolithisation other than a massive population movement."[15]

Ancient genomic evidence from the Neolithic of the British Isles has since revealed an almost complete replacement of Mesolithic populations around 4000 BCE, when Neolithic settlers arrived in Britain and

Ireland by sea from Iberia and Atlantic France.[16] Interbreeding between Neolithic settlers and Mesolithic hunter-gatherers was almost nonexistent at the start of the migration process. However, the genetic signature of the latter increased in significance across Europe over time, as former hunter-gatherers switched to food production, especially along the agricultural margins in northern Europe.[17]

It was also quickly realized by geneticists that the Neolithic migration into Europe originated mainly in Anatolia,[18] and that the ancient farmers of Anatolia were genetically distinct from those in the Jordan Valley and its hinterland in the southern Levant, and different yet again from those in Iran. Farming in the Middle East did not originate with a single genomic population. Not surprisingly, these differences suggested from the beginning a kind of radial scenario for the Fertile Crescent Neolithic, with populations moving outward like the spokes of a wheel:

> The impact of the Near Eastern farmers extended beyond the Near East: farmers related to those of Anatolia spread westward into Europe; farmers related to those of the Levant spread southward into East Africa; farmers related to those of Iran spread northward into the Eurasian steppe, and people related to both the early farmers of Iran and to the pastoralists of the Eurasian steppe spread eastward into South Asia.[19]

The pastoralists of the Eurasian steppe will feature later in this chapter.

Migrations from the Eastern Fertile Crescent

The eastern Fertile Crescent underwent developments similar to those in the Levant and Anatolia, albeit slightly later in time, perhaps because the Zagros foothills of southern Iran extended beyond the wild distributions of many of the domesticated Fertile Crescent cereals and legumes.[20] Nevertheless, soon after 10,000 BCE, hunters and gatherers in Iraqi Kurdistan were settling down into small settlements with circular stone house foundations, similar to those of the Natufians to their west. By 7000 BCE, villages of mud brick rectangular houses with cubicle-like small rooms, like those in many later Pre-Pottery Neolithic settlements

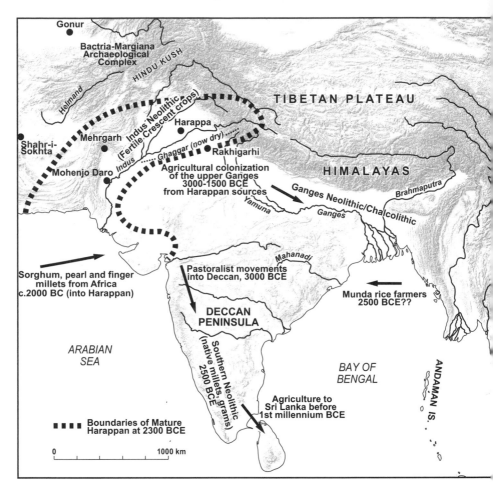

Figure 10.2. The Neolithic settlement of South Asia.

in the Levant and Anatolia (e.g., figure 8.2F), were coming into wide-spread existence down the western Zagros flanks.[21]

By at least 6500 BCE, this eastern Fertile Crescent cultural tradition had spread as far east as the archaeological site of Mehrgarh, in the Bolan Pass of Baluchistan province in Pakistan (see figures 10.1 and 10.2). Here, close to the Indus Valley, Pre-Pottery Neolithic farmers reached the eastern limit of the winter rainfall climate that had nurtured the Fertile Crescent agricultural revolution in the first place.[22] Beyond Baluchistan lay the summer monsoon climates of the Indus and Ganges Basins,

Peninsular India, and Sri Lanka. Not surprisingly, movement stopped at the Baluchistan limit, at least temporarily.

Food-producing populations also spread north and northeast of the Fertile Crescent around 6000 BCE, through the Caucasus, and around the southern Caspian Sea into the semidesert oases of southern Turkmenistan. These slightly younger movements were associated with pottery production, in parallel with the contemporary Pottery Neolithic settlement of southeastern Europe. Neolithic population movements from Iran and Armenia also continued beyond the Caucasus into the Pontic (Black Sea) Steppes, to contribute around half of the genomic ancestry of the developing pastoralist populations who lived beyond the Black Sea.[23]

Another observation of significance has been revealed by ancient genomic analysis in and around the Fertile Crescent. The Anatolian, southern Levant, and Iranian Neolithic populations were genomically different from each other at the start of the Neolithic, but during its course there was a great deal of mixing between them. After 5000 BCE, Anatolian Neolithic ancestry appeared in high percentages (30 to 50 percent) throughout Iran and the Caucasus, extending beyond the Caspian Sea into the Bactria-Margiana cultures of the upper Amu Darya (Oxus) River in central Asia (figure 10.3).[24] Likewise, Iranian Neolithic ancestry began to appear in the genomes of some contemporary populations in the Levant, and Levant ancestry spread north into Anatolia.[25] Farmers not only spread out of the Fertile Crescent but also mixed within it, eventually breaking down genetic boundaries that had been in place since Paleolithic times.

Early Farmers in South Asia

Beyond the site of Mehrgarh lies the Indus Valley, the gateway from the west into South Asia. The appearance of agriculture by 6500 BCE at Mehrgarh was clearly associated with a Fertile Crescent repertoire of crops and animals, including wheat, barley, sheep, and goats, with only the humped zebu cattle being domesticated locally. Farmers had presumably entered the region from Iran (although there is no ancient

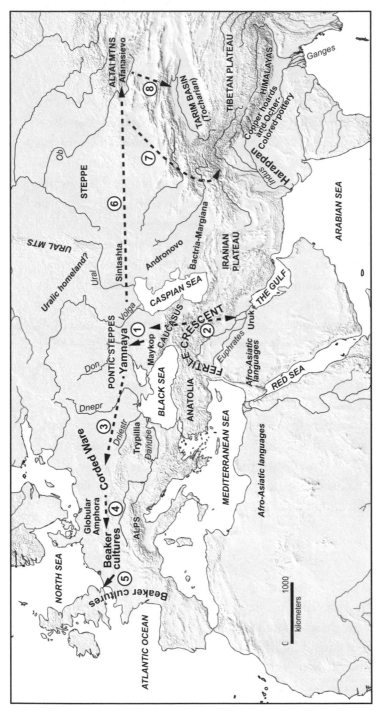

Figure 10.3. The Pontic Steppe migration hypothesis of the Copper and Early Bronze Ages. Arrow 1 represents movements into the Pontic Steppes from Iranian and Caucasian Neolithic sources, followed by Bronze Age contacts with the Maykop culture. Arrow 2 represents contacts between Maykop and Uruk Mesopotamia. Arrows 3, 4, and 5 represent the movements of Yamnaya genomic populations circa 2800–2300 BCE within the Corded Ware and Beaker cultures of central and western Europe. Arrow 6 represents movements from the Pontic Steppe eastward along the steppes toward the Altai Mountains circa 2700 BCE, followed by hypothetical movements south (arrows 7 and 8) into the Tarim Basin (Tocharian languages?) and South Asia around 2000–1500 BCE.

DNA information from Mehrgarh to demonstrate this), bringing in a tradition of mud-brick rectilinear architecture with small conjoined rooms, like that so common in later Pre-Pottery Neolithic settlements in the Fertile Crescent.

For perhaps two millennia, Mehrgarh represented the limit of Fertile Crescent Neolithic expansion to the east. Beyond the hills of Baluchistan lay the monsoonal (summer rainfall) climate of the Indus plains, where the Fertile Crescent cereals and legumes would have needed labor-intensive irrigation to grow in the dry winters. The first Indus Valley farmers therefore had to manage an agricultural system for which the climate presented certain disadvantages. As a result, the Indus plains lack definite Pre-Pottery Neolithic settlement contemporary with the foundation of Mehrgarh (although current archaeological evidence is admittedly limited). The oldest known Neolithic villages (apparently all pottery users) were established around 5000 or 4500 BCE (figure 10.2).[26]

Once farmers had settled the Indus Plains, population growth was fairly rapid. By 3200 BCE, small towns with defended citadels announced an early stage of the majestic Indus Valley Civilization, termed the Harappan culture by archaeologists. Dated during its Mature Harappan phase to between 2600 and 1900 BCE, features of the Indus civilization included large brick cities with raised fortified citadels and public architecture, house compounds with drainage systems, rectangular street grids, an undeciphered script, copper tools, fine painted pottery, and trading connections as far afield as Afghanistan and Mesopotamia.

Most importantly for what was to follow, the Harappans did not remain confined just to the Indus and its tributaries. They also spread eastward toward the upper Gangetic Plain. This population spread commenced as early as 3000 BCE, and it was magnified close to 2000 BCE, when many of the larger settlements such as Mohenjo Daro and Harappa in the Indus Basin itself were abandoned.

Whatever the reasons for this core-region Harappan decline (again, no one knows for certain—shifts in river courses and changing riverine water volumes are possibilities), the demographic center of gravity of the Late Harappan culture moved increasingly eastward, toward and into the upper Yamuna and Ganges valleys of northern India. The farming settlement of the Gangetic Plain had begun, with wheat and barley cultivation reaching the middle Ganges around 2400 BCE.[27]

The village-based archaeological cultures that were established on the Gangetic Plain (Copper Hoards and Ochre-Colored Pottery in Indian archaeological terminology) were ultimately derived from the Harappan in terms of their pottery styles, copper tools, and Fertile Crescent crops and animals. The Harappan script disappeared beyond the Indus, not to be replaced until at least a millennium later when alphabetic scripts of western Asian origin (including Brahmi and Kharoshthi) were adopted via the Persian Empire. The late Indus Valley civilization thus expanded, despite its apparent Indus core-region decline, to establish the oldest phases of the subsequent Gangetic civilization of Vedic (early Hindu) and Buddhist India.

At present, the archaeological record reveals a clear Fertile Crescent origin for the oldest food production in northern South Asia. But what about ancient genomics? There is ancient DNA so far from only one Harappan individual, who was buried in the Mature Harappan city of Rakhigarhi, on the Ghaggar River and close to the upper Gangetic Plain in Haryana State, not far from modern New Delhi. This person belonged to an indigenous Iranian Plateau population with close relatives traced in ancient DNA from a few individuals buried in the Bactria-Margiana city of Gonur in eastern Turkmenistan, and in the Bronze Age city of Shahr-i-Sokhta in Iran (both shown in figure 10.2).[28]

Gonur, occupied between 2250 and 1700 BCE, was in direct archaeological contact with contemporary Harappan cities. Indeed, the similarities in citadel-focused settlement plans between the Bactria-Margiana and Harappan Bronze Age urban civilizations suggests that their populations might have been closely related genetically and linguistically. And the Bactria-Margiana people, as noted above, carried up to 50 percent of their DNA from the Anatolian Neolithic source. However, the Harappan ancient genomic population sample is tiny, so resolving the issue of Harappan identity as a whole through genomics must await further discoveries in Pakistan and India, especially of human remains older than the Harappan. This is a crucial question for our further understanding of South Asian population history.

Europe and the Steppes

The previous sections have dealt with Neolithic migrations out of the Fertile Crescent, both to the west and to the east within Eurasia. Between 6500 and 4000 BCE, the Fertile Crescent repertoire of domesticated crops and animals, and the archaeological cultures associated with them, spread as far afield as Ireland, Iberia, western Russia, the Tien Shan Mountains, and Pakistan. This distribution overlaps closely with that of the Indo-European language family prior to the Colonial Era.

I am holding the origins of the Indo-European language family for more detailed consideration soon, but at this point it is necessary to examine another phase of considerable population movement in Europe and western Asia, that associated with steppe peoples during the Copper and Bronze Ages, mainly after 3000 BCE. The Eurasian steppe grasslands extend from Ukraine in the west to Mongolia in the east, generally between latitudes of 45° to 50°, with a break formed by the Altai Mountains in central Asia. Migrations of pastoralist peoples through and out of these steppe lands spread over the older Neolithic cultural and genetic foundations, but the courses they took were different from those of the older Neolithic populations.

The genomic reality of Early Bronze Age steppe migration from north of the Black Sea into Europe was first announced in 2015, side by side with that of the earlier Neolithic migration from Anatolia and the Aegean. Many archaeologists had favored the idea of steppe migration since the early twentieth century, especially through the writings of archaeologists Gordon Childe and Marija Gimbutas, and, more recently, David Anthony.[29] The genomics verified that a pastoralist population from the steppes of the Ukraine and western Russia, termed "Yamnaya" by archaeologists, spread across Europe to the north of the Alps about 2800 BCE, with onward repercussions eventually extending as far as the British Isles. Another Yamnaya expansion extended eastward to the Altai Mountains (figure 10.3). Yamnaya genomic ancestry was derived partly from an indigenous steppe hunter-gatherer background, with significant inputs from Neolithic eastern Europe and the Caucasus.[30] There was also influence on them from Copper and Bronze Age sources in the Caucasus, especially from the remarkable Maykop culture (circa 3500 BCE) with its contacts with Uruk period (early Sumerian) Mesopotamia, its elite shaft burials with wheeled vehicles under large mounds, and its fine metalworking in gold, silver, copper and bronze.[31]

By 2800 BCE, the Yamnaya were clearly eyeing the lands to their west, especially those occupied by Neolithic farmers of the later Trypillia culture that occupied the fertile black soils of the Dniester and Bug valleys to the northwest of the Black Sea. At that time the Trypillia culture was in apparent decline, with settlements being abandoned. The Yamnaya took advantage of the situation and rumbled west in their ox-drawn carts through the Carpathians and across the North European Plain, burying their elite dead with their carts in shafts beneath high earthen mounds, or *kurgans* in Russian and Ukrainian archaeological terminology. Although new genetic evidence indicates that the ancestors of modern horse breeds were domesticated in the Pontic Steppes soon after 2200 BCE, this relatively late date also implies that the earlier Yamnaya migrants into Europe at 2800 BCE did not have military chariotry or cavalry and were therefore not fast-moving equestrian conquerors.[32] Basically, they were mobile pastoralists seeking land for settlement.

Once they reached the Danube Basin and the North European Plain, the Yamnaya overran and absorbed indigenous populations to stimulate the formation of the Corded Ware and Bell Beaker cultures of European archaeology.[33] The Corded Ware and Bell Beaker populations eventually replaced the cultures of the Neolithic farmers of northern Europe, such as those of the Globular Amphora Culture (figure 10.3), in circumstances that might sometimes have involved lethal violence.[34]

The Yamnaya impact on the northern European population was massive, with the Yamnaya genomic signature accounting for between 50 and 80 percent of the DNA in many populations north of the Alps at the time of the migration. Even today, many northern Europeans still share between 30 and 50 percent of Yamnaya inheritance, the rest made up from Fertile Crescent Neolithic and indigenous Mesolithic hunter-gatherer sources.

However, the Yamnaya genomic spread differed from that of the Early Neolithic in one major respect. It was limited in impact south of the Alps and in the Balkans, and almost absent in Anatolia. Ancient DNA from

Greece and the Aegean dating between 3000 and 1000 BCE reveals no Yamnaya ancestry in the Minoans of Crete, and only a small quantity in the early Greek-speaking Mycenaeans, increasing in proportion after 2600 BCE among individuals from northern Greece.[35] Further west, in Mediterranean islands such as Sicily and Sardinia, a Yamnaya signature accounted for 25 percent or less in Bronze and Iron Age genomes, generally appearing only after 2200 BCE and becoming more important during and after the Roman Empire.[36] In France, steppe genomes appeared around 2650 BCE and contributed between 0 and 55 percent Yamnaya ancestry to the French Bell Beaker (Early Bronze Age) population.[37]

The Yamnaya, therefore, added to the existing genomic configuration in many regions of Europe rather than replacing all of it. In southern Scandinavia, for instance, people of the Battle Axe culture, a close relative of the Corded Ware culture, still carried up to half of their DNA from the earlier Fertile Crescent Neolithic farmers.[38] Certainly, there was a major population replacement in the British Isles during the Early Bronze Age, between 2450 and 1600 BCE, that perhaps introduced Celtic languages to these islands, but this did not arrive directly from the steppes; rather, it came from Bell Beaker populations in adjacent regions of Europe who had inherited large proportions of Yamnaya DNA.[39]

A major question now arises. What was the impact of the Yamnaya expansion on the distribution of languages in Europe, especially languages within the Indo-European family that occupies most of temperate Europe today? The time has come to examine the Indo-European language family itself.

The Contentious Prehistory of the Indo-European Language Family

In 1786, a British juror and oriental linguist living in Calcutta made some important observations about a number of languages now included in the Indo-European family that had played important roles in European and Asian history:

> The *Sanscrit* language, whatever be its antiquity, is of a wonderful structure; more perfect than the *Greek*, more copious than the *Latin*,

and more exquisitely refined than either, yet bearing to both of them a stronger affinity, both in the roots of verbs and in the forms of grammar, than could possibly have been produced by accident; so strong indeed, that no philologer could examine them all three, without believing them to have sprung from some common source, which, perhaps, no longer exists; there is a similar reason, though not quite so forcible, for supposing that both the *Gothick* and the *Celtick*, though blended with a very different idiom, had the same origin with the Sanscrit; and the old *Persian* might be added to the same family.[40]

So wrote Sir William Jones, who is sometimes credited with being one of the founding fathers of linguistics, although to be fair he was beaten to the post by Johann Reinhold Forster, who made similar comments in 1774 about a common source ("descended from the same original stem") for a number of Southeast Asian and Pacific Island languages that he had recorded on James Cook's second voyage across the Pacific.[41] Today this Oceanic family is called Austronesian, and I examine it in chapter 11.

In modern terms, Jones was referring to six of the ten living subgroups of the Indo-European languages (table 9.2 and figure 9.4): Indic, Greek, Italic (Romance), Germanic, Celtic, and Iranian. The others are Albanian, Armenian, Baltic, and Slavic. Jones was unaware of the existence of the Anatolian and Tocharian languages, both long extinct by the time he was writing.[42]

The Anatolian languages, represented most famously by second millennium BCE Hittite, and probably also including the unrecorded language of Troy, were lost to history until the early twentieth-century decipherment of clay tablets from the Hittite capital of Hattusa (Turkish Bogazköy) in central Anatolia. Most documents actually written in the Hittite language date between 1600 and 1200 BCE, so it was contemporary with two other famous Bronze Age Indo-European languages that have descendants today, these being Linear B Mycenaean Greek and early Vedic Sanskrit. Hittite personal names are also recorded on clay tablets found in the Old Assyrian trading city of Kanesh (modern Kultepe) in central Turkey (circa 1800 BCE), and perhaps also in the palace archive at Ebla (modern Tell Mardikh), near Aleppo in northern Syria (circa 2300 BCE). The Anatolian languages were replaced principally by Greek during the later first millennium BCE.

The Tocharian languages were brought to light early last century in collections of middle to late first millennium CE Buddhist texts and commercial documents, found in several oasis locations in the arid Tarim Basin in Xinjiang Province of western China. They present us with some intriguing mysteries, being located more than 1,000 kilometers from any other Indo-European languages and hidden from them by the massive bulk of the Tien Shan, Pamir, and Himalayan mountain ranges (see figure 10.3). The Tocharians also lack any clear historical identity, apart from their languages and their Buddhism. The Tarim Basin is well provided with Bronze Age mummies with western Eurasian biological features preserved in the desert sands, but none have so far been linked directly with Tocharian-speaking owners.[43] Today, the indigenous populations of this region speak Turkic languages, especially Uighur.

As we will see, the extinct Anatolian and Tocharian subgroups are of great importance in understanding the origins of the Indo-European-speaking peoples. In many ways they are lucky accidents, only known to scholarship because of the existence of contemporary written records. The fact that both of these subgroups were completely unknown until the twentieth century, apart from biblical mentions in the case of the Hittites, surely makes one wonder how many other Indo-European subgroups might have disappeared completely without trace, especially those without written records.

Celtic languages, for instance, were once widespread across much of central Europe and even into Anatolia, according to existing placenames and historical accounts from the Classical Era, but they survive today as minority languages only in Brittany (France) and parts of the British Isles. The Indo-European language family stands revealed as a potent eraser of the traces of its own past because of its many internal linguistic expansions. Greek replaced Anatolian, the Romance and Germanic languages replaced many Celtic languages, and the extensive expansion of Slavic languages in eastern Europe since Medieval times might have replaced many older Indo-European languages now lost without record, including some perhaps in the Pontic Steppes. As linguist Harry

Hoenigswald noted perceptively more than half a century ago: "Hittite and Tokharian . . . are now extinct; there are other splinters, barely known to us, of which the same is true, and we may conjecture, though with meagre profit, that there were many additional groups, now lost without a trace."[44]

Despite these problems, linguists who wish to trace the origins and expansion history of the Indo-European languages and their speakers can only work with the evidence that has survived. According to every Indo-European genealogy, or "family tree," created by linguists in the past twenty-five years, mostly through complex statistical analysis of vocabulary items, the Anatolian languages are always the first branch to emerge. This suggests that they developed in or close to the Indo-European homeland, which most logically must have been situated somewhere in or close to Anatolia, or at least within the northern Fertile Crescent. So far so good, but is it possible to state in which order the other Indo-European subgroups emerged? In most family tree reconstructions, Tocharian follows Anatolian, and then come Armenian and Greek. This is perhaps not surprising because the last two subgroups are geographically close to Anatolia, although no one knows where Tocharian was originally spoken before it arrived in Xinjiang. However, as one goes farther down the line of other subgroups, so the uncertainty and disagreements grow. The outer branches of the Indo-European language family do not have a genealogy that is agreed upon by all linguists.[45]

Calculations of a date for Indo-European origin that are derived from the observed rates of change in historical language vocabularies also disagree with each other, varying frustratingly between 6700 and 3000 BCE. My chronological preference is for the separation between Anatolian and the rest of Indo-European, a point in time labeled "Proto-Indo-European" by linguists, to have occurred between 6500 and 5500 BCE, in agreement with the calculations of a highly productive group of computational linguists.[46] An Indo-European genealogy reconstructed by these linguists in 2013 is shown in figure 10.4. In my view, it matches well with the Neolithic archaeological record that I regard as the most likely expression of the first Indo-European language dispersals.

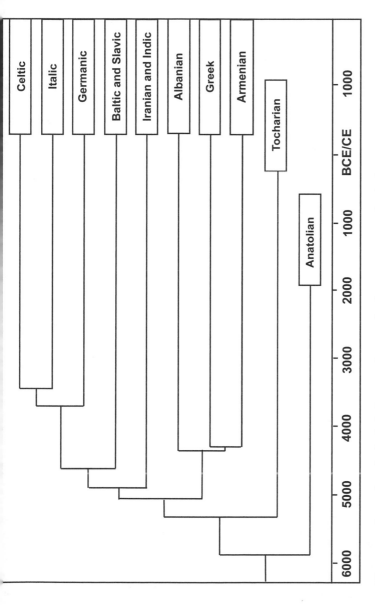

Figure 10.4. A genealogy for the Indo-European language family against a time scale of subgroup separation times. In reality, the separations between these subgroups were not as sharp as this diagram indicates. Like related hominin species gradually losing their ability to interbreed, languages with a common source also took time to lose mutual comprehension. Divergence rates between Polynesian and Romance languages suggest several centuries to over a millennium to reach this state, depending on the degree of continuing interaction. Modified by the author from Remco Bouckaert et al., "Mapping the origins and expansion of the Indo-European language family," *Science* 337 (2012): 957–960; Remco Bouckaert et al., "Corrections and clarifications," *Science* 342 (2013): 1446.

What can we conclude about Indo-European origins and expansion so far? The following observations appear to be justified:

- The replacement of Mesolithic populations with the arrival of Neolithic settlers in Europe was so widespread and thorough, and so unparalleled at any later time in European prehistory and history, that it must have been accompanied by a significant replacement of Mesolithic languages.

- The Anatolian languages are the oldest attested form of Indo-European and most probably are indigenous to central Anatolia. Suggestions that their speakers arrived after the Neolithic from some unidentified homeland are backed by no linguistic, archaeological, or genetic evidence.

- Because Anatolia has been identified as the archaeological and genomic source region for the Neolithic migration into Europe, it surely stands as a likely source region for the languages that were associated with that migration. We can never know for certain if those languages were indeed early Indo-European, but ancestral Anatolian speakers were in a geographical and chronological position to have spread Neolithic languages through much of Europe, as suggested many years ago by archaeologist Colin Renfrew.[47]

- It has sometimes been claimed that all the Neolithic languages of Europe were non-Indo-European, but the evidence is weak. Europe does have a few "substrate" languages represented in place-names, inscriptions, and comments by Classical authors. But except for Basque of the Pyrenees there is no certainty that they were non-Indo-European, even in the celebrated cases of Etruscan of Italy and Minoan of Crete.[48] Too little is known about these extinct languages to be certain of their language family affinities.

Did Yamnaya People from the Pontic Steppes Spread the Oldest Indo-European Languages?

The reader will realize that we are approaching a crucial but complex question. Did the oldest Indo-European languages and their speakers travel with the Neolithic migrations out of the Fertile Crescent between

6500 and 4000 BCE as I have suggested above, or, as others have suggested, did they travel with Bronze Age migrations from the Pontic Steppes north of the Black Sea between 2800 and 2300 BCE? Neither hypothesis is perfect, but which fits available evidence the best?

The Anatolian hypothesis, in terms of archaeology and ancient genomics, matches well with a hypothetical spread of early Indo-European languages with Neolithic populations through Europe. Anatolian Neolithic DNA trails that might have carried Indo-European languages also reached Iran, Armenia, Turkmenistan, and the Pontic Steppes. The genealogy of Indo-European languages discussed above also fits with a homeland in the Fertile Crescent.

South Asia offers some problems for the Anatolian hypothesis because the sample size of ancient DNA from there is currently so small, as discussed above. However, Fertile Crescent food production spread from Iran into the Indus and Ganges Basins, in the process blending eventually with indigenous monsoonal cropping systems involving millets and rice. Ultimately, cultures derived from the Harappan appear to have founded the Vedic (Indo-European-speaking) civilization of the first millennium BCE in the Ganges Basin. So a continuity in archaeological terms from the Harappan into the historical cultures of northern India is evident. There are no traces of a Pontic Steppes archaeological presence in the Ganges Basin.

The Yamnaya hypothesis fits well with the spreads of the Early Bronze Age Corded Ware and Beaker cultures across Europe north of the Alps, in the case of the Beaker people, to as far as the British Isles. Yamnaya people also gave rise to a number of major Bronze Age archaeological cultures and genetic populations east of the Caspian Sea, such the Sintashta, Andronovo, and Afanasievo complexes (see figure 10.3 for locations).

A Yamnaya spread becomes less convincing beyond northern Europe and central Asia because of the faintness of its DNA trail and the nonexistence of any Yamnaya-related archaeological trail, especially along the northern Mediterranean shoreline and in Anatolia, Iran, and South Asia. Those who favor a Yamnaya spread of Indo-European languages into all these regions need to explain how small numbers of migrants managed to impose their languages on much larger indigenous populations of Neolithic farmers on so many different occasions and across

such a vast extent of territory.[49] As I indicated in chapter 9, such a replacement definitely would go against the grain in terms of other trans-
continental situations of language change in recorded human history.

It should also be added that a Pontic Steppes homeland for Indo-
European is not supported by linguistic genealogies (e.g., figure 10.4).
Although some archaeologists and linguists favor a steppe origin based
on cognate words for wheels and carts, there is no particular reason why
this vocabulary (or the wheels and carts themselves) should have originated in the Pontic Steppes. Neither is the cognate vocabulary fully present in all subgroups; indeed, it is absent in Anatolian and Tocharian,
the two oldest subgroups that have survived for linguistic inspection.

An older version of the Pontic Steppes theory for Indo-European
origins was widely promulgated last century by archaeologist Marija
Gimbutas, who held that pastoralist and patriarchal Indo-European
societies migrated from the Pontic Steppes to replace more peaceful
matriarchal Neolithic societies who worshipped a mother goddess.[50]
However, this idea is not supported by new ancient DNA evidence
related to megalithic tombs in western Europe. This suggests that Neolithic farmers were just as patriarchal, status-conscious, warlike, and
even incestuous within high-ranked families as any of their Bronze Age
successors.[51]

In my view, the Yamnaya ancestors were originally an indigenous
hunter-gatherer population on the Pontic Steppes, close to the homeland region of the Uralic languages (these include modern Finnish,
Saami, Estonian, and Hungarian). During the Neolithic, the Pontic
Steppes were impacted by populations from eastern Europe and the
eastern Fertile Crescent who brought domesticated plants and animals,
including cattle, sheep, and goats, as far east as the Dnieper River in
Ukraine, beyond which crop farming became increasingly difficult in
the drier steppe climate, and pastoralism with dairying took over.[52] The
Slavic languages that occupy the Pontic Steppes today are of early me-
dieval origin and there are no traces of any language, whether Uralic or
Indo-European, that can convincingly be related to the Yamnaya.

By 2800 BCE, opportunities arose for the Yamnaya to migrate westward with their pastoralist and dairying economy, possibly taking

advantage of a temporary decline in the European population due to agricultural collapse. It is possible also that the Yamnaya spread an early variant of the plague, *Yersinia pestis*, to populations who lacked any resistance, although as I noted above it is also likely that *Yersinia* was present already in Neolithic and even hunter-gatherer populations in northern Europe.[53] I would suggest that some of these Yamnaya migrants might have replaced older Indo-European languages in subgroups that are now lost without record, although this is pure guesswork and there is no specific reason why the Yamnaya should have replaced any languages at all.

Meanwhile, other steppe populations living beyond the Yamnaya continued hunting and gathering, especially those who were speaking early Uralic languages ancestral to Finnish and Saami. After 3000 BCE, some of these ancestral Uralic-speakers headed north beyond farming limits via the Volga River to settle in Scandinavia and the Baltic region.[54] Indeed, the Yamnaya genomic signature is also widespread amongst Uralic speakers, so its correlation with the distribution of the Indo-European language family is not absolute.[55]

This scenario for the Yamnaya resembles that for the decline of the Roman Empire, when uninvited migrants from the north and east poured into a weakened western Europe. These post-Roman raiders and conquerors imposed few languages permanently, and I suspect neither did the Yamnaya across all the lands reached by their genes. Like the Vikings three millennia later, despite considerable Viking settlement in the British Isles and Normandy, there were probably not enough of them compared with the native population to redraw the western Eurasian linguistic map in any significant way.

South Asia beyond the Indus Valley

We still have the conundrum of South Asia to consider. A Fertile Crescent Neolithic economy was introduced through Iran into the northwest of the subcontinent, but so far the record from ancient DNA is too small to register the arrival of any significant quantity of Neolithic Fertile Crescent DNA. Because of this, it is currently suggested by many scholars that Indo-European languages were introduced into South Asia not from the Fertile Crescent, but from a Yamnaya source in the Pontic Steppes.[56]

The spread of Yamnaya genomic ancestry into South Asia was also little more than a trickle, like that in Mediterranean Europe and Anatolia, although some modern high-caste South Asians, especially Brahmins, carry up to 50 percent of it. This suggests to me that a few charismatic male individuals of central Asian origin entered South Asia during the second millennium BCE and founded high-status lineages among indigenous populations. Many Iranian and central Asian rulers of historical South Asian kingdoms did likewise, right down to the Mughals during the sixteenth century. So there has been a constant flow of Iranian and central Asian DNA into South Asia during its long history.

However, I see no particular reason why these Yamnaya migrants should have brought the first Indo-European languages to restart the whole linguistic prehistory of northern South Asia. It is more likely that they adopted existing Indic languages ancestral to the vernacular Prakrit languages of early historical times, lubricating their spreads through state formation and thereby creating the linguistic situation that exists within South Asia today.

There are two major archaeological problems with any major Yamnaya cultural impact within South Asia. First, south of the Himalayas, South Asia has no Neolithic or Bronze Age archaeology that relates to the Pontic Steppes. Archaeological cultures of steppe origin spread as far as southern Turkmenistan, the Andronovo culture coming the closest to South Asia. But Andronovo pastoralists never penetrated beyond the fringes of the urbanized Bactria-Margiana Archaeological Complex (figure 10.3).

Second, the Gangetic archaeological continuity from the Harappan that I discussed above suggests that the Indus civilization had at least some Indic speakers, even if no inscriptions have survived that can be translated. Indologist Asko Parpola has pointed out many aspects of Harappan art and iconography that overlap with those of later Hinduism.[57] Indeed, given that the Ganges civilization was Indic-speaking (Indo-European family, Indo-Iranian subgroup) during the first millennium BCE, the existence of an archaeological ancestry for it within the

Harappan but not within the Yamnaya must surely carry some weight in ultimate interpretations. From my perspective, an arrival of Indo-Iranian languages in Iran and South Asia with the Neolithic movement from the eastern Fertile Crescent between 5000 and 4500 BCE remains the most likely hypothesis.

South India and the Dravidian Language Family

Speakers of Indic languages were not the only inhabitants of prehistoric South Asia. How did the Dravidian language family of southern South Asia (figure 9.1) and its speakers originate? Dravidian languages are classified in a family separate from Indo-European, and are clearly of independent origin. They include Brahui of Pakistan, Kannada of Karnataka, Malayalam of Kerala, Tamil of Tamil Nadu and northern Sri Lanka, and Telugu and Gondi of Andhra Pradesh.

The Deccan Peninsula had a prehistoric record different from that of the Ganges Basin in that cattle-herding pastoralists and farmers spread through it with a use of pottery, apparently from the northwest, at about 3000 BCE (figure 10.2). At this time, according to new paleoclimatic research, it appears that a retreat of summer monsoon rainfall encouraged an expansion of savanna grassland suitable for pastoralism.[58] Within the peninsula these people domesticated native species of millet and mung bean (green gram), and later acquired rice from the Ganges and Brahmaputra basins.[59] African pearl millet and sorghum, both well adapted to monsoon climates, also arrived in South Asia around 1800 BCE as a result of Harappan contacts across the Arabian Sea. It is currently unclear when Sri Lanka received its first farmers, but a date before 1000 BCE seems likely.

Genetically, most Dravidian-speaking populations in South India have a higher proportion of South and Southeast Asian ancestry than do the Indo-Iranian speakers to their north. The simplest explanation here might be that Dravidian speakers are mainly indigenous to South Asia, but also share some ancestry with Indo-Iranian speakers due to Neolithic and historical contacts. These contacts culminated in the southward spreads of Hinduism and Buddhism from the Ganges Basin with the conquests of the Mauryan ruler Ashoka (circa 250 BCE) and his successors. As yet, the absence of an ancient DNA record from Peninsular India renders precise details elusive.

Dravidian languages today are confined to southern and eastern areas of Peninsular India, except for a single outlier called Brahui that is still spoken in parts of Baluchistan, Afghanistan, and eastern Iran. According to linguists Frank Southworth and David McAlpin, the Dravidian language family shares a distant connection with the Elamite language of early historical Iran.[60] This observation suggests to me that the survival of Brahui in its current location might be significant, even though this language now reveals a substantial linguistic influence from Iranian languages.

One explanatory possibility that I think deserves consideration is that Dravidian speakers originated in or close to the Indus region itself, perhaps even in Baluchistan. In an earlier section I referred to a Harappan individual from Rakhigarhi who carried a DNA profile found also on the Iranian Plateau. Even though no Dravidian languages are spoken in the Indus Valley proper today, could this ancient person have been a speaker of an early Dravidian language? This seems certainly possible, even if it will always remain unverifiable.

A Dravidian origin in the Indus region would suggest that some speakers within this language family began to move south through Gujarat and Maharashtra prior to 3000 BCE, perhaps after the first Indo-European populations in Pakistan arrived from Iran or Afghanistan. In Gujarat and Maharashtra they left a place-name trail, since overlain by the expansion of Indic languages. The immediate linguistic ancestry of the present-day Dravidian family in India, excluding Brahui, was established somewhere within the Deccan Peninsula through a proto-language that existed around 2500 BCE.[61]

Interestingly, the Dravidian language family has no ancient place-name presence in the Ganges Basin, and this circumstance hints at contemporary expansions of both Dravidian languages into the Deccan and Indic languages down the Ganges, each constraining the movements of the other in terms of the occupation of new territory.[62] This situation would have held until the eventual expansions of both Indic and

Dravidian languages into Sri Lanka, possibly during the first millennium BCE.

What Happened Next in Southwest Asia?

The Neolithic movements that emerged from the Fertile Crescent were some of the most important in recent human prehistory. But this does not mean that all migration stopped when these populations reached their eventual destinations. Including the Yamnaya, many smaller scale movements continued, some prehistoric, some revealed by historical records.

For instance, during the sixth millennium BCE, a 600-kilometer distribution of "Halafian" painted pottery appears to track the Semitic-speaking ancestors of the Akkadians, migrating eastward into northern Mesopotamia from their Levant homeland (Semitic peoples and the Afro-Asiatic language family are discussed in chapter 12). The Halafian culture spread from the environs of modern-day Aleppo in northern Syria to Mosul in northern Iraq, intruding in Iraq into territory that was probably inhabited by ancestral Sumerians, speakers of a mysterious language unrelated to Afro-Asiatic or Indo-European, or indeed to any other known language family. The Sumerians are famous in history as the founders of one of the world's oldest urban and literate civilizations, the other being that of the Afro-Asiatic Egyptians.

The Sumerians got their own back on the Halafians in the fourth millennium BCE, when a period of trading enterprise, possibly controlled by the powerful Sumerian city of Uruk in southern Mesopotamia, established Sumerian colonies more than 1,000 kilometers upriver in the middle and upper Euphrates Valley in northern Syria and Anatolia, as well as in regions of Elamite interest in western Iran. Some of these settlements existed as independent urban foundations with Sumerian-style tripartite temples, wheel-made Uruk pottery, clay tokens used in accounting, and engraved cylinder seals. Others were smaller enclaves within existing indigenous settlements.[63]

Sumerian city-states continued to control Mesopotamia during much of the third millennium BCE, until the Semitic-speaking Akkadians under a ruler named Sargon established a short-lived Mesopotamian empire around 2300 BCE. After this, the Sumerian city-states powered back again with the succeeding Third Dynasty of Ur, rendering Sargon like most other conquerors in history—lots of conquest but no linguistic replacement, at least not yet.

In actuality, however, the writing was on the wall for the Sumerians. After 2000 BCE their language gradually died out in both spoken and written form, although its cuneiform syllabic characters were adapted to write many non-Sumerian languages across the Middle East. As the major founders of Mesopotamian civilization, why did the Sumerians disappear so irrevocably, and where exactly did they fit within the linguistic and population kaleidoscope of the ancient Middle East, given that Sumerian was an isolated language with no known relatives? I would love to know.

With all of this coming and going revealed for us by the power of writing, it might come as no surprise that the Tower of Babel was a product of Middle Eastern imagination. After all, the Fertile Crescent flowering of people and language as a result of the development of food production at the beginning of the Holocene did not eliminate all linguistic and cultural diversity in the heart of the flower. As well as its Sumerian, Afro-Asiatic (Semitic), and Indo-European languages, the Middle East at 2000 BCE also contained Caucasian languages (Hurrian and Hatti of Anatolia), and powerful Elamite of southern Iran. These last, like Sumerian, are long extinct. Small Caucasian language families still exist in the Caucasus itself, but only Afro-Asiatic and Indo-European languages survived elsewhere in the Middle East through to Classical times, prior to the more recent expansions of Turkic peoples.

It is now time to head east. This chapter has taken us as far east as the Altai Mountains. Beyond them lies another amazing human tapestry.

11

Asia-Pacific Adventures

THE EASTERN HALF of the Asian continent, including the many islands that lie beyond its eastern shoreline and throughout the Pacific
Ocean, witnessed Neolithic developments and population expansions
that were equal in significance to those generated within the Fertile
Crescent. Three linked homelands of food production in northern and
central China gave rise to the initial expansions of no less than five
major language families and their speakers, including Transeurasian,
Sino-Tibetan, Hmong-Mien, Kra-Dai (Tai), and Austronesian (see figure 9.2 for their overall distributions today and table 11.1 for their contents). In particular, the origins and remarkable seaborne migrations of
the Polynesians, as the farthest-flung speakers of Austronesian languages, have been attracting the sometimes romanticized interests of
scholars for more than 250 years.

As discussed in chapter 7 and shown in figure 11.1, there are essentially
three main agricultural homelands in East Asia. Each nurtured one or
more different ancestral linguistic populations:

- The Liao River basin (Manchurian Plain) in northeastern China
  is widely regarded as the homeland for the Transeurasian languages and their speakers, including the ancestral Japanese,
  Korean, Tungusic, Mongolic, and Turkic languages.
- The Yellow River basin in north-central China is regarded as
  the homeland for the Sino-Tibetan languages and their speakers,

TABLE 11.1. The major language families of eastern Asia and the Pacific Islands.

Linguistic subgroup or geographical description	Important languages spoken now or historically recorded
Transeurasian	
Japonic	Japanese, Okinawan
Koreanic	Korean (now a single state language)
Mongolic	Mongolian, Oirat, Buryat
Tungusic	Evenki, Manchu (rulers of Qing Dynasty China, 1644–1912 CE)
Turkic	Uzbek, Uighur, Kazakh, Kirgiz, Turkmen, Turkish, Yakut
Sino-Tibetan	
Sinitic	Eleven major Sinitic languages, including Mandarin, Cantonese (Yue), and Taiwanese
Tibeto-Burman	Tibetan, Burmese, and many other languages of the southern Himalayas, from Assam to Himachal Pradesh
Karenic	Karen (Myanmar)
Hmong-Mien	
Scattered distribution in southern China, northern Vietnam, Laos and Thailand	
Kra-Dai (Tai, Tai-Kadai, Daic)	
Thai (Thailand), Lao, Shan (Myanmar), Zhuang (southern China), many northern Vietnam languages, Li (Hainan Island), Ahom (Assam and Arunachal Pradesh, India)	
Austroasiatic	
Munda	Languages in Northeast Peninsular India
"Mon-Khmer" (not a unified subgroup)	Mon (Myanmar), Khmer (Cambodia), and many Mainland Southeast Asian languages such as Khasi of Assam, Khmu of Laos
Vietic	Vietnamese (Kinh), Muong (northern Vietnam)
Aslian	Many interior Malay Peninsula languages
Nicobarese	Nicobar Islands
Austronesian	
Formosan	Fifteen Indigenous Formosan languages in Taiwan
Malayo-Polynesian (i.e., all Austronesian languages apart from Formosan)	Philippine languages
	Mariana Islands (Chamorro of Guam), Palau Islands
	Indonesian languages (except for Papuan languages in and around New Guinea)
	Peninsular Malaysia (Malay) and central Vietnam (Chamic languages)
	Malagasy (Madagascar)
Oceanic (a subgroup of Malayo-Polynesian)	Melanesian languages (coastal New Guinea through Solomons, Vanuatu, and New Caledonia to Fiji)
	Central and eastern Micronesian languages (including Caroline Islands, Kiribati, Marshall Islands)
	Polynesian languages (including Tuvalu, Tonga, Samoa, the "Polynesian Outliers" (see text), Tahiti, Hawaii, Easter Island [Rapa Nui], New Zealand)
Papuan	
Trans-New Guinea	Much of the New Guinea mainland, plus eastern Timor and Alor, and some Bismarck Archipelago and Solomon Island languages
Sepik-Ramu	Sepik and Ramu basins of northern Papua New Guinea
West Papuan	Bird's Head and northern Moluccas (Halmahera Island)

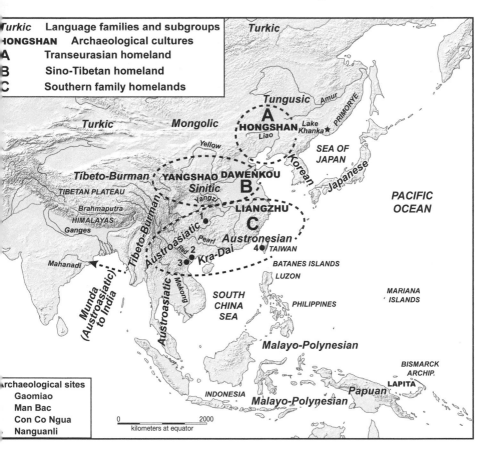

Figure 11.1. Estimated linguistic homelands and population movements consequent on the developments of food production in eastern Asia.

including the Sinitic-speaking Chinese and the Tibeto-Burman sector of the family that extends into Tibet, Myanmar, and around the southern Himalayas into the northern fringes of South Asia.

- The Yangzi Valley in south-central China, together with south-eastern coastal China and Taiwan, is regarded as the homeland region for the separate Hmong-Mien, Kra-Dai, and Austronesian language families and their speakers, the latter ultimately traveling via Taiwan across more than 13,000 kilometers of sea to reach Easter Island (Rapa Nui), and even the coast of South America.

- The widespread Austroasiatic language family that extends from northeastern India to Peninsular Malaysia and the Nicobar Islands is less certain in terms of its homeland, probably because older phases of its linguistic prehistory have been masked by the more recent expansions of Sinitic and Kra-Dai languages.

Ancient Human Populations of East Asia and Sahul

Previously, I described how two Paleolithic populations of *Homo sapiens* developed in Late Pleistocene and early Holocene East Asia. The northern one (see chapter 6) became ancestral to modern East Asians and Native Americans. The southern one (see chapter 5), more tropically adapted, was ancestral to Late Pleistocene Southeast Asians and modern Indigenous Australians, Papuans, and Melanesians, plus a few Negrito hunter-gatherer populations in the Philippines, Malaysia, and the Andaman Islands.

It is clear from craniofacial features of ancient skulls that major elements of the southern population extended quite far north in late Paleolithic times, at least to the Yangzi River and Jomon Japan, although few traces of its former presence survive in these northern regions today. Genetically, this Paleolithic population was quite varied, as revealed by its ancient DNA, containing many localized hunter-gatherer groups who had differentiated from each other following the initial expansion of *Homo sapiens* into the region perhaps 40,000 years beforehand.[1] During the Neolithic, however, substantial migrations with a general north to south trend pushed the boundary zone of intermixture between these two populations southward and eastward, toward the islands of eastern Indonesia where the boundary exists today.[2]

These Neolithic migrations carried farming populations of northerly East Asian origin into southern China, throughout both Mainland and Island Southeast Asia, around the southern Himalayan foothills, and right through the islands of Oceania. This immense shift in the human geography of eastern Asia and the Pacific occurred mostly after 6,000 years ago, and it involved populations with food production—especially of *japonica* (East Asian) rice, foxtail and broomcorn millet, many fruits and tubers, pigs, chickens, and dogs—and ultimately the most remarkable tradition of ocean voyaging in the prehistory of humanity, which gave rise to the Polynesians.

In regions south of the Yangzi Basin, the migrating farmers left a skeletal record, especially in their faces and crania, that allows them to be distinguished from the indigenous populations who preceded them.[3] Similar migrations by early food-producing populations also occurred in northeastern Asia out of the Yellow and Liao River basins, but in these higher latitudes the skeletal and genomic differences between the indigenous and immigrant populations were less marked than in the tropical south.

These skeletal observations are enhanced by observations about burial behavior. Over most of the region south of the Yangzi, including Jomon Japan, pre-Neolithic burials of late Pleistocene and early Holocene date have seated or tightly flexed skeletal postures, usually without grave goods. Neolithic burials, on the contrary, have East Asian cranial and facial affinities and normally lie supine on their backs with offerings of pottery vessels, body ornaments, and stone adzes.

Given the biological and cultural differences between these two cranial and burial categories in the archaeological record, it is usually not difficult in well-preserved situations to attribute them either to late Paleolithic indigenous populations or to Neolithic immigrants, and sometimes to both in situations of population mixing. Examples of such mixing, both genetic and cultural, in seemingly peaceful circumstances in Vietnam and central China are discussed below. In the last few years, this reconstruction of East Asian population history has been amply reinforced by studies of ancient DNA, although samples are still limited in coverage and many details remain to be clarified.[4]

Plotting the Course of Transeurasian Dispersal

Within those northerly temperate latitudes where agriculture and pastoralism are possible, the course of East Asian Neolithic prehistory has been closely bound up with the expansionary histories of the speakers of two major language families, known to linguists as Transeurasian (figure 11.2) and Sino-Tibetan (figure 9.1).

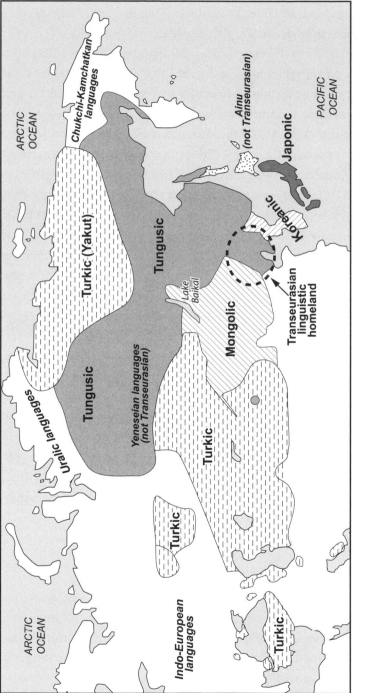

Figure 11.2. The Transeurasian languages at their greatest precolonial extents. Background map by CartoGIS Services, Australian National University, with linguistic boundaries redrawn from M. Ruhlen, *A Guide to the World's Languages* (Stanford University Press, 1987).

The former, which I deal with first, ultimately occupied a vast area from Turkey to Japan. During the past few years a productive research project led by linguist Martine Robbeets of the Max Planck Institute for the Science of Human History has brought together archaeologists, linguists, and geneticists to uncover the Transeurasian family's origins and routes of expansion.[5] Its main populations of speakers today are the Japanese, Koreans, Mongolians, the Turkic-speaking peoples, and the Tungusic peoples of Manchuria and the Russian Far East. These Transeurasian-speaking peoples have shared a complex prehistory across an enormous area of the Asian continent.

Linguistic analysis suggests that the Transeurasian homeland was located in northeastern China, in or close to the Liao River basin within the Manchurian Plain, where millets were domesticated by Xinglongwa people around 6000 BCE, as I described in chapter 8. They also practiced weaving and kept domesticated pigs and dogs. By soon after 5000 BCE, during the time span of the following Hongshan culture in Liaoning, millet farmers from these riverine plains were searching for new territory.

At the same time, the Yellow River basin to the south contained people of the densely populated Yangshao and Dawenkou cultures and their predecessors, currently regarded as the originators of the Sino-Tibetan language family. Early Transeurasian and Sino-Tibetan speakers evidently began their expansions in close proximity to each other, but they expanded in different directions. The early Transeurasian speakers kept generally to the north, beyond the range of Neolithic rice cultivation. The early Sino-Tibetan speakers looked to the west and southwest, partly because China south of the Yangzi was occupied at that time by ancestral Southeast Asian populations, including the ancestral Hmong-Mien, Kra-Dai, and Austronesian peoples.

Ancient DNA research implies that the initial Liao and Yellow River farmers (figure 8.3) belonged to separate genetic populations, but intermixture between them by 4000 BCE contributed to the ancestry of many Transeurasian-and Sino-Tibetan-speaking populations today, especially Koreans, Japanese, and Chinese.[6] Culturally, for instance, the Hongshan and Yangshao cultures at 4000 BCE shared pottery manufacture in closed kilns, large villages, millet agriculture with pigs and dogs (but no cattle or dairy production), and weaving technology, presumably for native fibers such as hemp and ramie.

By 3500 BCE, millet cultivation, especially of broomcorn millet, had spread from northern China in several directions—west to the Altai Mountains and to the edge of the Tibetan Plateau, northeast to the Primorye region of the Russian Far East, and into Korea, possibly with ancestral Korean languages.[7] The ancestral Japanese took longer to penetrate the Japanese islands, which during most of the Holocene were occupied by Jomon populations—indigenous people with Neolithic technology who lived mainly by hunting and gathering, with some cultivation of native plants.[8] The Yayoi culture, considered to be ancestral to the modern Japanese (excluding the non-Japonic Ainu of Hokkaido), was brought into the island of Kyushu from Korea during the early first millennium BCE. With it came irrigated rice cultivation, introduced to Korea around 1300 BCE after hardy temperate races of the crop had been developed in China, and transmitted onward into Japan with bronze working and weaving technology.[9] Modern Japanese are mainly of Yayoi Bronze Age origin but still share about 20 percent of their DNA with the older Jomon population.

The expansions of the Mongolian, Turkic, and Tungusic populations into inner Asian regions more suitable for pastoralism than crop production occurred later in time than the initial Transeurasian farming movements in northeastern China, the Russian Far East, and Korea. After the Neolithic, contact with the Pontic and West Asian steppes around 2700 BCE introduced sheep and cattle dairying into Mongolia through the Afanasievo peoples of the Altai Mountains,[10] and Mongolian and Tungusic populations undertook their main migrations after this time. They spread eventually to form many of the pastoralist peoples who existed widely in northeastern Asia prior to the recent movements of Russian and Chinese settlers. The main expansions of the Tungusic-speaking peoples, who later included the Manchu rulers of Qing Dynasty China (1644–1912 CE), commenced late in the first millennium BCE from the vicinity of Lake Khanka in the southern Amur Basin of the Russian Far East.[11]

So far, little ancient DNA research has been carried out on the deeper ancestry of the Turkic-speaking peoples. Linguistically, however, the first millennium BCE witnessed their main expansions as pastoralists and millet farmers through large parts of central Asia, possibly replacing earlier populations of Indo-European speakers and eventually reaching the Arctic Circle in the case of the Yakut horse, cattle, and reindeer herders.[12]

The subsequent adventures of the Transeurasian speakers take us forward into historical times, culminating in the migration into Anatolia by the Seljuk Turks in the eleventh century CE, and the Mongol conquests of Genghis Khan and his grandson Kublai Khan in the twelfth and thirteenth. In table 9.1 I rank the Transeurasian language family as the third most extensive in the world prior to the year 1492, although it must be said that Genghis Khan and his Mongol Empire did not have a great deal to do with this linguistic achievement, and neither did his grandson Kublai Khan when he founded the Yuan dynasty (1279–1368 CE) in Beijing. The Mongols did not replace the indigenous languages throughout the lands they conquered.

The Yellow River and the Sino-Tibetan Language Family

Between 5000 and 3000 BCE, the Yellow River loess (glacial era wind-blown dust) soils supported the Yangshao and Dawenkou Neolithic cultures and their predecessors, eventually with some of the greatest population densities of any period in prehistoric China.[13] While the early Transeurasian developments were taking place in and beyond the Liao Valley to the north, early speakers of Sino-Tibetan languages were increasing rapidly in number on these Yellow River loess soils. Outward population movements soon began, but not all opted to migrate far, which is precisely why the Sinitic-speaking Chinese have become such a large population today. They stayed put on the fertile riverine plains of their Yellow River homeland, an excellent environment for their millet and rice agriculture and pig husbandry.

At present, linguistic evidence provides the clearest information on the overall expansion history of the Sino-Tibetan-speaking peoples. Three independent lexical analyses of the Sino-Tibetan family pinpoint the middle Yellow River as its Neolithic homeland, between 6000 and 4000 BCE.[14] The dates correspond with the Yangshao culture and its immediate antecedents, and the linguistic genealogies presented in each case reveal initial separations into stay-at-home ancestral Sinitic languages on the one hand, and migratory Tibeto-Burman languages on the other.

Those who opted to move out of the Yellow River homeland perhaps tried initially to move downriver around 5000 BCE into the Shandong Peninsula.[15] But southward beyond the Yangzi they would have soon found themselves checked by the presence of other Neolithic populations, the rice-cultivating progenitors of the Hmong-Mien, Kra-Dai, and Austronesian linguistic populations whose descendants today occupy much of Southeast Asia. Chinese settlers were only able to occupy the southerly regions of China intensively around 2,200 to 1,900 years ago, when several million people from central China followed conquering Han Dynasty armies south to as far as central Vietnam.[16]

In geographical terms, the ancestral Tibeto-Burmans were the most widespread branch of Sino-Tibetan language speakers. They channeled their energies in directions of least resistance from other food-producing populations such as the ancestral Chinese; they migrated west and southwest toward the Tibetan Plateau and the fringes of the Himalayas, especially into Bhutan, Nepal, and northern India. Heading farther south, they moved through southwestern China into Myanmar and western Thailand, with the ancestral Karen penetrating as far as the top of the Malay Peninsula.

Western Yangshao millet farmers with domesticated pigs, dogs, and occasional sheep appear to have begun the migration process of the Tibeto-Burmans, arriving along the eastern edge of the Tibetan plateau and in northern Sichuan Province by 3500 BCE.[17] However, permanent agricultural settlement of the Tibetan Plateau only took place around 1600 BCE, assisted by the arrival of cold-adapted wheat and barley from western Asia.[18] Presumably, these early migrants were the direct ancestors of the modern Tibetans.

Archaeological and genomic evidence for the movement beyond southwestern China into Myanmar and South Asia during the Neolithic is currently lacking, although rice cultivation had reached the upper Yangzi in Sichuan by 2600 BCE. The remarkable and stylistically unique culture of Sanxingdui on the Chengdu Plain in Sichuan (late second millennium BCE), with its striking bronze human statues and masks, stands as a mysterious counterpart to the contemporary Sinitic civilization of the Bronze Age Shang Dynasty along the Yellow River. Sanxingdui is surely a likely candidate for an early Tibeto-Burman linguistic association.

Southern China and the Neolithic Settlement of Mainland Southeast Asia

With the establishment of a full and transportable agricultural economy with domesticated *japonica* rice in the Yangzi Basin after 4500 BCE, a number of Neolithic population movements headed out from this general region toward southern China, in parallel with but to the east of those of the early Tibeto-Burmans. Some followed the coastline toward Fujian and Taiwan, although many of the rice-growing lowlands that exist in deltas and along rivers and coastlines nowadays had not yet been built up by Holocene sediment deposition.[19] Others moved up the southern tributaries of the Yangzi toward Guangxi and Guangdong.[20] These movements involved populations with East Asian Neolithic cranial and facial characteristics, known from large cemeteries in several central Chinese archaeological sites that date between 6000 and 3000 BCE.[21]

These Neolithic populations either replaced, or lived alongside, indigenous hunter-gatherer groups with indigenous Pleistocene DNA ancestry. In southern China and Mainland Southeast Asia, these indigenous populations were associated with Hoabinhian pebble tools (named after sites in Hoa Binh Province in northern Vietnam), to which were added polished stone axes and pottery during the early Holocene. The remarkable cemetery site of Con Co Ngua in northern Vietnam (4500 BCE) belongs to this late hunter-gatherer phase, with pottery and polished stone axes. Con Co Ngua had almost 300 squatting and flexed burials, but they lacked intentionally placed grave goods.[22]

Ancient genetic data are not yet enough to say if the expanding Neo-lithic populations coexisted with their indigenous predecessors, or if replacement was rather more abrupt. One example of coexistence comes from the site of Gaomiao in the middle Yangzi Basin, where an indigenous hunter-gatherer population adopted the decorated pottery style of adjacent rice farmers. Another comes from Man Bac in northern Vietnam, where skeletons from both an immigrant and an indigenous population (the latter of presumed Hoabinhian ancestry) were buried side by side with similar decorated earthenware pots, nephrite beads, and nephrite bracelets.[23]

Indeed, the population history of southern China is likely to have been complex. For example, some archaeologists have recently raised the possibility that indigenous Hoabinhian hunter-gatherers could have been cultivating taro, bananas, and sago palms prior to the arrival of Neolithic cultures from farther north.[24] Some of the pre-Neolithic and craniofacially indigenous people in Fujian also had DNA profiles that link them with Chinese Neolithic rather than Hoabinhian populations elsewhere, at least according to current genomic interpretations.[25] A great deal of future research is necessary to resolve these issues.

Turning now to the migrations onward from southern China into the mainland of Southeast Asia, the archaeological record currently situates them mainly during the third millennium BCE.[26] Clusters of well-dated archaeological sites exist in coastal Vietnam and in central and north-eastern Thailand. Many are associated with rice cultivation, but remains of foxtail millet have also been found widely in southern China, Vietnam, and Thailand in recent years.[27] One important addition to the roster of domesticated animals in this region might have been the chicken, bred from a wild jungle fowl ancestor.[28] Most strikingly, the oldest Neolithic sites in central Thailand, which date to about 2300 BCE, occur only with foxtail millet. This is a crop of Yellow River rather than Yangzi origin, thus raising the possibility of an association with Tibeto-Burman mi-gration southward from the Sino-Tibetan homeland via southwestern China.[29]

As with many early agricultural dispersals in other parts of the world, the oldest Neolithic archaeological cultures across Southeast Asia, in both the mainland and the islands, reveal close connections in terms of stylistic features, especially in pottery. One specific feature of pottery decoration that occurred from the Yangzi Basin southward into the Malay Peninsula and eastward from the Philippines to Tonga and Samoa in Polynesia involved the stamping of patterns created from parallel rows of small indentations into the surface of the pot before firing, usually in motifs enclosed within incised lines (this pottery is termed "Lapita" in the Melanesian and western Polynesian islands—an example from Vanuatu is shown in figure 11.3). These similarities indicate a community of shared culture, within which widespread populations were exchanging information and ideas. It will come as no surprise to find that these related pottery styles appear to have originated in the middle part of the Yangzi Basin, commencing around 5000 BCE.[30]

As for the populations who were spreading these pottery styles, the comparative linguistic situation indicates that they are likely to have been ancestral to present day Hmong-Mien, Austroasiatic, Kra-Dai, and Austronesian speakers. The potential homelands of these language families, as understood from archaeological, linguistic, and genomic information, are indicated in figure 11.1. The Kra-Dai speakers originated close to early Austronesian-speakers, with whom they shared ancestral linguistic and genetic connections, in southeastern China and spread into Hainan Island and northern Vietnam in Neolithic times.[31] Their historical spreads southward to found states in what are today Thailand and Laos occurred only during the last millennium.

Neolithic migrations around 4,500 to 4,000 years ago in southern China and Southeast Asia were prolific and multidirectional. However, there is no whole-genome ancient DNA yet reported from Neolithic cemeteries in the Yangzi Basin, although analysis of Y chromosomes in ancient DNA from the remarkable Liangzhu culture (circa 3000 BCE) of the Shanghai region has revealed a high frequency of haplogroups associated today with Kra-Dai- and Austronesian-speaking populations far to the south.[32] This linkage is reinforced by new observations on the whole genomes of living southern Chinese populations. The study of ancient DNA in China is developing rapidly as I write, and exciting new perspectives from Chinese scholars are eagerly awaited.

Figure 11.3. A Lapita ceramic design from the Teouma burial site on Efate Island, Vanuatu. The indentations on this 3,000-year-old pottery sherd are made with toothed and circular stamping implements. Similar pottery is found widely in other Melanesian and western Polynesian islands, plus the Philippines, northern Indonesia, and the Mariana Islands. Photo courtesy Matthew Spriggs.

The Austroasiatic Mystery

The most widely dispersed language family in Mainland Southeast Asia is Austroasiatic, of which the two most spoken languages today are Khmer and Vietnamese. The former was installed as a state language during the Kingdom of Angkor (802–1431 CE); the latter was spread from the Red River, eventually to the Mekong Delta, by migrating northern Vietnamese (who call themselves Kinh) after their independence from a millennium of Chinese rule in the tenth century CE.[33] Today, it is obvious from the interrupted distribution of Austroasiatic speakers that much of the former extent of this language family has been overlain by Kra-Dai languages in Thailand and Laos, by Malay in the Malay Peninsula, by Indic languages in northeastern India, and possibly by Sinitic languages in southern China.

Austroasiatic speakers stand as the most likely candidates for the first Neolithic dispersal through much of Mainland Southeast Asia. But they had competitors, especially among the Tibeto-Burmans. Above I raised the possibility of early Tibeto-Burman migration from western China into central Thailand with foxtail millet farming at around 2300 BCE. This possibility links well with a suggestion by linguists Felix Rau and Paul Sidwell that Tibeto-Burman speakers were already in the western mainland of Southeast Asia prior to the arrival of Austroasiatic speakers. As a result, they suggest that the Neolithic migration of the Munda speakers (within the Austroasiatic family) from Mainland Southeast Asia to the Mahanadi Delta in the Indian state of Odisha had to cross the Bay of Bengal by boat rather than using a land route (figure 11.1).[34]

The Munda-speaking peoples today are a minority Austroasiatic population in northeastern India who are surrounded by Indic speakers. Genomic analysis of living Munda populations in India supports an arrival of their ancestors from Mainland Southeast Asia, possibly commencing from the Andaman Sea coastline of Myanmar about 2000 BCE.[35] Presumably, the Mundas met both Tibeto-Burman and Indic speakers during their traverse into India, and this may be why they were only able to settle in relatively infertile regions such as the dry Chota Nagpur Plateau of Jharkand State (formerly southern Bihar), away from major rivers such as the Ganges and the Brahmaputra. The people of the Nicobar Islands also speak Austroasiatic languages, and their ancestors must also have arrived there by crossing at least one sea passage.

The source of the whole Austroasiatic language family is still a mystery because it is unclear whether it once extended into southern China, and, if so, how far. Early claims that it once existed as far north as the Yangzi prior to the Kra-Dai and Sino-Tibetan expansions are unproven, although the ancient DNA affinities of the Neolithic populations of mainland Southeast Asia, including most speakers of Austroasiatic languages, lay with Chinese Neolithic populations and not with Hoabinhian hunter-gatherers. This renders an origin for them entirely in pre-Neolithic (Hoabinhan) Mainland Southeast Asia rather unlikely. The Aslian-speaking peoples (Orang Asli) of the interior Malay Peninsula are an exception here, in having considerable Hoabinhian ancestry. This suggests adoption of ancestral Aslian languages (a subgroup of Austroasiatic) by them in Neolithic times or later.

Despite the origin mystery for the Austroasiatic family, I would suggest that early Tibeto-Burman and Austroasiatic speakers were spreading through Mainland Southeast Asia at much the same time, between 2500 and 2000 BCE, adjusting to each other's presence as necessary. Likewise, Indic (Indo-Iranian) speakers were also settling in the lower Ganges and Brahmaputra basins at the same time, extending the Indo-European language family to its ultimate pre-1492 eastern limit in Bangladesh, a limit shared with the Tocharian languages far to the north.

The Austronesians

We now come to the most dramatic episode of expansion by a food-producing population at any time in human prehistory, an expansion that has occupied my interest during most of my research career. Today, there are over 1,000 Austronesian languages, with over 300 million speakers. As with all of the major populations discussed in this agricultural section of the Odyssey, we can study their prehistory through their languages, their genes, and their archaeology.

Between 2000 BCE and 1250 CE, the Austronesians spread across more than 210 degrees of longitude from their southern China and Taiwan homeland, clocking up an incredible 25,000 kilometers from Madagascar in the west to Polynesia in the east, with further contacts (but no major settlement) reaching the western coast of South America about one millennium ago (figure 11.4). Three outstanding cultural elements

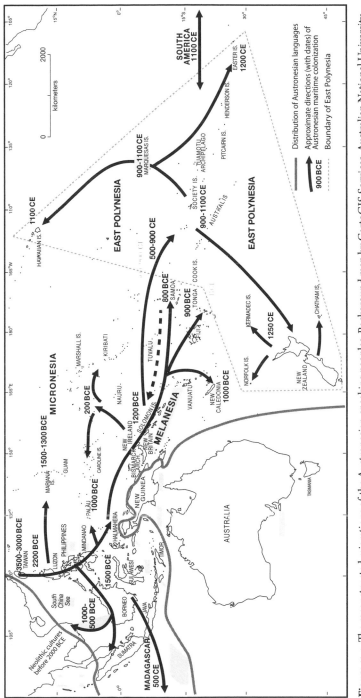

Figure 11.4. The reconstructed migrations of the Austronesian-speaking peoples. Background map by CartoGIS Services, Australian National University.

drew them into the open Pacific and westward to Madagascar: their transportable food-producing economies, their sailing outrigger and double canoes (ancestors of modern catamarans), and their enthusiasm to found new communities in which their concepts of founder-associated and inherited rank could propagate anew.[36]

Linguistically, the Austronesians can be traced to a homeland in Taiwan, the location for Proto-Austronesian as it can be reconstructed from the Austronesian languages that are spoken today.[37] Older ancestral stages of the language family once existed in southern China; however, their traces have been completely erased by the expansions of the Sinitic and Kra-Dai languages, so the details will perhaps never be known to us. Perhaps the same fate was met by the early Austroasiatic languages, as I discussed above.

The first Austronesian speakers to arrive in Taiwan crossed the Taiwan Strait from the Fujian and Guangdong coastlines of southern China between 3500 and 3000 BCE. They brought in a subsistence lifestyle of food production, reconstructed from both archaeology (the Dabenkeng culture in Taiwan; see figure 11.1) and comparative linguistics, which included rice and millet agriculture, domesticated pigs, and dogs (chickens were perhaps introduced later from Mainland Southeast Asia). Taiwan was also the source of a female clone of the paper mulberry tree, the bark of which was used across many regions of Southeast Asia and Oceania for making bark cloth.[38]

These people constructed village settlements, wove fibers (but not yet with the ethnographically widespread backstrap loom[39]), and used pottery, polished stone tools, and shell body decorations that included beads, bracelets, and anklets. Taiwan has the oldest Neolithic cultures in Southeast Asia, befitting their direct origins from the southern Chinese mainland. In terms of stylistic features of pottery and polished stone tools, the oldest Neolithic cultures to the south, especially in the Philippines, Indonesia, and the Pacific islands, carried clear traces of an ultimate southern Chinese and Taiwan origin.[40]

For about a millennium, the Neolithic inhabitants of Taiwan occupied all coastal regions of the island, but apparently went no farther. This pause might have reflected the strength of the north-flowing

Kuroshio Current, which would have inhibited them from heading south to the Philippines until someone developed the use of sails, outriggers, and side-planks lashed on top of dug-out hulls to keep out seawater. By 2200 BCE, they were ready to travel in their boats through the Batanes islands to reach the island of Luzon in the northern Philippines. This movement led to the emergence, through shared linguistic innovation, of the Malayo-Polynesian subgroup of Austronesian languages, which includes all Austronesian languages spoken outside Taiwan.[41] The Malayo-Polynesians were to spread, eventually, from Island Southeast Asia to Madagascar in the west and Easter Island in the east—more than halfway around the world.

Lessons from the Austronesians

The Austronesian world highlights a number of important themes in human prehistory. The languages share a common origin in Taiwan, and all still reflect that heritage today, notwithstanding episodes of contact in the past with speakers within other language families, especially Austroasiatic and Papuan.

As a result of these contacts, not all of the people who speak Austronesian languages are necessarily of identical genetic origin, even though a great majority of them reflect fairly direct descent from ancestral Neolithic populations with food production, pottery, and sailing canoes who were once located in southern China, Taiwan, and the Philippines.[42] A subsequent Papuan genetic migration into the islands of Melanesia at around 500 BCE (to be described below) illustrates this point clearly. In this instance, Papuan genes moved, but Papuan languages did not because the migrants adopted Malayo-Polynesian languages. Languages and genes do not, and need not, always match with 100 percent accuracy.

During my earlier career, such observations of disjunction between languages and biology posed great uncertainty for students of prehistory, with some prompted to claim that languages, cultures, and biological populations always evolved independently of each other. Such views, from a modern perspective, and especially one informed by ancient

DNA analysis, are misguided. Within the Austronesian world, genes, languages, and cultures might have had separate prehistories on some occasions, but such disjunctions were not necessarily the norm.

In this regard, it is necessary to remember that the Austronesian world is not a unity in cultural or genetic terms today, even if it started close to being one around 5,000 years ago during the Dabenkeng Neolithic period in Taiwan. Not only were more than 1,000 separate Austronesian languages spoken at European contact, but modern Austronesian societies range from the majority Islamic nation of Indonesia, with 270 million people, to the tiny Christian nation of Tuvalu, a Polynesian island group in the central Pacific with only 12,000 people. There are also genetic differences between the people who speak Austronesian languages today. This observation informs us that it is essential to distinguish the primary spreads of Austronesian languages, Austronesian-speaking people, and their archaeological cultures from the impacts of the many subsequent population movements that have taken place during the past 5,000 years. The past is a palimpsest; its creation cannot be understood simply from what exists today.

My understanding of the human past tells me that human behavior, especially behavior related to migration, has generally been predictable rather than random and chaotic. The challenge for those who seek to interpret human population history is to identify the predictions that best fit the available multidisciplinary evidence. Malayo-Polynesian expansion across the Pacific provides many excellent examples.

Malayo-Polynesians and Papuans

The Philippines, as the first stop beyond Taiwan, offered to Malayo-Polynesian voyagers a remarkable inland seascape protected by surrounding islands, excellent for improving their sailing skills, and quite different as a geological configuration compared with the large islands of Indonesia (Sumatra, Java, Borneo). By 1500 BCE, having spread through the Philippines, they launched movements onward into Borneo, Sulawesi, the Moluccas, and the islands of the western Pacific, in the latter case perhaps first to the Mariana Islands, potentially a 2,300-kilometer open sea crossing from the Philippines unless they traveled via the Palau Islands.[43] At around 1200 BCE, Malayo-Polynesians sailed southeastward from the Philippines toward the Admiralty Islands and the Bismarck Archipelago in Melanesia, to commence the settlement of the islands of the open Pacific.

The remarkable colonization of the previously unsettled islands of Oceania culminated in the settlement of the Polynesian Triangle, with apices 8,000 kilometers apart in Hawaii, New Zealand, and Easter Island. This occurred in two phases separated by a long pause of almost 2,000 years. First of all, the remarkable decorated pottery of the Lapita culture (figure 11.3) recorded the movement of people through the islands of Melanesia from the Bismarcks to as far east as Tonga and Samoa, located on the western border of Polynesia. This movement is now tracked through ancient DNA, especially from the Lapita site of Teouma in Vanuatu, and it took place mainly between 1200 and 800 BCE. After this, the Lapita culture lost its coherence; in archaeological terms, it was replaced by different pottery styles in the Melanesian islands.[44]

For many years, the reason for this change in the archaeological record of Melanesia was not clear to archaeologists, even though European explorers and anthropologists had commented for almost two centuries on the differences in physical appearance between Island Melanesian populations with their New Guinea biological affinities, and Polynesian populations with their Indonesian and Filipino biological affinities. Johann Reinhold Forster, a scientist who traveled on James Cook's second voyage through the Pacific in the 1770s, was an early commentator on this issue, and he collected lists of common words from the languages encountered across the Pacific. Cook and Forster quickly noted that, despite their physical differences, Polynesians and Melanesians were speaking languages that were closely related. They wondered why.

Now we know why, and the answer involves New Guinea, where there was an indigenous development of agriculture that I discussed in chapter 8. We might know far less about the rise of agriculture in New Guinea than we do in the Fertile Crescent or China, but this island was obviously of tremendous importance in the prehistory of the Melanesian Islands. Two important observations can be made.

First, most of the interior languages of New Guinea, apart from a number at the western end of the island and in the Sepik and Ramu valleys in northern Papua New Guinea, belong to a single language family called Trans-New Guinea, the largest family within the Papuan grouping. Did ancestral speakers within this family spread with the development of agriculture in the New Guinea Highlands? Many linguists think this is possible, although the situation is by no means certain. The Trans-New Guinea languages are not as closely related as the Malayo-Polynesian languages; if they spread with farming, it would probably have occurred more than 4,000 years ago, long before the arrival of Malayo-Polynesian-speaking peoples in the coastal regions of the island.[45] The archaeological record of early agriculture in the New Guinea Highlands (see chapter 8) clearly allows for such a time scale.

The second observation comes from the study of both ancient and living DNA. Soon after the dissolution of the Lapita culture in archaeological terms (around 800–500 BCE), Papuan populations of Bismarck Archipelago genetic origin, mainly from New Britain, migrated in large numbers through the islands of Melanesia, especially to Vanuatu and New Caledonia, overlying the earlier occupation by the Lapita people, who were of Taiwan/Philippine Malayo-Polynesian ancestry. But, and this is one of the most remarkable observations, these Bismarck migrants did not carry their original Papuan languages, at least not beyond the Solomon Islands.[46] Instead, they adopted the Malayo-Polynesian languages (Oceanic subgroup) of their Lapita predecessors.

Linguists have pointed out that the languages of Vanuatu, for instance, have many features derived from Papuan languages that must have been acquired through episodes of linguistic mixing between the original Lapita (Malayo-Polynesian-speaking) settlers and the later Papuan-speaking immigrants. In my view, it is possible that the Malayo-Polynesian languages were adopted by these Papuan populations because they were widely spoken. Regional Malayo-Polynesian dialects would have been closely related to each other at this early stage in their migratory history by virtue of their recent common source in what linguists term Proto-Malayo-Polynesian. Papuan languages were far more diverse and thus not intercomprehensible, except through multilingualism, so their speakers might have been eager to identify and adopt useful lingua francas when they arrived in the Oceanic islands beyond New Guinea.

This Papuan expansion, which replaced the Lapita culture in Melanesia, appears not to have spread farther east than Fiji. Although Polynesian people carry a significant percentage of Melanesian genes, especially in the male-inherited Y chromosome, they did not acquire these genes in the initial Lapita phase of expansion but through later contacts within the western Pacific prior to the settlement of the eastern islands of Polynesia. The Lapita populations who reached Tonga and Samoa in West Polynesia soon after 1000 BCE continued to live there without further significant immigration from Melanesia, despite continuing genetic exchange with Melanesian island groups, especially through Fiji.

The Settlement of Polynesia

The West Polynesians who arrived in Tonga and Samoa with Lapita pottery about 3,000 years ago became the main founders of the Polynesian cultures and societies that existed at the time of European contact. However, the two millennia between 1000 BCE and 1000 CE that followed the settlements of Tonga, Samoa, and other islands in West Polynesia still present a degree of mystery. There was clearly a standstill before people sailed farther east across the wider sea gaps in the middle of the Pacific to reach the islands of East Polynesia. This final Malayo-Polynesian migration took place as a series of dramatic sailing events across thousands of kilometers of open sea between 800 and 1250 CE, a date range confirmed both by the archaeological record and by genomic dating using the DNA of living populations (figure 11.4).[47]

Perhaps the standstill occurred because the coralline atolls of the western Pacific had not yet grown upward with the rapid postglacial rise in sea level to create land surfaces upon which humans could survive. Many Pacific atolls only became sufficiently emergent above high tide to support human occupation between 3,000 and 1,000 years ago, these dates becoming younger toward the eastern Pacific. Before the numerous atolls that dot the vast Pacific Ocean had emerged, the distances between the main volcanic islands would have been much greater, hence more difficult to cross.

It is also apparent from the archaeological record that the oldest cultures in East Polynesia were not founded directly from Tonga or Samoa but from an intermediate staging location that has long been hidden from view. Where was that location? Linguists have come to the rescue by suggesting that the languages of East Polynesia are most closely related to some of the "Polynesian Outlier" languages located on small islands within Melanesia. These Polynesian Outlier languages are thought to be of relatively recent West Polynesian origin, spreading back to the west with the prevailing trade winds from Samoa or the Wallis and Futuna Islands. Whatever their ages, they were certainly widespread among the small islands and atolls of Melanesia prior to 1,000 years ago, when the migrations to East Polynesia commenced.

Many of the Polynesian Outliers are atolls, and a cluster of them in the northern Solomons, perhaps settled by Polynesians between 500 and 1000 CE, are now identified as likely linguistic homelands for the first eastern Polynesians.[48] It is difficult to derive the material culture of the first eastern Polynesians entirely from these coralline atolls, which lack extensive agricultural soil deposits (aside from deliberately composted soils) and volcanic rocks for making stone tools. However, the obvious mobility of these people with their efficient boats makes the idea of a single-island homeland for all aspects of culture a little superfluous.

There is another important social factor that might have been significant in stimulating the first migrations toward the far horizons of eastern Polynesia. Ancient DNA research has demonstrated the Polynesian genetic affinities of three of the forty or so members of the burial entourage of a chief—remembered by modern Vanuatu people as Roi Mata—who were buried on a small island in central Vanuatu around 1600 CE.[49] The region still has Polynesian Outlier linguistic populations who live among larger Melanesian linguistic populations today. Twenty-two of the Roi Mata burials were male and female side-by-side pairs, thought to have been buried alive by the original excavator of the site.

Roi Mata's thought-provoking burial ritual implies a presence in central Melanesia of socially stratified societies, like those that met the first

European explorers in Polynesian archipelagoes such as the Tongan, Society (including Tahiti), and Hawaiian Islands. Roi Mata was evidently of Polynesian genetic ancestry. Ranked societies with charismatic chiefs often have schisms in their higher social echelons, schisms that can lead to persons with chiefly ambitions departing with followers to seek a better life elsewhere. Perhaps similar-minded individuals in the Solomon Island Polynesian Outliers became the first East Polynesians.

In the islands of East Polynesia, the first human settlements occurred around 1,200 years ago. Possibly some lucky navigators managed to reach some of them and then get back to their homelands in the western Pacific in order to tell the tale. If they did so, we can expect that many expeditions would have sailed east toward the rising sun, settling all the islands throughout the enormous East Polynesian triangle shown in figure 11.4 within a period of perhaps 500 years. We can never know how many lives were lost in these incredible feats of voyaging over seemingly endless horizons, but we do know that some of the most recent to arrive were the ancestors of the New Zealand Maori, who sailed far south into temperate latitudes some 3,000 kilometers from their tropical Polynesian homeland around 1250 CE. The Maori were creators of one of the most remarkable and concentrated 500 years of prehistory, from initial settlement to the eighteenth century arrival of European explorers, anywhere in the world.

Recent DNA research has also suggested that there was a small population exchange between South America and East Polynesia at about 1,000 years ago.[50] It is possible that Polynesian contact with Native Americans along the Pacific coastline of Colombia or Ecuador stimulated this exchange, although it must be emphasised that Polynesians never settled in large numbers in South America. Neither were Native American cultures and languages permanently established in the Pacific Islands. Trans-Pacific contact occurred, but not on a level significant for the movement of whole populations, languages, or archaeological cultures.

This topic of American contact is of historical interest within Pacific archaeology, however, in view of the Kon-Tiki sailing raft expedition of 1947 and the hypotheses about Pacific prehistory presented by

Norwegian explorer Thor Heyerdahl. He believed that Native Americans had first settled the islands of eastern Polynesia, especially Easter Island, prior to the arrival of the Polynesians from Southeast Asia. Heyerdahl's overall views (which did not, in fairness, claim that the Polynesians themselves originated in South America) are no longer taken seriously by modern prehistorians because there is no evidence for any actual colonization of Polynesia by South Americans nor any former presence there of South American languages. However, this new genetic research certainly rejuvenates certain aspects of the debate.

Rice versus Yams?

Where does the debate over Austronesian origins and dispersal patterns stand now? Genetic evidence leaves little doubt that New Guineans and Island Melanesians have always been, for the most part, indigenous to the western Pacific. The majority of the speakers of Malayo-Polynesian languages have undeniable genetic ancestry in southern China, followed by ancestral migrations through Taiwan and the Philippines. The archaeological record of Neolithic sites and material cultures from southern China to eastern Polynesia makes this clear.

However, there are some obvious complications in what might appear to be a continuous pattern of migration in one direction out of Taiwan by one people. One complication involved the Papuan migration around 500 BCE into the islands of Melanesia that I discussed above. This undoubtedly led to a great deal of language shift and genetic admixture.

Another complication was related to some significant changes in subsistence economies. As Malayo-Polynesians settled the small islands of Oceania, so their systems of food production were adapted to the decreasing opportunities for cereal cultivation. Pigs, dogs, and chickens traveled almost everywhere (except to some distant Oceanic islands), but rice and the millets did not, probably because they were unsuited to the nonseasonal climates that existed close to the equator and through which people had to travel to reach Oceania. Many small Oceanic islands also have unreliable surface water supplies, especially the atolls. Oceanic peoples subsisted mainly on fruits and tubers, many (such as coconuts and breadfruit) of Indonesian and New Guinea origin, together with pigs, chickens, and, above all, fish and other marine resources.

When did rice cultivation become important in Island Southeast Asia? The hot, wet climates of Island Southeast Asia do not encourage good preservation of botanical materials, and many former village sites along rivers and coastlines are now buried so deeply under sediment that they will only be recovered through massive bulldozing, normally only available with deep construction projects. Such deeply buried but archaeologically rich waterlogged sites have so far only been excavated in southwestern Taiwan, a nation that has a developed industrial economy and protective heritage legislation.[51] Elsewhere in Island Southeast Asia, evidence for ancient food production has so far been virtually absent or perhaps ignored.

In the Karama Valley on Sulawesi Island in central Indonesia, the recent discovery of evidence for intensive rice cultivation dating between 1500 and 1200 BCE therefore indicates what might survive if only it can be located and recovered. The Neolithic settlement of Minanga Sipakko was located about 100 kilometers inland by river from the western coastline of Sulawesi—thus, far from coastal sea level changes and downstream deep soil deposition. Its inhabitants were growing and processing rice close to their river terrace settlement for several centuries, according to new evidence derived from analysis of microscopic phytoliths, silica bodies that exist in plants.[52]

This evidence from Minanga Sipakko is currently unique in Island Southeast Asia outside Taiwan, although this situation surely reflects a lack of discovery in difficult conditions rather than any real absence. Rice cultivation was widespread in Taiwan well before 2000 BCE, and its presence in the Proto-Austronesian linguistic environment is strongly supported by linguistic reconstructions. The source of this Sulawesi rice cultivation, arriving from southern China via Taiwan and the Philippines, is quite apparent, even if many of the modern varieties of rice grown across Island Southeast Asia reflect later spreads as the rice plant evolved to suit varying climatic conditions.[53] The first Malayo-Polynesians migrated with rice from the Philippines into Indonesia but then modified their food production repertoires to adapt better to conditions among the small islands of Oceania.

12

Africa, Australia, and the Americas

THREE MAJOR PORTIONS of the earth's surface remain for investigation—Africa, Australia, and the Americas. Africa was impacted in the north by Neolithic migrations from the Fertile Crescent, but its Sub-Saharan peoples also made their own transitions to food production and pastoralism, between latitudes 5° and 15° north, followed by extensive migrations southward in the cases of the ancestral Khoisan and Bantu-language speakers. Australians were impacted by developments involving food production in neighboring Indonesia and New Guinea, and major archaeological and language changes occurred across much of the continent during the mid-Holocene as possible reflections of these contacts. Native Americans underwent several episodes of migration from their major agricultural homelands, especially with developments in maize production after 4,000 years ago, to give rise to many of the food-producing linguistic populations that were in place by 1492 CE.

The African Continent

In Africa, the Holocene human population history of the past 12,000 years has been intimately connected with the histories of four major language families: Afro-Asiatic, Nilo-Saharan, Niger-Congo, and Khoisan. I commence with Afro-Asiatic, the source of the oldest subsistence economy of food production in Africa. This was introduced from the Levant into Egypt more than 7,000 years ago, and also at an uncertain date through Yemen and across the Red Sea into East Africa.

Afro-Asiatic Migrations from the Southern Levant into North Africa

In chapter 10, I focused on the spread of food production out of the Fertile Crescent, especially with the expansion of early Indo-European speakers from its northern regions into Europe and Asia. It might be wondered why Indo-Europeans never spread into Africa. There is one very good reason: another important food-producing population stood in the way, that which gave rise to the Afro-Asiatic-speaking populations of the Middle East and northern Africa (figure 9.2 and table 12.1).

The migrations of Afro-Asiatic agricultural and pastoral peoples took Fertile Crescent food production from the southern Levant into northern Africa by two separate routes. One, which traveled with crops and domesticated animals, went through Sinai into Egypt. From here, one branch traveled up the Nile Valley toward Sudan, while another went along the Mediterranean coastline toward fertile farmlands in Tunisia, Algeria, and Morocco. The other route, which perhaps brought sheep and goats but apparently not Fertile Crescent crops into East Africa, went south through Arabia and Yemen (more humid than now during the early Holocene) and probably across the Bab el Mandeb. This pastoral migration entered the monsoonal climate regions of Sudan, Ethiopia and the Horn of Africa, although rather less is known about it than about the movement into the Nile Valley. These population and language spreads out of the Levant date archaeologically from 6000 BCE onward, although outer limits between 8000 and 5000 BCE are quite possible, as I discussed in chapter 8.

Where was the homeland of the Afro-Asiatic language family? Some linguists prefer northeastern Africa because more linguistic subgroups are located there than in the Levant, although there is no linguistic genealogy that conclusively demonstrates an African origin.[1] Others

TABLE 12.1. The Afro-Asiatic subgroups and major languages.

(Proto-Afro-Asiatic was located in the Early Holocene Levant)	
Semitic	Extinct: Akkadian (ancestor of Babylonian and Assyrian), Eblaite, Amorite, Phoenician (including Punic/Carthaginian)
	Living: Arabic (many regional dialects), Hebrew (reinstated from a nonspoken status in Israel), Aramaic, Amharic (Ethiopia)
Egyptian	Ancient Egyptian and its Coptic successor in southern Egypt
Chadic	Hausa (Niger and northern Nigeria)
Berber	Berber, Tuareg (Niger and Mali)
Cushitic	Oromo and Somali (Sudan and the Horn of Africa/Ethiopia)
Omotic	Ethiopia, but the status of Omotic as an Afro-Asiatic language is debated

prefer the Levant, partly on the grounds that the reconstructed Proto-Afro-Asiatic vocabulary for subsistence economy and material culture matches that of the Natufian and Pre-Pottery Neolithic cultures of the Fertile Crescent.[2] One recent suggestion has placed the proto-language in the Levant at about 10,000 BCE.[3] Other linguists have pointed to early word-borrowing connections that suggest that Proto-Afro-Asiatic was a near-neighbor to Proto-Indo-European, thus supporting a southwest Asian origin.[4]

However, the main clues to an ultimate Levant origin for the speakers of Afro-Asiatic languages come not from linguistics but from archaeology and ancient genomics. In early Holocene times, the flow of both farming cultures and human genes was positively out of the Levant, rather than the other way around. In chapter 8, I discussed the archaeological evidence for the spread of Fertile Crescent crops and domesticated animals into Egypt. None spread the other way, apart from the donkey. Sheep, goats, and cattle moved much farther than the Fertile Crescent crops, doubtless with some of their human owners, and eventually contributed greatly to the prehistory of Sub-Saharan Africa.

Some linguistic reconstructions also support these separate Afro-Asiatic population movements. According to linguist Vaclav Blazek, the movement through Sinai brought the ancestral Ancient Egyptian and Berber languages to North Africa, although the main spread of the Berbers through the northern and western Sahara appears to have occurred only in the first millennium BCE. Prior to this time, indigenous Nilo-Saharan and Niger-Congo populations must have been more widespread in the Sahara than they are today.

The Bab el Mandeb movement spread ancestral Cushitic and Omotic speakers to Ethiopia and the Horn, whereas ancestral Chadic speakers traveled most probably along the southern edge of the Sahara (the Sahel zone) to arrive in their current locations in Niger, Nigeria, and Chad. Semitic languages, especially Arabic, now dominate much of North Africa and have replaced many preexisting Afro-Asiatic languages. The spread of Arabic occurred with Islamic conquests and settlement from Arabia in and after the seventh century CE.

Genetically, the Neolithic population that brought the Levant genomic profile to North Africa must have contributed to the ancestry of many Afro-Asiatic-speaking peoples today. However, Chadic, Cushitic, and Omotic speakers have high percentages of indigenous Sub-Saharan African DNA derived from ancient mixture with Niger-Congo and Nilo-Saharan peoples.[5] The genetic link with the Levant is much stronger in the Mediterranean populations of North Africa, and recently it has been shown that Egyptian mummies dating between 1200 BCE and Roman times have genomic ancestries most closely paralleled in Neolithic populations in the Levant, Anatolia, and even Europe.[6] Sub-Saharan African DNA only made a significant appearance in Pharaonic Egypt during and after Roman times. Ancient DNA from Neolithic skeletons found in Morocco has also revealed Fertile Crescent Natufian and Neolithic affinity mixed with indigenous Paleolithic ancestry, plus a small component introduced from Neolithic Iberia.[7]

The Transformation of Sub-Saharan Africa

As discussed in chapter 8, Nilo-Saharan populations presumably occupied a much larger area of the Sahara during its early Holocene wet phase (the Green Sahara), before they retreated southward into their current distribution with its return to desert around 5,000 years ago. By the time they did so, pastoralism with sheep, goats, and cattle had been introduced along with Afro-Asiatic languages from the Levant. Both Nilo-Saharan and Afro-Asiatic (especially Cushitic) peoples then spread south with pastoral economies to reach the equator and the Rift Valley of Kenya by 2500 BCE.[8] Here, their incursions appear to have had a major impact on a deeply indigenous Khoisan-speaking population of hunter-gatherers.

As I noted in chapter 5, the southern regions of Africa were originally occupied by hunter-gatherers whose surviving descendants today include the San of southwestern Africa, together with their Khoekhoen pastoralist relatives.[9] Related groups include the Sandawe and Hadza of Tanzania, and the rainforest hunter-gatherers of the Congo Basin in central Africa. These groups exist nowadays as hunter-gatherer minorities, or as pastoralists in the case of the Khoekhoen, among a majority population of Bantu-speaking farmers.

These Sub-Saharan hunter-gatherer populations are today far-flung and varied in lifestyle, but many (except for the Congo hunter-gatherers) share an important linguistic characteristic in having click consonants in their languages. These languages are often grouped together linguistically as "Khoisan," although it is generally agreed that there is more than one Khoisan language family. The Congo hunters ("Pygmies") have adopted Niger-Congo and Nilo-Saharan languages.

After herding of cattle, sheep, and goats had been established in northeastern Africa, it appears that hunter-gatherer populations in the vicinity of Kenya and the Rift Valley, including the ancestors of the Khoekhoen, adopted a pastoralist lifestyle. Some of them then migrated to reach their historical territories in southwestern Africa by 2,000–1,500 years ago, before the spread of the Bantu-speaking farmers. By the first few centuries CE, pastoralists with sheep, cattle and pottery-making skills had reached the Cape of Good Hope (figure 12.1).

Genetic evidence for a proportion of northeast African and Levant DNA among the living Khoekhoen pastoralists of Namibia, Botswana, and South Africa strongly favors such a migration by ancestral Khoisan-speaking peoples, even if the details are still unclear. According to geneticist Carina Schlebusch, these migrating Khoisan populations mixed with the southern African San hunter-gatherers, thus contributing genes to both of the non-Bantu (San and Khoekhoen) populations who inhabit southwestern Africa today.[10]

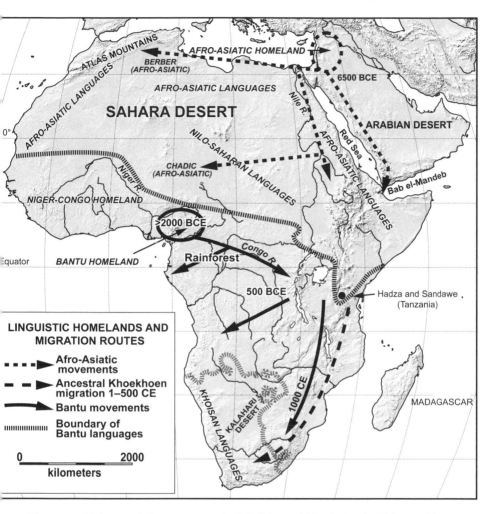

Figure 12.1. Major population movements in Sub-Saharan Africa during the Holocene. The dates refer to likely times of Bantu arrival, and the arrows for Bantu dispersal are derived (approximately) from Ezequiel Koile et al., "Phylogeographic analysis of the Bantu language expansion supports a rainforest route," forthcoming in *PNAS*.

The Bantu Diaspora

The fourth African linguistic population, comprising the Niger-Congo-speaking peoples, includes most of the Sub-Saharan African population today. Niger-Congo peoples are indigenous to tropical West Africa, and their ancestors must have overlapped with the distribution of

Nilo-Saharans in Mali and Niger during the Saharan humid phase. Indeed, Nilo-Saharan languages are still spoken in parts of the Niger Valley today.[11]

The Bantu languages form only one branch of the much larger Niger-Congo language family. In the past 4,000 years they have spread through a huge area of Africa that extends from their homeland in eastern Nigeria and Cameroon in the northwest to the eastern coastline of South Africa, a distance of about 4,500 kilometers. Today, these languages are spoken by about 250 million people. According to one group of geneticists, the Bantu dispersal was "one of the most dramatic demographic events in human history."[12]

In chapter 8, I discussed the food-producing economy that developed in Sub-Saharan Africa, with its millets, legumes, and Fertile Crescent domesticated animals. Domestication of pearl and sorghum millet had begun by 2500 BCE in the Sahel and Sudan savanna and parkland vegetation zones, and the Bantu homeland is agreed by virtually all linguists to have been within this region, located especially in eastern Nigeria and Cameroon. Food production and herding were present here by 2000 BCE, and a new linguistic analysis of Bantu languages suggests that expansion into the rainforest to the immediate south, probably following rivers and natural zones of relatively open vegetation, was already starting by this time.[13]

During the last few centuries BCE the ancestral Bantu had also adopted ironworking, although whether through independent invention or adoption from elsewhere is uncertain. Perhaps as a result of this adoption of iron tools, a full north-to-south open passageway through the western side of the West African rainforest was opened by about 500 BCE, allowing the Bantu to emerge on the southern side of the rainforest with their cereals, legumes, domestic animals (cattle, sheep and goats), and iron tools. From this point, they began their expansion to the east, toward Lake Victoria and the drier seasonal landscapes of eastern Africa. From the general vicinity of Lake Victoria they continued southward through 4,000 kilometers of mostly savanna terrain to reach their prehistoric limit in Natal by around 1000 CE (figure 12.1).

The Bantu migrations were relatively recent compared with the other farmer migrations discussed in these chapters, and we might wonder why. By the time the expansion through the rainforest was underway, the regions north and northeast of the rainforest had already become settled by other non-Bantu farmers and pastoralists, especially in East Africa. This is perhaps why the Bantu essentially headed south in the first instance, into and through the rainforest with its indigenous populations of hunters and gatherers. Furthermore, the previous Khoisan migrations that I discussed above were pastoralist, and it is possible that much of Sub-Saharan Africa to the south of the rainforest was also occupied by these people before the Bantu arrived. Perhaps the Bantu migrations were successful in part because of their acquisition of iron tools and in part because of the arrivals of taro, greater yam, banana, sugar cane, and chickens from Indonesia. These important domesticated food resources arrived in the early centuries CE, when Madagascar was in the process of being settled by seaborne Malayo-Polynesians from Borneo. Thus, they arrived on the eastern African mainland in good time to play an important role in the Bantu expansion.

By 500 CE, Bantu settlement had reached the vicinity of Swaziland, and by 1000 CE the limit of monsoonal climate in the far southeastern portion of South Africa. There is a considerable record of radiocarbon-dated archaeology that plots this progress, despite an apparent resettlement of the Congo rainforest around 1000 CE following a regional population decline, still of uncertain causation.[14] All recent genetic discussions of the Bantu migration process concur that it caused a major genetic shift in the human landscape of Sub-Saharan Africa, although mixture with Khoisan populations increased as the migration progressed southward toward the limits of the monsoonal summer rainfall system of food production.[15] Interestingly, it is likely that the ancestral Bantu, like most living Bantu speakers, also carried Duffy antigens and abnormal hemoglobins in their blood that gave them resistance to malaria.[16]

The Bantu migrations led to one of the most extensive spreads of an early farming population in world prehistory. Their possession of iron tools and their dependence on a broad range of crops as well as domestic animals gave the Bantu the necessary demographic edge to migrate through areas inhabited earlier by pastoralist as well as hunter-gatherer populations, until they reached the Kalahari Desert and the Mediterranean winter rainfall climate of southwestern Africa. Neither of these environments was suitable for their monsoonal crops, so here their expansion stopped. When the Dutch began to settle in South Africa in the mid-seventeenth century, they met both Bantu and Khoisan peoples.

The Australian Continent

An idea that the Australian continent was isolated from the rest of the world until Europeans arrived has been circulating among students of the human past for a long time. I suspect that this idea has never been entirely correct. It seems unthinkable that a land mass so large, flanked to its north for the past 3,000 years by maritime populations, could have been entirely isolated. Ancient Australia changed during its 55,000 or more years of prehistory, just like any other major region of the world. It simply changed in its own hunter-gatherer way, and that way was not identical to other ways that were followed during the Holocene by food-producing populations elsewhere. Nowadays, we know that some remarkable cultural and linguistic changes unfolded over much of the continent during the past 3,000 years, on a geographical scale equivalent to the expansions of some of the most widespread archaeological cultures and language families in other parts of the world.

Let us first examine some archaeology. Australian populations across most of the central and southern continental mainland, particularly between 1500 and 500 BCE, focused heavily on the manufacture of small backed stone artifacts, coincidentally similar to those made during much earlier Paleolithic times in South Africa, India, and Sri Lanka (see figure 12.2). Such items are rare outside this date range in Australia, although examples do exist. However, the remarkable focus upon them across most of the continent during this date range suggests a cultural change of some magnitude.

The only region in nearby Southeast Asia that contained similar tools at the same time as Australia was the southwestern arm of the island of Sulawesi, in central Indonesia, where they occurred in pre-Neolithic "Toalian" archaeological contexts after about 5000 BCE, prior to the arrival of Malayo-Polynesian-speaking rice farmers. I return to Sulawesi

Figure 12.2. Virtually identical mid-Holocene backed artifacts from South Sulawesi (*top*) and southeastern Australia (*bottom*). Lake Illawarra is in coastal New South Wales. The specimen at bottom left is 4.5 centimeters long. Collections of the Australian Museum on loan to the School of Archaeology and Anthropology, Australian National University, photo by the author. Published originally in Peter Bellwood, *First Migrants* (Wiley-Blackwell, 2013), figure 5.5.

below, but add here that Arnhem Land and the Kimberley Plateau in the tropical north of Australia had different Holocene stone tool industries, without backed artifacts, that were focused on the manufacture of bifacial points and edge-ground axes, as indicated on the map in figure 12.3.[17] Backed tools were also absent in New Guinea and in the eastern islands of Indonesia, as well as in Tasmania, which had become cut off from the Australian mainland during the early Holocene by the rising postglacial sea level in Bass Strait.

Whether these backed tools were invented independently in Australia or introduced from South Sulawesi is unknown. It is the high

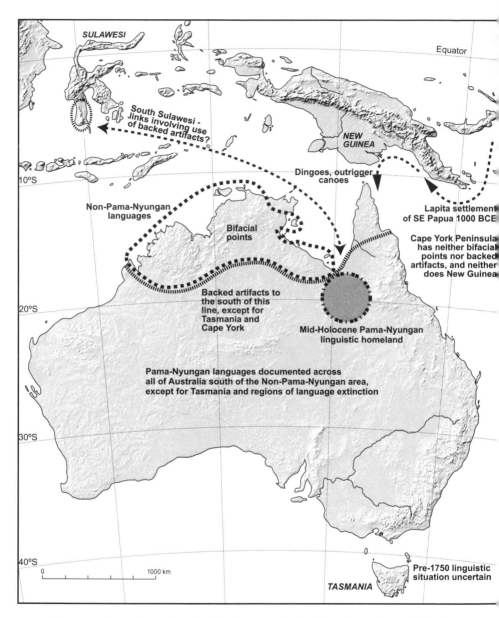

Within the figure:

SULAWESI

Equator

South Sulawesi - links involving use of backed artifacts?

NEW GUINEA

10°S

Non-Pama-Nyungan languages

Bifacial points

Dingoes, outrigger canoes

Lapita settlement of SE Papua 1000 BCE

Cape York Peninsula has neither bifacial points nor backed artifacts, and neither does New Guinea

Backed artifacts to the south of this line, except for Tasmania and Cape York

20°S

Mid-Holocene Pama-Nyungan linguistic homeland

Pama-Nyungan languages documented across all of Australia south of the Non-Pama-Nyungan area, except for Tasmania and regions of language extinction

30°S

40°S

0   1000 km

Pre-1750 linguistic situation uncertain

TASMANIA

Figure 12.3. Australia, to show the distributions of archaeological complexes and the Pama-Nyungan languages.

intensity of their use at around 3,500 years ago in Australia that is so striking because this was a time of major change in both Australia and in the archipelagoes to its north, regardless of where the idea of backing a stone tool might have arisen in the first place.[18]

Linguistically, the past few millennia have also seen the unfolding of a major language family across the same region of mainland Australia as that occupied by the backed stone implements. Termed the Pama-Nyungan language family by linguists, it is believed to have originated close to the base of Cape York Peninsula. Its date of initial expansion is uncertain, but recent estimates tend to fall somewhere between 5,000 and 3,000 years ago.[19] Linguistic dating is unable to be precise on this issue. However, a recent linguistic appraisal of the Pama-Nyungan language family traces its origins to a "cataclysmic spread."[20] Like the backed stone tools, Pama-Nyungan languages also do not occur in Arnhem Land, the Kimberley Plateau, or Tasmania, although the Aboriginal languages of Tasmania are poorly recorded. Those of Arnhem Land and the Kimberley belong to many small and deeply diversified families; these languages cannot be connected with Pama-Nyungan and might have been in place since Pleistocene times.

At first sight, this conjunction of archaeological and linguistic data seems almost too good to be true. It implies that the greater part of the Australian continent witnessed major changes within the past 3,500 years, in both stone tools and languages, that might have been directly associated. Furthermore, the stone tools and languages are not the end of the story.

Dingoes, canoes, and possibly even agriculture also enter the fray. Genetic evidence from modern dogs suggests that the dingo, originally a domesticated Asian dog, might have arrived in Cape York Peninsula via southeastern New Guinea, presumably connected with the Lapita migration toward Polynesia around 1000 BCE (see chapter 11).[21] Recent analysis of historical records suggests that dingoes might have played a role in Aboriginal hunting. If so, they would have been helpful aids for people who were hunting kangaroos, wallabies, and emus for meat.[22]

The outrigger canoes used at European contact along Cape York Peninsula were also introduced from the southeastern coastline of New

Guinea. The precise date of introduction is unknown, but canoe terms borrowed from Malayo-Polynesian (Oceanic subgroup) languages suggest that it was before the arrival of Europeans, possibly as much as 3,000 years ago, when Malayo-Polynesian navigators appeared off the eastern end of New Guinea.[23] Furthermore, there have been suggestions that some crop plants, such as yams, bananas, and taro, were occasionally introduced from New Guinea into northern Australia during the Holocene.[24] All of this implies that Australia was not entirely isolated from some fundamental changes that were happening to its north.

As we saw in the previous chapter, the time span between 1500 and 500 BCE was a particularly important one in the prehistory of Island Southeast Asia and the Melanesian islands. It witnessed the migration of Malayo-Polynesian-speaking peoples, of both Asian Neolithic and New Guinea Papuan genetic origin, into the Nusa Tenggara Islands in southeastern Indonesia, Melanesia, and Polynesia. These people traveled with economies of food production, advanced canoe-construction skills, dogs, and a set of closely related languages into many islands that lay close to Australia. Any idea that these Malayo-Polynesians somehow avoided contact with Australia is surely misguided.

However, the Pama-Nyungan languages are not related to Malayo-Polynesian or Papuan languages, so they can hardly have originated from them. Indeed, had Pama-Nyungan languages originated from either of these overseas linguistic sources we might expect them to occur throughout northern Australia, rather than just in central and southern Australia. The Pama-Nyungan languages do not have such a distribution. Furthermore, neither Malayo-Polynesians nor Papuans ever used backed stone tools according to their associated archaeological records, and the Toalians, who did use such tools, can hardly have belonged to either linguistic grouping in terms of their archaeological date and geographical location. Put simply, there is no evidence for Malayo-Polynesian or Papuan settlement in Australia on any significant scale at any time in the past. Another explanation is needed.

What about a connection between Sulawesi and Australia before Austronesians arrived, to explain the stone tool similarities? In my book *First Migrants* I thought through these issues and came up with the idea that some members of the Pre-Austronesian Toalian population of South Sulawesi might have migrated into the Gulf of Carpentaria as a reaction to the settlement of their hunting territories by immigrant Malayo-Polynesian-speaking farmers.[25] Did some Toalians adopt canoe technology and dogs from their Malayo-Polynesian-speaking neighbors and sail away toward Australia with a collection of backed stone artifacts, on a summer-season northwest monsoon wind?

When I wrote *First Migrants* I was not aware of the information relating to dingoes and canoes that also pointed to contacts between Cape York and southeastern New Guinea, so the situation might have been more complex than I realized then. Even so, if Toalian people did come from Sulawesi, they would have been following the sea routes used by the Macassans of South Sulawesi during the eighteenth and nineteenth centuries. These people traveled to the northern coastline of Australia each year until the British government in Australia stopped them; they would collect there the black sea slug known in French as *bêche-de-mer*, a constituent of Chinese cuisine.

However, we have absolutely no idea what languages the Toalians might have spoken 3,500 years ago because any descendants of the pre-Austronesian languages once spoken in western and central Indonesia are now all extinct. Presumably, Toalian languages were not ancestral to Pama-Nyungan, but the main grounds for suggesting this are that the latter are considered by linguists to be indigenously Australian. A hypothetical Toalian linguistic source for Pama-Nyungan languages thus fails to be convincing, but it has to be admitted that there are still many loose ends in this debate.

To summarize, we clearly have a situation of significant potential cultural impact in much of late prehistoric Australia, especially some of its northern regions. Might we face a scenario in which the Pama-Nyungans were migrants from somewhere within northeastern Australia, thus indigenous Australians, who had somehow acquired some useful advantages in stone tool technology, canine support, and boat and fishing technology as a result of connections with the outside world? It is of interest that small shell fishhooks, like those made in pre-Neolithic eastern Indonesia, especially Timor and nearby islands (figure 6.4), were made as far south as coastal New South Wales in Holocene times. Did all of this new knowledge encourage Pama-Nyungans to spread with their languages across and around the coasts of most of the central and southern Australian mainland?

Can genetic evidence assist a decision? Unfortunately, there have been no studies of ancient DNA across Australia that can throw direct light on questions of Pama-Nyungan expansion, although a recent study of mitochondrial DNA in hair collected from ethnographic Pama-Nyungan-speaking populations in Queensland and South Australia yielded no evidence for secondary migration within the continent at any time in the past 50,000 years.[26] This is not conclusive evidence, however; an earlier study dating from 2001 did claim that a specific Y-chromosome haplotype became widespread across mainland Australia from about 4,000 years ago.[27] The authors pointed to a possible association with the spread of backed tools and the dingo. Recent whole-genome research on living Pama-Nyungan populations has also supported a generalized Holocene population spread within Australia from northeast to southwest.[28]

What to conclude? The Pama-Nyungan language family and the archaeology that overlaps with its distribution would be sufficient to raise the likelihood of a major continental-scale population expansion in any situation elsewhere that involved food production and consequent population increase. However, the Australian Holocene archaeological record appears not to reveal convincing evidence for either. Because food production requires the management of domesticated plants and animals, Australia cannot be claimed to have fully developed this way of life in prehistory.[29] The dingo is not relevant in this regard because it was introduced as a domesticated dog. At the moment, we seem to have a stalemate.

I await future developments in Australia with great anticipation. For ancient Australians, the fundamental hunting and gathering economy that fueled virtually all of the Five-Million-Year Odyssey worked perfectly well throughout their prehistory. According to anthropologist Peter Sutton and archaeologist Keryn Walshe, Australians were spiritual managers of their lands rather than food producers with domesticated resources.[30] But could contacts from the north have promoted

Holocene population movements within the southern two-thirds of the continent? I am unsure, but I would love to know.

The American Continents

Holocene Migrations in the Americas

In the Old World, food-producing populations with ever-increasing numbers spread outward when investment in a transportable economy of domesticated resources allowed and encouraged them to do so. It should come as no surprise, therefore, to find that exactly the same occurred in the New World, but with the catastrophic end result that we remember as the year 1492 and its consequences. Massive upheavals of indigenous populations, their cultures, and their languages took place after that fateful year. Geneticists and historians estimate that up to 90 percent of Native Americans died from introduced diseases during the sixteenth century, with the result that precise details of the linguistic and cultural landscape that existed in 1492 can sometimes be difficult to reconstruct. This is especially true in those middle temperate latitudes of both North and South America where Europeans transplanted their own systems of food production and dispossessed the indigenous people.

Native Americans did survive, of course, and patterns from the prehistoric past are still readily identifiable in the distributions of the major language families, even if there are many unfortunate gaps in the data. However, prehistorians researching in the Americas have been slow in recent years to question how the populations and languages present at the time of European contact originated and spread. The archaeological record is often difficult to relate to linguistically defined populations in the past, and genetic research, especially the study of ancient DNA, is only just beginning. As in Australia, living Indigenous populations have not always welcomed archaeological or genetic research into their ancestry by outsiders, although it is gratifying to see that desires for accurate and informative research into such issues are now gaining in strength.

Despite this relatively slow start, many interesting observations can still be made. The farming populations of the Americas occupied most of the tropical and temperate latitudes of the two continents at European contact. They are listed by language family in table 12.2 and plotted as they were when Europeans arrived in figure 9.3. All those language families listed in table 12.2, with the possible exception of Algonquian, overlapped in their suggested linguistic homelands with the American agricultural homelands, as reconstructed in figure 12.4.

North American Hunter-Gatherers on the Move

As in the Holocene Old World, New World peoples prior to the Colonial Era included both hunter-gatherers and food producers. The hunters and gatherers, as might be expected, were mostly situated in regions where farming was not possible, but there were exceptions. California, for instance, a fertile region for agriculture today, had no prehistoric farmers with domesticated plants and animals. This region was too well protected by deserts and mountains for any significant immigration by maize farmers from the US Southwest to have occurred. Indigenous hunters and gatherers, some with intensive methods of resource management, occupied the region until the arrival of the Spanish.

Some large and widespread linguistic populations also straddled both farming and nonfarming landscapes, ranging in the case of the Uto-Aztecans from the mighty maize-farming Aztecs of Mexico to the Paiute and Shoshone hunters and collectors in the semiarid Great Basin of the western United States. The Algonquians likewise occupied agricultural terrain in the northeastern United States as well as much colder hunting and gathering territory on the Canadian Shield. These populations had mixed subsistence lifestyles that raise interesting questions about origins. Were their initial expansions tied to hunting and gathering or to food production? I discuss the Uto-Aztecan and Algonquian cases further below.

In linguistic terms, wholly hunter-gatherer populations in the Americas at European contact included the Eskimo-Aleuts in the Arctic and the Na-Dené peoples who occupied much of northwestern North America beyond the range of prehistoric farming. Both were attached to substantial and recent episodes of hunter-gatherer migration, in both cases well recorded.

TABLE 12.2. Major agriculturalist and mixed-economy language families of the New World.

Language family	Surviving distribution	Approximate origin region	Economic and cultural orientation
North America			
Siouan	Carolina coast to eastern Great Plains	Ohio Valley?[1]	Eastern Woodlands agriculture, with maize after 200 CE. Possible association with the Mississippian culture 800–1600 CE.
Iroquoian	Appalachians to St. Lawrence River	Western New York state?[2] Appalachians?	Eastern Woodlands agriculture, with maize after 200 CE.
Muskogean	Southern US states—Atlantic coast to Mississippi	Southern US states	Eastern Woodlands agriculture and hunting-gathering (Gulf and Atlantic coasts), with maize after 200 CE.
Algonquian	Northeast United States, surrounding the Iroquoians; Eastern Canada	Great Lakes?	Hunting and gathering in Canada. Eastern Woodlands and maize agriculture in the United States.
Mesoamerica[3]			
Otomanguean	Parts of central Mexico	Oaxaca	Agriculture, and probably associated with the initial Balsas Basin domestication of maize. Associated with the Zapotec and Mixtec Formative, Classic and Postclassic cultures.
Uto-Aztecan	El Salvador to Idaho	Central Mexico, close to Otomanguean homeland	Agriculture, with maize terminology borrowed from Otomanguean. Hunting and gathering in the Great Basin (Paiute, Shoshone) and southern Great Plains (Comanche). Associated in Mesoamerica with the Aztecs (speakers of the Nahua language at Spanish contact), and possibly the earlier civilizations of Teotihuacan and the Toltecs.
Mixe-Zoque	Isthmus of Tehuantepec, Mexico	Isthmus of Tehuantepec	Agriculture, and probably associated with the Formative Olmec civilization.
Mayan	Guatemala, Yucatan, Chiapas, with a Huastecan enclave in northeastern Mexico	Guatemala Highlands	Agriculture, and associated through inscriptions with Indigenous America's most significant literary tradition.

(continued)

TABLE 12.2. (*continued*)

Language family	Surviving distribution	Approximate origin region	Economic and cultural orientation
South America			
Arawakan	Amazonia and the Caribbean Islands, except for western Cuba	Southwestern Amazonia (upper Madeira and Purus valleys)[4]	Agriculture, especially manioc and maize. Ranked societies with circular villages around plazas, associated with Saladoid and Barrancoid decorated pottery.
Tupian	Southern Amazonia and east coastal South America down to Argentina and Paraguay	Southwestern Amazonia (upper Madeira and Guapore valleys)[5]	Agriculture, especially manioc and maize, associated with polychrome pottery. Southwards expansion (Tupi-Guarani) after 500 BCE.
Cariban	Much of northern South America, but no certain prehistoric presence in the Caribbean Islands	Eastern Amazonia[6]	Agriculture, especially of manioc and maize.
Quechuan and Aymaran	Central Andes	Central Andes[7]	Main expansions after 1 CE with Andean Formative and Classic civilizations, continuing with Incas after 1438 CE.

[1] Robert Rankin, "Siouan tribal contacts and dispersions," in John Staller et al., eds., *Histories of Maize: Multidisciplinary Approaches to the Prehistory, Linguistics, Biogeography, Domestication, and Evolution of Maize* (Elsevier, 2006), 564–577.

[2] Michael Schillaci et al., "Linguistic clues to Iroquoian prehistory," *Journal of Anthropological Research* 73 (2017): 448–485.

[3] Jane Hill, "Mesoamerica and the southwestern United States: Linguistic history," in Peter Bellwood, ed., *The Global Prehistory of Human Migration* (Wiley, 2015), 327–332.

[4] Robert Walker and Lincoln Ribeiro, "Bayesian phylogeography of the Arawak expansion in lowland South America," *Proceedings of the Royal Society of London, Series B: Biological Sciences* 278 (2011): 2562–2577.

[5] Jose Iriarte et al., "Out of Amazonia: Late-Holocene climate change and the Tupi-Guarani trans-continental expansion," *The Holocene* 27 (2017): 967–975; Thiago Chacon, "Migration and trade as drivers of language spread and contact in indigenous Latin America," in S. Mufwene and A. Escobar, eds., *The Cambridge Handbook of Language Contact* (online prepublication, 2019).

[6] Alexandra Aikhenvald, "Amazonia: Linguistic history," in Peter Bellwood, ed., *The Global Prehistory of Human Migration* (Wiley, 2015), 384–391.

[7] Paul Heggarty and David Beresford-Jones, eds., *Archaeology and Language in the Andes* (published for the British Academy by Oxford University Press, 2012); Nicholas Emlan and Willem Adelaar, "Proto-Quechua and Proto-Aymara agropastoral terms," in Martine Robbeets and Alexander Savelyev, eds., *Language Dispersal beyond Farming* (John Benjamins, 2017), 25–46.

Figure 12.4. Map to show how widespread New World agriculturalist language families overlap in origin with agricultural homelands.

The Inuit (Eskimo-Aleut speakers) migrated along the Arctic coastlines of Canada as far as Greenland during warm climatic conditions between 800 and 1300 CE, as I discussed at the end of chapter 6. Further south, ancestral Apache and Navajo buffalo hunters speaking Athabaskan languages (the major language family within the Na-Dené macrofamily) migrated 3,000 kilometers southward from Canada into the US Southwest after 1350 CE, taking advantage of a catastrophic population decline within the maize-growing Pueblo cultures of Arizona and New Mexico.

The Inuit and Apachean migrations were the two most significant to be undertaken by hunter-gatherers in recent American prehistory, and both expanded into territories that had been virtually or entirely abandoned by previous populations. Had there not been these prior population withdrawals, it is unlikely that the Inuit and Apache/Navajo migrations could have been so successful. Hunter-gatherer migrations into regions already occupied by other hunter-gatherer populations of similar density and technological capacity are not recorded elsewhere in anthropological or historical circumstances, as perceptively noted in 1976 by biological anthropologist Grover Krantz: "Hunting peoples did not make significant migrations . . . except into unoccupied or underutilized territories. . . . Anyone who postulates a greater movement of hunting peoples at the expense of other hunters ought at least to suggest some means by which this could have been accomplished."[31] With farmers, the situation was clearly different.

Farming Spreads in the Americas: Some Examples

As in the Old World, populations of American farmers were frequently able to expand into territories held by hunter-gatherers and to occupy them permanently. There is little doubt that these migrating populations already practiced food production with domesticated resources before their expansions began. Many American language families that consisted entirely of farming populations at European contact have estimated linguistic time depths of less than 6,000 years, mostly much less, bringing them well within the time period of agricultural development.

From this perspective within North America, the Siouan, Iroquoian, and Muskogean language family distributions can be read as the results of prehistoric agricultural enterprise in the Eastern Woodlands of the United States, focused along the many tributary rivers of the Mississippi Basin. Domestication of indigenous plants began perhaps 5,000 years ago in this region, but it is unlikely that the ultimate extents of these language families and their speakers were reached as long ago as this. Their main expansions appear to have occurred after the arrival of maize farming during the first millennium CE. The Siouans might have been associated with the celebrated Mississippian culture, centered after 800 CE on the massive complex of earthen mounds at Cahokia on the eastern bank of that river, near St. Louis, although it has to be admitted that the ancestral Muskogeans to their south might have equal claims. All indigenous languages once spoken along the middle Mississippi died without record after European contact, so we may never know.

The Chibchan, Mayan, Otomanguean, Mixe-Zoque, and Uto-Aztecan populations were the products of similar agricultural enterprise in Central America, albeit tightly constrained by the narrow isthmian geography of the region, except for the spectacular breakout of the Uto-Aztecans from Mexico almost to the Canadian border. With so many agriculturalist populations packed cheek by jowl into such a narrow strip of land, it will come as no surprise to find that the prehistory of the past 6,000 years in this region has been complex.

And here is one example. New research on ancient DNA and dietary traces in human bones recovered from caves in Belize suggests that agricultural populations migrated here from Costa Rica and Panama around 3500 BCE, bringing improved varieties of maize from South America back into Mesoamerica. This is a reverse direction to what might be expected, given that Mesoamerica was the original homeland of maize. It reveals that human migration can often go back on itself in circumstances of improved food production, beyond the actual origin region of the crop in question (in this case maize). The DNA of these early agricultural migrants in Belize matches that of Chibchan-speaking peoples in Costa Rica and Panama today (Chibchans are located in figure 12.4), and modern Mayans also carry around 50 percent of this DNA ancestry.[32]

There is a fairly obvious complication here, however, in that the Mayan languages are not closely related to Chibchan languages and appear instead to have originated around 2000 BCE from a source in the Guatemalan Highlands. Modern Mayan DNA also reveals significant links to people in the Mexican and Guatemalan Highlands. This situation is still difficult to explain, but it suggests that the narrow and contested strip of isthmian land that occupies much of Mesoamerica witnessed many population movements and mixtures during the past four millennia.

In South America, the Arawakan, Tupian, Panoan, and Tacanan language families were the products of prehistoric agricultural enterprise in southwestern Amazonia. They originated within the Madeira, Purus, and other upper Amazon drainage systems where manioc and other important crops were first domesticated (figure 12.4).[33] The Quechuan and Aymaran families likewise originated among early farming populations in the nearby central Andes, although their main linguistic expansions occurred with the development of Andean empires during and after the first millennium CE.[34] The Caribbean Islands were settled by Arawakan farmers (the Taino) migrating via the Orinoco Valley and the Lesser Antilles around 800 BCE, who replaced (according to genetic evidence) an older population of hunters and gatherers who had occupied Cuba and Hispaniola since about 4000 BCE.[35]

It is striking that so many major language families appear to have developed in these adjacent sectors of the Andes and southwestern Amazonia, despite the obvious environmental differences between these two regions. Contacts between them were clearly of significance, especially in the circulation of knowledge and successful domesticates.

Some other South American agriculturalist families, such as Jê (or "Macro-Ge") and Cariban, both located in the eastern tropics (figure 9.3), were also extensive, but their origins are difficult to relate to agricultural homelands. Perhaps they belonged to populations who adopted agriculture from elsewhere. However, completely independent switches to food production by hunter-gatherers were rare in prehistory, and it is difficult to demonstrate examples without access to ancient DNA. This research still lies ahead.

Algonquians and Uto-Aztecans

Apart from those examples discussed above, there are two major American language families that had separate and widespread hunter-gatherer and food producer populations, thus raising interesting questions about how they developed.

The first is Algonquian. Many northeastern regions of the United States and all of central and eastern subarctic Canada were populated by Algonquian-speaking peoples at European contact. Those living around and south of the Great Lakes and down the eastern seaboard of the United States had access to maize agriculture since about 1,500 years ago, while those of the Canadian Shield, such as the Cree, were hunters and Colonial Era fur trappers. Readers of James Fennimore Cooper's novel *The Last of the Mohicans* (1826) will know that the Mohicans of upper New York state and Vermont were Algonquians, whereas their Mohawk neighbors and enemies to the immediate west were Iroquoians.

The Algonquian language family raises an interesting question. Were the early Algonquians farmers or hunters? If farmers, did they switch to hunting when they moved north? If hunters, did they switch to farming when they moved south? Or did they originally combine both hunting and farming, emphasizing either as necessary in new environments?

There seem to be no clear answers to these questions from archaeology or linguistics, except that linguists have suggested an origin for the Proto-Algonquians somewhere close to the Great Lakes. Proto-Algonquian had no reconstructed maize-related vocabulary, so the expansion of the family cannot be related to a presence of this crop. Nevertheless, ancestral Algonquian-speaking populations undoubtedly spread on a large scale, mostly within the past 1,500 years, according to linguistic and archaeological dating estimates.[36] Did they begin as Eastern Woodland pre-maize farmers? If so, did their spread as hunter-gatherers into Canada owe its success to the same warm medieval weather conditions that allowed the Inuit to settle the Canadian Arctic coast as the sea ice retreated? Interesting questions indeed.

In the case of the Uto-Aztecan language family, which spread from Mesoamerica to the northern Great Basin and almost to the Canadian border, the answers are clearer than they are for Algonquian. Linguist Jane Hill has located the origin of this language family among early maize farmers in central Mesoamerica, between 5,000 and 4,000 years ago.[37] Her results are based on comparisons of Uto-Aztecan languages in the US Southwest (including Hopi in Arizona) with their relatives in Mexico, including Nahua (the Aztec language). They countermand, in my opinion decisively, an older linguistic view that the Proto-Uto-Aztecans began as hunter-gatherers in the US Southwest, migrated into Mesoamerica, adopted maize farming, and then migrated back into the US Southwest and started building pueblo villages. There is satisfaction in an economical explanation, and the archaeological record in Arizona and northern Mexico is in strong support of Jane Hill's views.

Until the late 1990s, it was also believed by many archaeologists that maize farming spread from Mesoamerica into the US Southwest only around 400 BCE, becoming adopted there by hunter-gatherers with no incoming population movement. As a result of recent excavations in northern Mexico and southern Arizona, many in connection with road and urban construction projects, this picture has changed dramatically in recent years.

A number of archaeological sites located on flood plains in the Tucson region of southern Arizona indicate an arrival of maize from Mesoamerica before 2000 BCE, in association by 1500 BCE with multihectare landscapes of small square fields fed by irrigation channels linked to nearby rivers. The associated settlements have sunken circular house floors and occasional sherds of pottery, the latter becoming more common after 1500 BCE.[38] Contemporary sites in northern Mexico have extensive areas of terracing for housing and cultivation, and both complexes have ample evidence for maize cultivation and its storage in underground bell-shaped pits. Remains of squash, tobacco, and cotton reinforce the agricultural nature of these sites. There was also a decrease in the number of large hunted animals over time, suggesting increasing pressure on the environment by a growing human population.

By 1500 BCE, maize cultivation had spread over much of the US Southwest, including high plateau country in Colorado and New Mexico. It is likely that its introduction came with the arrival of a Uto-Aztecan-speaking population from northern Mesoamerica, one ancestral to the ancient Pueblo-building peoples of the Southwest, including the modern Hopi. Because much of lowland northern Mexico is arid, the most likely route of maize introduction would have been through the better-watered regions of the Sierra Madre Occidental, inland from the Gulf of California.[39]

What is so surprising about the introduction of maize agriculture into the US Southwest is its early date, coming not long after the beginnings of village life in Mesoamerica itself. Indeed, it is likely that the irrigation channels excavated in Arizona are older than any excavated so far in Mesoamerica, as also may be the sherds of pottery, although this need not mean that the ideas of irrigation and pottery were invented first of all in Arizona. Much older pottery, for instance, occurs in hunter-gatherer contexts in Florida and Venezuela. Over most of northern Mexico, except for sites located fairly close to the US border, the archaeological record prior to 2000 BCE is poorly known, so new discoveries are always likely.

The prehistoric Algonquians and Uto-Aztecans were internally unified by their possession of closely related languages yet quite diverse in their economies, which ranged from hunting and gathering to food production with maize. Does this mean that they switched between both modes of subsistence with ease during prehistory? I suspect not.

The Uto-Aztecans of the Great Basin and southeastern California, people such as the Paiute, Shoshone, and Chemehuevi, migrated into regions where there was insufficient rainfall to support agriculture. They had little choice but to become hunter-gatherers unless they returned to where they had come from. They made their choices around 2,000 years ago, at least according to modern perspectives from both linguistics and ancient DNA.[40]

Similar choices (not always free ones) were made by many other North American peoples, especially after European contact, when so much agricultural land in the Eastern Woodlands became occupied by invading European settlers and the original inhabitants had to travel west to escape the onslaught. Like the Great Basin, the Great Plains were too dry for prehistoric North American agriculture because they occupy a rain shadow zone east of the Rocky Mountains. However, many Native American groups moved onto them during the eighteenth and nineteenth centuries as their farmlands east of the Mississippi were expropriated. Here, they were able to hunt buffalo on horseback, horses having been reintroduced into the Americas by the Spanish after their extinction at the end of the Pleistocene.

Many Siouan, Algonquian, and Caddoan peoples moved on to the central Plains in this way, likewise the Uto-Aztecan-speaking Comanche who occupied the southern Plains. The Hollywood Western movie genre owes a great deal to them, even if many film connoisseurs perhaps do not realize that this equestrian way of life does not truly represent Native American prehistory.

13

Ape to Agriculture

I HAVE NOW FINISHED my treatment of the Five-Million-Year Odyssey. The world that we inhabit today reflects events that occurred *not only* during the last few thousand years of recorded history. It was created on a more fundamental level by the achievements of people who lived long before that history was even dreamed of. These people remained hidden from knowledge until recently, but they and their creations are now being brought back into the spotlight through research by thousands of committed scholars across the world.

If we look back at the hominin and human past as a whole, we recognize points within its progression that appear to have seen decisive change. They include the primary separation of hominins from ancestral chimpanzees and bonobos, the first hominin migrations out of Africa, the emergence of the genus *Homo*, the emergence of the species *Homo sapiens* and its own eventual migration out of Africa, and, not least, the several developments of food production and the migrations consequent upon them. Each of these developments stands out as a point of no return, propelling hominins and their human successors onward in new directions.

Can we use knowledge of the hominin and human past to project our common future? I suspect that we cannot in any precise way, although perspectives for the future that take account of past failures can sometimes be comforting, especially in a world that remains as unstable today as it has ever been, if not more so. My own perspective on this is not that knowledge of our past will help us determine our future in any meaningful way, but rather that it will make us aware of our common humanity. One result of such realization should be that humans will control their worst outbursts of ethnic and racial suspicion and hatred, and work to ensure equal access by all peoples to the world's ample resources.

Most relevant as a conclusion for this book is a recapitulation of some of the major issues that arise in our understanding of human prehistory. I cannot resolve these issues to the satisfaction of everyone, and all of them involve research in multiple fields of study. I suspect we will never know the answers to many of them, a situation that undoubtedly increases their attraction as sources of debate. If we knew everything about the human past, curiosity would die, and boredom would reign supreme.

The first issue is that of the separation between hominins and panins. Where and how did it happen, and when? The exposed Rift Valley sediments and South African caves that have produced so much of the early hominin fossil record do not contain accessible deposits of the required antiquity, this being between 5 and 10 million years ago. Paleoanthropologists cannot simply rush out to known potential field sites and expect to hit the jackpot. Nothing can be done to increase the speed of recovery; indeed, there may never be any recovery of the smoking gun of fossil remains, a perfect common ancestor for both hominins and panins.

A second question asks which Pliocene hominins in Africa provided the ancestry for the first members of the genus *Homo* between 2.5 and 2 million years ago. Ancient DNA and protein research could, in theory, answer this question, but with fossils in which the original bone has been completely mineralized the prospects of extracting ancient DNA and proteins are zero. So far, the oldest ancient DNA recovered from hominin bone only dates to about 400,000 years ago, from Sima de los Huesos in Spain, although the protein record now extends back much earlier, even to Dmanisi at 1.7 million years ago in the case of its animal species. We can only wait and see what the future holds.

Once early Pleistocene *Homo* had come into existence, its first transformative activity was to seek an exit from Africa. Exactly when this auspicious event occurred is debatable because the available dates from scientific analysis do not come directly from the stone tools or fossils in question but from the sediments that contain them, and here there can be major issues over depositional context. The oldest hominin remains outside Africa are those from Dmanisi in Georgia, at possibly 1.7 million years ago. The oldest stone tools are claimed from China, at slightly over two million years ago. Are these dates "correct"? I suspect there will never be a simple answer.

Regardless of date, the hominin exit from Africa was probably one of the most decisive events in the evolution of the emergent genus *Homo*, unless one takes seriously the Eurasian alternative for hominin origins that I mentioned in chapter 2 (and I find myself unable to do this with full confidence, in terms of present evidence). Suddenly, the extent of the hominin playground was almost trebled. Africa covers about 30 million square kilometers and Eurasia about 55 million square kilometers, although not all of it is easy for human life, even today. The eventual result in both Africa and Eurasia was a proliferation of species, some evolving into descendants with larger brains, others remaining small-brained in their isolation.

What was the real status as separate breeding populations of those several Pleistocene hominin species that are named and recognized as distinct by paleoanthropologists? Did they become increasingly bounded with the passage of time and eventually unable to reproduce fertile offspring between each other? The evolutionary sciences suggest that this would often have been the case. In other circumstances, they might have maintained mutual fertility through gene flow, triggered especially when members of one species migrated deep into the territory of another.

The fossil and genetic perspectives inform us that both possibilities were probably in play, meaning, for example, that *Homo sapiens* could both interbreed with Neanderthals and Denisovans, yet also replace them as independent species some 700,000 years after their shared genetic origin. Unfortunately, such observations are only possible at present for hominins of the Middle and Late Pleistocene because of the absence of an ancient DNA record from older time periods.

From the perspective presented in this book, the course of evolution within the genus *Homo* during the past two million years suggests an initial expansion out of Africa of small-bodied and small-brained hominins, as identified in Georgia, China, and Java, with migratory offshoots eventually entering states of genetic isolation in Flores and Luzon. Possibly, the same happened in parts of southern Africa with ancestral *Homo naledi*, although here the isolation is harder to explain.

After one million years ago, another burst of migration carried larger-brained Middle Pleistocene populations throughout Africa and much of Eurasia, impinging on the territories of the earlier hominin populations, especially *Homo erectus*. By 300,000 years ago, this Middle Pleistocene ancestor had differentiated into early *Homo sapiens* in Africa, and Neanderthals and Denisovans in Eurasia. All three maintained an ability to interbreed until the latter two finally became extinct in the Late Pleistocene. The shadowy possibility arises that Denisovans were also able to reach New Guinea and Australia prior to the arrival of the *sapiens* ancestors of the modern populations of these regions.

So far so good, but then we reach *Homo sapiens*. Here, the picture becomes confused because of the apparent conflict between the paleoanthropological record, which favors movement out of Africa into Europe and the Middle East more than 100,000 years ago, and the genetic record, which dates this movement to under 70,000 years ago. This implies that the first *sapiens* populations to leave Africa were not successful in contributing their genes to living human populations today. Herein lies a degree of mystery. Perhaps Neanderthals provided more resistance to the immigrants than their popular image among the general public might suggest.

The settlements of Australia and Eurasia in combination, however, do clarify one observation about early *Homo sapiens*. The oldest stone tools that are definitely associated with *Homo sapiens* skeletal remains outside Africa, for instance in the Levant and Australia, carried Middle Paleolithic technological characteristics, implying that the archaeological category of "Upper Paleolithic," at least in terms of its stone tools, developed after the *sapiens* dispersal into Eurasia. Whether the Upper Paleolithic stone tool forms, such as backed blades and bifacial points, spread from one or more than one region of origin remains debated.

However, in Eurasian chronological terms, the Upper Paleolithic only appeared after 47,000 years ago, and its stone tool repertoire appears to have been an adaptation to the requirements of life in colder climates. There is only one occurrence of Upper Paleolithic tool types that is definitely older than 50,000 years and that occurred in southern Africa, where it overlapped with a Sub-Saharan genetic and skeletal origin for *Homo sapiens* as a separate species. Whether the African and Eurasian occurrences of Upper Paleolithic technology were connected through cultural contact is uncertain; most archaeologists, I suspect, would prefer to keep an open mind on the matter until more evidence from northern Africa is available.

As I have stressed, however, there is far more to the Upper Paleolithic than just a collection of backed and bifacial stone tools. Art, purposeful burial, personal vanity (body ornaments), and ocean crossings out of sight of land all made early appearances at around this time, although there are debates about whether these were purely *sapiens* attributes or whether they were also present among older hominin species such as Neanderthals.

The Madjedbebe edge-ground axes from northern Australia, stated to be between 65,000 and 53,000 years old, also raise some puzzling questions. Edge-ground axes are not found anywhere else in the world at such an early date. Are they correctly dated? And, if so, do they herald an arrival of *Homo sapiens* into Australia that was technologically independent of both the Middle and the Upper Paleolithic as formulated by archaeologists elsewhere? Was this a kind of third stream, perhaps, with edge grinding but otherwise Middle Paleolithic cores and flakes? I will have to keep an open mind on this possibility.

Whatever was going on with the initial dispersal of *Homo sapiens* out of Africa, one result is clear. They had succeeded in settling all habitable landscapes in the Old World by 40,000 years ago, in the process removing older hominin species from the scene, with or without interbreeding. Modern human ancestors had won the battle, if it is wise to see it in that light, and had taken over the Old World as the sole species of hominin left in existence. And they did not stop there.

By about 15,500 years ago, the Americas were actively under settlement from northeastern Asia, as the western coastline of Canada became released from its ice cover by the retreat of the last Ice Age. Another 42.5 million square kilometers of land quickly felt the passage of human feet, soon with a few canine paws in tow. By 14,500 years ago, all of the continents of the earth apart from Antarctica were feeling the impact of a human presence.

There is one final point to make about the Paleolithic. To reach Australia, Japan, and the Americas, people traveled in boats. But, in this regard, let us not forget the tiny hominins of Luzon and Flores. They remind us that crossing the sea was not just an achievement of the First Australians, the First Americans, or Neolithic farmers, although it was to be Neolithic peoples who eventually developed the technology to sail right across the Pacific.

Did Food Production Change the Rules?

Between 10,000 and 2000 BCE, different peoples in many parts of the world developed food production. The results were increasing human numbers and increasing desires for productive arable and pasture lands. The lid was lifted thereby on the constraints that had kept hominins and humans relatively low in numbers during the previous five million years.

Indeed, these increasing population numbers, as observed in the archaeological record, are in themselves a useful clue to the successful development of food production, especially useful in situations with poor preservation of organic remains of plants and animals. Other clues to the existence and success of ancient food production lie in the expansion histories of the languages spoken and the language families created by early farmers in many parts of the world.

With food production, migrations began that were eventually to distribute the major agriculturalist populations of the world to their pre-1492 CE limits, together with the languages and material cultures that were attached to them. However, lest there be any misunderstanding, no claim is made that the inceptions of food production necessarily led directly to the structure of the human population that existed in 1492 CE.

Many population and language family expansions occurred after those inceptions, and older human landscapes will have been wholly or partially erased on many occasions.

Furthermore, major migrations tend to occur in cycles, with periods of quiescence followed by sometimes rapid and long-distance movement. For instance, Polynesians did not settle the vast expanses of East Polynesia until 3,000 years after their ancestors began to move out of Taiwan. Sinitic populations did not settle southern China until the Han Dynasty, this being perhaps 4,000 years after their languages began to diversify close to the Yellow River. Turkic and Tungusic peoples in Asia began their expansions about 2,000 years ago, as also did the ancestral speakers of the Romance languages in Europe under the Roman Empire, many millennia after the likely origins of the Transeurasian and Indo-European families to which their languages belonged.

These expansions were all outer branches on much larger genealogies of human population and language, and sometimes they obscured or erased what went before. Older landscapes of humanity sometimes disappeared altogether, as in the cases of the Anatolian and Tocharian languages within the Indo-European family. Such population replacements occurred on their greatest recorded scales during the European Colonial Era, when background causes are not difficult to identify because of the existence of a detailed historical record.

Many archaeologists have questioned the significance of agriculture in human prehistory, suggesting that it was not as significant as is often claimed. I suspect they overlook the significance of the vast population and linguistic movements that I have documented in this book, especially in its second half. Did humans really make a bad mistake by entering a state of food production, unknowingly choosing disease, oppression, and hard labor, as opposed to the free and easy life of an affluent hunter-gatherer? We need to be fairly dispassionate about this because it is clear that both lifestyles had their ups and downs.

From my perspective, the demographic potential of food production drove the creation of much of the human pattern that occupies our world today. That pattern is visible to us through the distribution of the world's major biological populations and language families. It exists on a scale never created at any other time in the prehistory of *Homo sapiens*, at least not since the initial expansion of our species through and out of Africa.

Overall, therefore, food production was a success, at least demographically. Farmers had a definite edge over hunter-gatherers with their transportable economies of domesticated plants and animals, and also an economy that could be intensified by artificial means to increase production.

However, we must remember that hunter-gatherers sometimes also undertook significant migrations, but for the most part into territories that were not previously occupied or that had been previously depopulated. This is clear in the cases of the relatively recent Inuit and Apachean migrations in North America. The Pama-Nyungan language expansion in Australia is puzzling in this regard because there is no evidence for depopulation prior to its occurrence or for significant population movement attached to it. A full explanation is still lacking, especially in the absence of a sufficient record from ancient DNA.

What to conclude? By the time Alexander the Great had conquered his way from Macedonia to the Indus River, just before 320 BCE, the "big" patterns within humanity, which continued to exist until 1492 CE and the Columbian Exchange, had been in place for many millennia, except in remote regions of recent human settlement such as Madagascar and the outer islands of Polynesia. Neither Alexander nor the Romans changed much in terms of the distributions of peoples and languages across the surface of the earth, and neither did Genghis Khan with his Mongol armies.

In other words, all of us alive today have inherited a world created in its most basic foundations by our prehistoric ancestors, people who lived long before writing, statehood, and empires, and in most cases long before great cities, metallurgy, and wheeled vehicles. These hunter-gatherers and subsistence farmers, with their lifestyles of kinship-based tribalism sometimes mixed with instances of striking inequality, contributed far more to our humanity and diversity today than many of us perhaps realize.

So here comes a final question. Has there been one factor above all others that can stand as the most significant cause, among many, for the current state of humanity and for our fraught relationship with our rather subjugated natural world? Some factors are obvious: becoming human in the first place, becoming sentient, learning how to migrate into new environments, and learning how to produce food. But there is one factor that, for me, has always stood out: the ability of our species to increase its numbers, often on a remarkable scale for a large-brained species such as *Homo sapiens*.

Humans in certain circumstances are capable of procreative exploits, when opportunities arise, that rank us with, dare I say it, some of our fabled leporine friends. This is not to say that human movement has always been a reflection of overpopulation. Ancient migrations, I suspect, often began with small numbers of people but then roared into action as these people found themselves in new environments with access to greatly increased resources. The population histories of continents such as Australia and the Americas since 1492 CE make such abilities rather obvious.

In chapter 2, I referred to the recent book by Paul Morland, *The Human Tide: How Population Shaped the Modern World*. On the back cover of the book I found a clear and unambiguous statement: "Modern history is the story of global population change." From my deep-past perspective, I would conclude, "The whole of human history has been a story of population change." Of course, population change was not always "global" in the way that it has been since 1492, but the message still comes through with clarity.

Our world has limited resources, and those resources are currently under pressure. We must share them with each other more evenly. We must also escape from our twentieth-century belief, at least among capitalist nations, that the human future must be one of constant growth. As we proceed into the twenty-first century, global population growth may well be slowing down. But we still have much to do if future generations are to respect us.

ACKNOWLEDGMENTS

FIRST OF ALL, I thank my wife, Claudia Morris. She has been my main editor throughout the preparation of this manuscript and has kept me on track to the finish. I also thank the following colleagues who have discussed specific issues with me: Debbie Argue, Katharine Balolia, Murray Cox, Noreen Cramon-Taubadel, Norman Hammond, Paul Heggarty, Mark Hudson, Philip Piper, Cosimo Posth, Keith Prufer, David Reich, Martine Robbeets, Paul Sidwell, Pontus Skoglund, Chris Stringer, and Peter Sutton. Not all necessarily agree with what I say here, and I look forward to much further discussion. Philip Piper also kindly read the whole manuscript, and Katharine Balolia the first four chapters. Maggie Otto at the Australian National University helped with studio photography of artifacts and crania. Much of this manuscript has been written in the era of COVID-19, when face-to-face discussion with colleagues has been difficult.

It would be churlish of me not to thank the various online sources that I have used for bibliographic checking and quick facts: Academia, Google Scholar, and Wikipedia bulk quite large in this respect. Needless to say, however, my primary research has used other sources, available to me through the library of my home institution, the Australian National University. I list these sources in the notes for each chapter. Most of the illustrations were prepared by me in Adobe Illustrator 2020, using world base maps provided to me in 2012 by CartoGIS Services in the School of Pacific and Asian Studies at the Australian National University. The photos of archaeological sites were mostly taken by me, except where otherwise indicated.

I also wish to thank the team of people in the United States who have brought this book to fruition: my literary agent, James A. Levine of Levine Greenberg Rostan Literary Agency in New York; Alison Kalett, Hallie Schaeffer, Elizabeth Byrd, and Dimitri Karetnikov at Princeton University Press; and John Donohue and Vickie West at Westchester Publishing Services.

NOTES

Journal Abbreviations

AAS	*Archaeological and Anthropological Sciences*
AJPA	*American Journal of Physical Anthropology*
JAS	*Journal of Archaeological Science*
JHE	*Journal of Human Evolution*
JICA	*Journal of Island and Coastal Archaeology*
PNAS	*Proceedings of the National Academy of Sciences of the United States of America*
PTRSB	*Philosophical Transactions of the Royal Society B: Biological Sciences*
QSR	*Quaternary Science Reviews*

Preface

1. Clare Goff, *An Archaeologist in the Making: Six Seasons in Iran* (Constable, 1980).

2. For instance, New Zealand Maori genealogies that recorded the arrival of most of their ancestors from "Hawaiki" around 1250 CE are strongly supported by modern archaeology and radiocarbon dating.

Chapter 1

1. Charles Darwin, *The Descent of Man, and Selection in Relation to Sex* (John Murray, 1870), 689.

2. Use of the term "panin" for these two closely related species is derived from the name of their common genus, *Pan*. It was suggested to me by my colleague at the Australian National University, Katharine Balolia.

3. Jared Diamond, *The Rise and Fall of the Third Chimpanzee* (Vintage Books, 1992).

4. Simon Lewis and Mark Maslin, *The Human Planet: How We Created the Anthropocene* (Yale University Press, 2019), 3.

5. Paul Morland, *The Human Tide: How Population Shaped the Modern World* (John Murray, 2019).

Chapter 2

1. Sergio Almécija et al., "Fossil apes and human evolution," *Science* 372 (2021): eabb4363.

2. Thomas Mailund et al., "A new isolation with migration model along complete genomes infers very different divergences processes among closely related great ape species," *PLoS*

Genetics 8 (2012): e1003125; Yafei Mao et al., "A high-quality bonobo genome refines the analysis of human evolution," *Nature* 594 (2021): 77–81.

3. Frederick Engels, *The Origin of the Family, Private Property, and the State* (1884; repr., Pathfinder Press, 1972).

4. Graeme Ruxton and David Wilkinson, "Avoidance of overheating and selection for both hair loss and bipedality in humans," *PNAS* 108 (2011): 20965–20969.

5. Milford Wolpoff, *Paleoanthropology*, 2nd ed. (McGraw-Hill College, 1999), 222. See also Ian Tattersall, *Masters of the Planet: The Search for Our Human Origins* (Palgrave Macmillan, 2012), for a good discussion of this topic.

6. Madelaine Böhme et al., "A new Miocene ape and locomotion in the ancestor of great apes and humans," *Nature* 575 (2019): 489–493.

7. Madelaine Böhme et al., *Ancient Bones: Unearthing the Astonishing New Story of How We Became Human* (Scribe, 2020).

8. Scott Williams et al., "Reevaluating bipedalism in *Danuvius*," *Nature* 586 (2020): E1–E3; see also Almécija et al., "Fossil apes."

9. Mao et al., "High-quality bonobo genome."

10. A mutation is a change in DNA sequence that normally occurs during reproduction through errors in DNA replication. Selection is basically the differential contribution to future generations of those who, for whatever reason (environmental, sexual, cultural), reproduce offspring the most prolifically. Genetic drift occurs through random fluctuations in gene frequencies from one generation to the next.

11. Trenton Holliday et al., "Right for the wrong reasons: Reflections on modern human origins in the post-Neanderthal genome era," *Current Anthropology* 55 (2014): 696–724.

12. Mailund et al., "New isolation."

13. For a review of recent early hominin finds in Africa, see Lauren Schroeder, "Revolutionary fossils, ancient biomolecules," *American Anthropologist* 122 (2020): 306–320.

14. As suggested by Clive Finlayson, *The Improbable Primate: How Water Shaped Human Evolution* (Oxford University Press, 2014).

15. Yohannes Haile-Selassie et al., "A 3.8-million-year-old hominin cranium from Woranso-Mille, Ethiopia," *Nature* 573 (2019): 214–219; Dean Falk, "Hominin brain evolution," in Sally Reynolds and Andrew Gallagher, eds., *African Genesis* (Cambridge University Press, 2012), 145–162.

16. Amélie Beaudet et al., "The endocast of StW 573 ('Little Foot') and hominin brain evolution," *JHE* 126 (2019): 112–123.

17. Jeremy DeSilva et al., "One small step: A review of Plio-Pleistocene hominin foot evolution," *AJPA* 168, suppl. 67 (2019): 63–140.

18. Matthew Skinner et al., "Human-like hand use in *Australopithecus africanus*," *Science* 347 (2015): 395–399.

19. Mark Grabowski et al., "Body mass estimates of hominin fossils and the evolution of human body size," *JHE* 85 (2015): 75–93.

20. Ignacio De la Torre, "Searching for the emergence of stone tool making in eastern Africa," *PNAS* 116 (2019): 11567–11569.

21. For Lokalelei, see H. Roche et al., "Early hominid stone tool production and technical skill 2.34 mya in West Turkana, Kenya," *Nature* 399 (1999): 57–60. For teacher-learner

interactions, see Dietrich Stout et al., "Archaeology and the origins of human cumulative culture," *Current Anthropology* 60 (2019): 309–340.

22. Jessica Thompson et al., "Origins of the human predatory pattern," *Current Anthropology* 60 (2019): 1–23.

23. Julio Mercader et al., "4,300-year-old chimpanzee sites and the origins of percussive stone technology," *PNAS* 104 (2007): 3043–3048.

24. Kenneth Oakley, *Man the Tool-Maker* (British Museum, 1949), is the classic text on this theme.

25. Thibaut Caley et al., "A two-million-year-long hydroclimatic context for hominin evolution in southeastern Africa," *Nature* 560 (2018): 76–79.

26. Brian Villmoare, "Early *Homo* at 2.8 ma from Ledi-Geraru, Afar, Ethiopia," *Science* 347 (2015): 1352–1355.

27. Frank Brown et al., "Early *Homo erectus* skeleton from west Lake Turkana, Kenya," *Nature* 316 (1985): 788–792. The skull of Nariokotome Boy was found in seventy fragments. Together with his young age at death, this makes a precise estimate of adult cranial capacity difficult. The skeleton was found in a layer of siltstone above a volcanic tuff dated by the potassium-argon method to 1.65 million years ago.

28. Paul Manger et al., "The mass of the human brain," in Sally Reynolds and Andrew Gallagher, eds., *African Genesis* (Cambridge University Press, 2012), 181–204.

29. Leslie Aiello and Peter Wheeler, "The expensive-tissue hypothesis," *Current Anthropology* 36 (1995): 199–223.

30. Engels, *Origin of the Family*, 176.

31. Robin Dunbar, *Human Evolution: Our Brains and Behavior* (Pelican, 2014).

32. Donna Hart and Robert Sussman, *Man the Hunted: Primates, Predators, and Human Evolution* (Westview, 2005).

33. Timothy Taylor, *The Artificial Ape: How Technology Changed the Course of Human Evolution* (Palgrave Macmillan, 2010).

34. Richard Wrangham, *Catching Fire: How Cooking Made Us Human* (Profile, 2009); R. Wrangham, "Control of fire in the Paleolithic," *Current Anthropology* 58, suppl. 16 (2017): S303–313. Burned bone and ash indicate a use of fire in Wonderwerk Cave, South Africa, at about one million years ago. This is probably the oldest evidence for human-associated fire known at present.

35. John Gowlett, "Deep roots of kin: Developing the evolutionary perspective from prehistory," in N. J. Allen et al., eds., *Early Human Kinship: From Sex to Social Reproduction* (Blackwell, 2008), 48.

36. Henry Bunn et al., "FxJj50: An Early Pleistocene site in northern Kenya," *World Archaeology* 12 (1980): 109–136.

37. Glynn Isaac, "Emergence of human behaviour patterns," *PTRSB* 292 (1981): 187.

Chapter 3

1. For detailed accounts of the climatic and sea level changes that occurred between glacials and interglacials, see Kurt Lambeck et al., "Sea level and global ice volumes from the Last Glacial Maximum to the Holocene," *PNAS* 111 (2014): 15296–15303; Andrea Dutton et al., "Sea-level rise due to polar ice-sheet mass loss during warm wet periods," *Science* 349, no. 6244 (2015): aaa4019;

Yusuke Yokoyama et al., "Rapid glaciation and a two-step sea level plunge into the Last Glacial Maximum," *Nature* 559 (2018): 603–607; Simon Lewis and Mark Maslin, *The Human Planet: How We Created the Anthropocene* (Yale University Press, 2019).

2. As suggested for the final Ice Age at 20,000 years ago by Geoff Bailey et al., "Coastlines, submerged landscapes, and human evolution: The Red Sea Basin and the Farasan Islands," *JICA* 2 (2007): 140.

3. A suggestion of back migration into Africa is made forcefully by Madelaine Böhme et al., *Ancient Bones: Unearthing the Astonishing New Story of How We Became Human* (Scribe, 2020), although in this case with a preference for an actual genesis of hominins in Europe rather than Africa.

4. As shown for phases of high rainfall over the past 400,000 years in Arabia by Huw Groucutt et al., "Multiple hominin dispersals into Southwest Asia over the past 400,000 years," *Nature* 597 (2021): 376–380.

5. As implied by Ofer Bar-Yosef and Miriam Belmaker, "Early and Middle Pleistocene faunal and hominin dispersals through southwestern Asia," *QSR* 30 (2011): 1318–1337; Hannah O'Regan et al., "Hominins without fellow travellers?," *QSR* 30 (2011): 1343–1352.

6. The dates come from magnetostratigraphy, electron-spin resonance, and mammalian biostratigraphy. See Mohamed Sahnouni et al., "1.9-million- and 2.4-million-year-old artifacts and stone tool-cutmarked bones from Ain Boucherit, Algeria," *Science* 362 (2018): 1297–1301.

7. Zhaoyu Zhu et al., "Hominin occupation of the Chinese Loess Plateau since about 2.1 million years ago," *Nature* 559 (2018): 608–612.

8. Rob Hosfield, "Walking in a winter wonderland," *Current Anthropology* 57 (2016): 653–682.

9. See the discussions of naive faunas by Margaret E. Lewis, "Carnivore guilds and the impact of hominin dispersals," and Robin Dennell, "Pleistocene hominin dispersals, naive faunas and social networks," in Nicole Boivin et al., eds., *Human Dispersal and Species Movement: From Prehistory to Present* (Cambridge University Press, 2017), 29–61 and 62–89.

10. Ann Gibbons, "The wanderers," *Science* 354 (2016): 959–961; David Lordkipanidze et al., "A complete skull from Dmanisi, Georgia," *Science* 342 (2013): 326–331.

11. The Indonesian Pleistocene is discussed in detail in Peter Bellwood, *First Islanders: Prehistory and Human Migration in Island Southeast Asia* (Wiley Blackwell, 2017).

12. Julien Louys and Patrick Roberts, "Environmental drivers of megafauna and hominin extinction in Southeast Asia," *Nature* 586 (2020): 402–406.

13. Shuji Matsu'ura et al., "Age control of the first appearance datum for Javanese *Homo erectus* in the Sangiran area," *Science* 367 (2020): 210–214. These authors claim an *erectus* arrival at Sangiran in central Java only 1.3 million years ago.

14. Marcia Ponce de Leon et al., "The primitive brain of early *Homo*," *Science* 372 (2021): 165–171.

15. Thomas Sutikna et al., "Revised stratigraphy and chronology for *Homo floresiensis* at Liang Bua in Indonesia," *Nature* 532 (2016): 366–369.

16. Gert van den Bergh et al., "*Homo floresiensis*–like fossils from the early Middle Pleistocene of Flores," *Nature* 534 (2016): 245–248.

17. Debbie Argue et al., "*Homo floresiensis*: A cladistic analysis," *JHE* 57 (2009): 623–629. See also the discussion by Debbie Argue, "The enigma of *Homo floresiensis*," in Bellwood, *First Islanders*, 60–64.

18. Jeremy DeSilva et al., "One small step: A review of Plio-Pleistocene foot evolution," *AJPA* 168 (2019): S67.

19. J. Tyler Faith et al., "Plio-Pleistocene decline of African megaherbivores: No evidence for ancient hominin impacts," *Science* 362 (2018): 938–941; and Louys and Roberts, "Environmental drivers," 402–406.

20. Böhme et al., *Ancient Bones*, 221.

21. Böhme et al., *Ancient Bones*, 226, suggests that both *floresiensis* and *luzonensis* were of Eurasian rather than African ultimate origin.

22. Florent Détroit et al., "A new species of *Homo* from the Late Pleistocene of the Philippines," *Nature* 568 (2019): 181–186. One tooth from Callao Cave is more than 130,000 years old according to recent uranium series dating (Rainer Grün, communication at International Conference on *Homo luzonensis* and the Hominin Record of Southeast Asia, Manila, February 2020).

23. Thomas Ingicco et al., "Oldest known hominin activity in the Philippines by 709,000 years ago," *Nature* 557 (2018): 232–237. The dating was by electron-spin resonance on tooth enamel and quartz.

24. It is likely also that pre-*sapiens* hominins were able to reach some of the non-landbridged islands of the Mediterranean, such as Crete and Naxos. See Andrew Lawler, "Searching for a Stone Age Odysseus," *Science* 360 (2018): 362–363.

25. A. P. Derevianko, *Three Global Human Migrations in Eurasia*, vol. 2, *The Original Peopling of Northern, Central and Western Central Asia* (Russian Academy of Sciences, 2017), 802.

Chapter 4

1. See the phylogenetic trees in Xijun Ni et al., "Massive cranium from Harbin in northeastern China," *The Innovation* 2, no. 3 (2021), 100130; Anders Bergström et al., "Origins of modern human ancestry," *Nature* 590 (2021): 229–237; Elena Zavala et al., "Pleistocene sediment DNA reveals hominin and faunal turnovers at Denisova Cave," *Nature* 595 (2021): 399–403.

2. Alan Thorne and Milford Wolpoff, "The multiregional evolution of humans," *Scientific American* 266, no. 4 (1992): 76–79, 82–83; Milford Wolpoff, *Paleoanthropology*, 2nd ed. (McGraw-Hill College, 1999). See also Sang-hee Lee, *Close Encounters with Humankind: A Paleoanthropologist Investigates Our Evolving Species* (W. W. Norton, 2018).

3. Svante Paabo, "The human condition—a molecular approach," *Cell* 157 (2014): 216–226.

4. For nongeneticists like me, mitochondria can be thought of as small engines in cells that contain enzymes for energy production. Only females normally transmit them to their male and female offspring (transmission by males apparently has been recorded but is rare), and only the daughters of those females transmit them further. Every now and again, nucleotides along the mitochondrial genes mutate during conception, forming new lineages (or haplotypes).

5. Rodrigo Lacruz et al., "The evolutionary history of the human face," *Nature Ecology and Evolution* 3 (2019): 726–736.

6. Frido Welker et al., "The dental proteome of *Homo antecessor*," *Nature* 580 (2020): 235–238.

7. José Maria Bermudez de Castro and Maria Martinon-Torres, "A new model for the evolution of the human Pleistocene populations of Europe," *Quaternary International* 295 (2013): 102–112; David Reich, *Who We Are and How We Got Here: Ancient DNA and the New Science of the Human Past* (Oxford University Press, 2018), 70. See also Madelaine Böhme et al., *Ancient Bones: Unearthing the Astonishing New Story of How We Became Human* (Scribe, 2020), for a suggestion that hominin evolution commenced in Eurasia rather than Africa.

8. Rainer Grün et al., "Dating the skull from Broken Hill, Zambia," *Nature* 580 (2020): 372–375.

9. A. P. Derevianko, *Three Global Human Migrations in Eurasia*, vol. 4, *The Acheulean and Bifacial Lithic Industries* (Russian Academy of Sciences, 2019).

10. Ceri Shipton, "The unity of Acheulean culture," in Huw Groucutt, ed., *Culture History and Convergent Evolution: Can We Detect Populations in Prehistory?* (Springer, 2020), 13–28.

11. Also sometimes spelled "Neandertal," after the modern German spelling of the Neander Valley.

12. Chris Stringer, "Evolution of early humans," in Steve Jones et al., eds., *The Cambridge Encyclopedia of Human Evolution* (Cambridge University Press, 1992), 248.

13. Martin Petr et al., "The evolutionary history of Neanderthal and Denisovan Y chromosomes," *Science* 369 (2020): 1653–1656. The Sima mitochondrial DNA lineages were replaced in Classic Neanderthals by lineages derived from interbreeding with *Homo sapiens*, as shown in figure 4.1, bottom diagram.

14. Nohemi Sala et al., "Lethal interpersonal violence in the middle Pleistocene," *PLoS One* 10, no. 5 (2015): e0126589.

15. Lu Chen et al., "Identifying and interpreting apparent Neanderthal ancestry in African individuals," *Cell* 180 (2020): 677–687.

16. Bergström et al., "Origins of modern human ancestry."

17. Benjamin Vernot et al., "Excavating Neanderthal and Denisovan DNA from the genomes of Melanesian individuals," *Science* 352 (2016): 235–239; João Teixeira et al., "Widespread Denisovan ancestry in Island Southeast Asia," *Nature Ecology and Evolution* 5 (2021): 616–624; Maximilian Larena et al., "Philippine Ayta possess the highest level of Denisovan ancestry in the world," *Current Biology* 31 (2021): 1–12.

18. The most readable report of this research can be found in Tom Higham, *The World before Us* (Viking, 2021), and technical details can be found in Zenobia Jacobs et al., "Timing of archaic hominin occupation of Denisova Cave in southern Siberia," *Nature* 565 (2019): 594–599; Katerina Douka et al., "Age estimates for hominin fossils and the onset of the Upper Palaeolithic at Denisova Cave," *Nature* 565 (2019): 640–644.

19. Viviane Slon et al., "The genome of the offspring of a Neanderthal mother and a Denisovan father," *Nature* 561 (2018): 113–116.

20. Fahu Chen et al., "A late Middle Pleistocene Denisovan mandible from the Tibetan Plateau," *Nature* 569 (2019): 409–412. The age comes from uranium series dating on an adhering carbonate matrix. See also D. Zhang et al., "Denisovan DNA in Late Pleistocene sediments from Baishiya Karst Cave," *Science* 370 (2020): 584–587.

21. Guy Jacobs et al., "Multiple deeply divergent Denisovan ancestries in Papuans," *Cell* 177 (2019): 1010–1021; Anders Bergström et al., "Insights into human genetic variation and population

history," *Science* 367 (2020): eaay5012; Diyendo Massilani et al., "Denisovan ancestry and population history of early East Asians," *Science* 370 (2020): 579–583; Larena et al., "Philippine Ayta."

22. Ni et al., "Massive cranium"; Qiang Ji et al., "Late Middle Pleistocene Harbin cranium represents a new *Homo* species," *The Innovation* 2, no. 3 (2021): 100132.

23. Ann Gibbons, "'Dragon Man' may be an elusive Denisovan," *Science* 373 (2021), 11–12; Bergström et al., "Origins of modern human ancestry."

24. Marie Soressi et al., "Neandertals made the first specialized bone tools in Europe," *PNAS* 110 (2013): 14186–14190; Jacques Jaubert et al., "Early Neanderthal constructions deep in Bruniquel Cave in southwestern France," *Nature* 534 (2016): 111–114; Tim Appenzeller, "Europe's first artists were Neandertals," *Science* 359 (2018): 853–853; Clive Finlayson, *The Smart Neanderthal: Bird Catching, Cave Art, and the Cognitive Revolution* (Oxford University Press, 2019).

25. Rebecca Wragg Sykes, *Kindred: Neanderthal Life, Love, Death and Art* (Bloomsbury, 2020).

26. Emma Pomeroy et al., "New Neanderthal remains associated with the 'flower burial' at Shanidar Cave," *Antiquity* 94 (2020): 11–26; Avraham Ronen, "The oldest burials and their significance," in Sally Reynolds and Andrew Gallagher, eds., *African Genesis: Perspectives on Hominin Evolution* (Cambridge University Press, 2012), 554–570.

27. Judith Beier et al., "Similar cranial trauma prevalence among Neanderthals and Upper Palaeolithic modern humans," *Nature* 563 (2018): 686–690.

28. Michael Balter, "The killing ground," *Science* 344 (2014): 1080–1083.

29. Kumar Akhilesh et al., "Early Middle Palaeolithic culture in India around 385–172 ka reframes Out of Africa models," *Nature* 554 (2018): 97–101.

30. Yan Rizal et al., "Last appearance of *Homo erectus* at Ngandong, Java, 117,000–108,000 years ago," *Nature* 577 (2020): 381–385. Note that this is not an extinction date for *erectus* but simply a youngest dated presence in Java.

31. Peter Bellwood, *First Islanders: Prehistory and Human Migration in Island Southeast Asia* (Wiley Blackwell, 2017).

32. According to uranium series and electron spin resonance dating. See Paul Dirks et al., "The age of *Homo naledi* and associated sediments in the Rising Star Cave, South Africa," *eLife* 6 (2017): e24231.

33. Lee Berger and John Hawks, *Almost Human: The Astonishing Tale of* Homo naledi *and the Discovery That Changed Our Human Story* (National Geographic, 2017).

Chapter 5

1. Paul Pettitt, "The rise of modern humans," in Chris Scarre, ed., *The Human Past: World Prehistory and the Development of Human Societies*, 4th ed. (Thames and Hudson, 2018), 117; John Hoffecker, *Modern Humans: Their African Origin and Global Dispersal* (Columbia University Press, 2017), table 4.3.

2. Carina Schlebusch et al., "Khoe-San genomes reveal unique variation and confirm the deepest population divergence in *Homo sapiens*," *Molecular Biology and Evolution* 37 (2020): 2944–2954.

3. Rebecca Cann, Mark Stoneking, and Alan Wilson, "Mitochondrial DNA and human evolution," *Nature* 325 (1987): 31–36.

4. Swapan Mallick et al., "The Simons Genome Diversity Project," *Nature* 538 (2016): 201–206; David Reich, *Who We Are and How We Got Here: Ancient DNA and the New Science of the Human Past* (Oxford University Press, 2018); Schlebusch et al., "Khoe-San genomes," 2944–2954.

5. Eva Chan et al., "Human origins in a southern African palaeo-wetland and first migrations," *Nature* 575 (2019): 185–189.

6. Carina Schlebusch et al., "Human origins in southern African palaeo-wetlands? Strong claims from weak evidence," *JAS* 130 (2021): 105374.

7. Mark Lipson et al., "Ancient West African foragers in the context of African population history," *Nature* 577 (2020): 665–670. For a similar chronological conclusion, see Anders Bergström et al., "Insights into human genetic variation and population history," *Science* 367 (2020): eaay5012.

8. Martin Petr et al., "The evolutionary history of Neanderthal and Denisovan Y chromosomes," *Science* 369 (2020): 1653–1656.

9. Reich, *Who We Are*, 52, 88. A *Homo sapiens* leg bone from Ust'-Ishim in Siberia provides radiocarbon dating and genetic evidence for an episode of *sapiens*-Neanderthal mixture at around 55,000 years ago; see Qiaomei Fu et al., "Genome sequence of a 45,000-year-old modern human from western Siberia," *Nature* 514 (2014): 445–451.

10. Katerina Harvati et al., "Apidima Cave fossils provide earliest evidence of *Homo sapiens* in Eurasia," *Nature* 571 (2019): 500–504.

11. Israel Herschkovitz et al., "The earliest modern humans outside Africa," *Science* 359 (2018): 456–459.

12. Bernard Vandermeersch and Ofer Bar-Yosef, "The Paleolithic burials at Qafzeh Cave, Israel," *Paleo* 30, no. 1 (2019): 236–275.

13. Avraham Ronen, "The oldest burials and their significance," in Sally Reynolds and Andrew Gallagher, eds., *African Genesis: Perspectives on Hominin Evolution* (Cambridge University Press, 2012), 554–570.

14. Maria Martinon-Torres et al., "Earliest known human burial in Africa," *Nature* 593 (2021): 95–100.

15. The Nesher Ramla cranial remains come from a sediment-filled limestone sinkhole at Nesher Ramla in central Israel. They date to about 140,000 years ago and were stated originally to be neither Neanderthal nor *sapiens*, even though they overlap in date with both of these species and were likewise associated with Levallois cores and flakes. See Israel Herschkovitz et al., "A Middle Pleistocene *Homo* from Nesher Ramla, Israel," *Science* 372 (2021): 1424–1428; Yossi Zaidner et al., "Middle Pleistocene *Homo* behavior and culture at 140,000 to 120,000 years ago and interactions with *Homo sapiens*," *Science* 372 (2021): 1429–1433. A more recent suggestion that the remains are unequivocally Neanderthal comes from Assaf Marom and Yoel Yak, Comment on "A Middle Pleistocene *Homo* from Nesher Ramla, Israel," *Science* 374, no. 6572 (2021), doi: 10.1126/science.abl4336.

16. Xue-feng Sun et al., "Ancient DNA and multimethod dating confirm the late arrival of anatomically modern humans in southern China," *PNAS* 118 (2021): e2019158118.

17. Petr et al., "Evolutionary history," 1653–1656.

18. Linda Schroeder, "Revolutionary fossils, ancient biomolecules," *American Anthropologist* 122, no. 2 (2020): 306–320. A similar view is presented by Aurélien Mounier and Marta Lahr, "Deciphering African late middle Pleistocene hominin diversity and the origin of our species," *Nature Communications* 10 (2019): 3406. See also Reich, *Who We Are*, chapter 9.

19. Sally McBrearty and Alison Brooks, "The revolution that wasn't: A new interpretation of the origin of modern human behavior," *JHE* (2000) 39: 453–563; Christopher Henshilwood et al., "An abstract drawing from the 73,000-year-old levels at Blombos Cave, South Africa," *Nature* 562 (2018): 115–118; Manuel Will et al., "Human teeth from securely stratified Middle Stone Age contexts at Sibudu, South Africa," *AAS* 11 (2019): 3491–3501.

20. Manuel Will et al., "Timing and trajectory of cultural evolution on the African continent 200,000–30,000 years ago," in Yonatan Sahle et al., eds., *Modern Human Origins and Dispersal* (Kerns Verlag, 2019), 25–72.

21. This point is also made by Robin Dennell, *From Arabia to the Pacific: How Our Species Colonised Asia* (Taylor and Francis, 2020).

22. Ian Gilligan, *Climate, Clothing, and Agriculture in Prehistory: Linking Evidence, Causes, and Effects* (Cambridge University Press, 2019); see also Peter Frost, "The original Industrial Revolution: Did cold winters select for cognitive ability?," *Psych* 1 (2019): 161–181; Hoffecker, *Modern Humans*, 66–67.

23. Stanley Ambrose, "Chronological calibration of Late Pleistocene modern human dispersals, climate change and Archaeology with geochemical isochrons," in Sahle et al., *Modern Human Origins*, 171–213.

24. See, for instance, Eugene Smith et al., "Humans thrived in South Africa through the Toba eruption about 74,000 years ago," *Nature* 555 (2018): 511–515; Michael Petraglia and Ravi Korisettar, "The Toba volcanic super-eruption," *Quaternary International* 258 (2012): 119–134. These authors favor only minor impact on human populations from the Toba eruption.

25. Alex Mackay et al., "Coalescence and fragmentation in the Pleistocene archaeology of southernmost Africa," *JHE* 72 (2014): 26–51.

26. Jean-Jacques Hublin et al., "Initial Upper Palaeolithic *Homo sapiens* from Bacho Kiro Cave, Bulgaria," *Nature* 581 (2020): 299–302.

27. Reich, *Who We Are*; Iñigo Olalde and Cosimo Posth, "Latest trends in archaeogenetic research of west Eurasians," *Current Research in Genetics and Development* 62 (2020): 36–43.

28. Michael Petraglia et al., "Middle Paleolithic assemblages from the Indian Subcontinent before and after the Toba eruption," *Science* 417 (2012): 114–116; Laura Lewis et al., "First technological comparison of southern African Howieson's Poort and South Asian microlithic industries," *Quaternary International* 350 (2014): 7–24.

29. Maxime Aubert et al., "Earliest hunting scene in prehistoric art," *Nature* 576 (2019): 442–445; Adam Brumm et al., "Oldest cave art found in Sulawesi," *Science Advances* 7 (2021): eabd4648.

30. Paul Mellars and Jennifer French, "Tenfold population increase in western Europe at the Neandertal-to-Modern human transition," *Science* 333 (2011): 623–637; Jennifer French, "Demography and the Palaeolithic archaeological record," *Journal of Archeological Method and Theory* 23 (2016): 150–199; M. Bolus, "The late Middle Palaeolithic and the Aurignacian of the Swabian Jura, southwestern Germany," in A. P. Derevianko and M. Shunkov, eds., *Characteristic Features of the Middle to Upper Palaeolithic Transition in Eurasia: Proceedings of the International Symposium "Characteristic Features of the Middle to Upper Paleolithic Transition in Eurasia— Development of Culture and Evolution of Homo Genus," July 4–10, 2011, Denisova Cave, Altai* (Department of the Institute of Archaeology and Ethnography [Novosibirsk], 2011), 3–10.

31. Tim Flannery, *Europe: A Natural History* (Grove, 2020).

32. Matthias Currat and Laurent Excoffier, "Strong reproductive isolation between humans and Neanderthals inferred from observed patterns of introgression," *PNAS* 108 (2011): 15129–15134. See also Reich, *Who We Are*, chapter 2, for a good explanation of Neanderthal and modern human hybridization issues.

33. Qiaomei Fu et al., "An early modern human from Romania with a recent Neanderthal ancestor," *Nature* 524 (2014): 445–449.

34. Hugo Zeberg and Svante Pääbo, "The major genetic risk factor for severe COVID-19 is inherited from Neanderthals," *Nature* 587 (2020): 610–612.

35. Anna Goldfield et al., "Modeling the role of fire and cooking in the competitive exclusion of Neanderthals," *JHE* 124 (2018): 91–104.

36. Kay Prufer, "The complete genome sequence of a Neanderthal from the Altai Mountains," *Nature* 505 (2014): 43–49; Sriram Sankararaman, "The genomic landscape of Neanderthal ancestry in present-day humans," *Nature* 507 (2014): 354–357; L. Rios et al., "Skeletal anomalies in the Neanderthal family of El Sidron (Spain) support a role of inbreeding in Neanderthal extinction," *Scientific Reports* 9 (2019): 1697.

37. Thomas Higham et al., "The timing and spatiotemporal patterning of Neanderthal disappearance," *Nature* 512 (2014): 306–309.

38. Dennell, *From Arabia to the Pacific*, 201.

39. For discussions of Paleolithic populations in eastern Asia, see Hirofumi Matsumura et al., "Craniometrics reveal two layers of prehistoric human dispersal in eastern Eurasia," *Scientific Reports* 9 (2019): 1451; Hirofumi Matsumura et al., "Female craniometrics support the 'two-later model' of human dispersal in eastern Eurasia," *Scientific Reports* 11 (2021): 20830; Melinda Yang et al., "Ancient DNA indicates human population shifts and admixture in northern and southern China," *Science* 369, no. 6501 (2020): 282–288; Chuan-chou Wang et al., "Genomic insights into the formation of human populations in East Asia," *Nature* 591 (2021): 413–419.

40. James O'Connell et al., "When did *Homo sapiens* first reach Southeast Asia and Sahul?," *PNAS* 115 (2018): 8482–8490.

41. Hugh Groucutt et al., "Stone tool assemblages and models for the dispersal of *Homo sapiens* out of Africa," *Quaternary International* 382 (2015): 8–30.

42. Milford Wolpoff and Sang-hee Lee, "WLH 50: How Australia informs the worldwide pattern of Pleistocene human evolution," *PaleoAnthropology* (2014): 505–564.

43. João Teixeira et al., "Widespread Denisovan ancestry in Island Southeast Asia," *Nature Ecology and Evolution* 5 (2021): 616–624.

44. Chris Clarkson et al., "Human occupation of northern Australia by 65,000 years ago," *Nature* 547 (2017): 306–310.

45. O'Connell et al., "When did *Homo sapiens*."

46. Jeremy Choin et al., "Genomic insights into population history and biological adaptation in Oceania," *Nature* 592 (2021): 583–589. They suggest a date of only 45,000 to 30,000 years ago for the first *Homo sapiens* settlement of New Guinea.

47. Nicole Pedro et al., "Papuan mitochondrial genomes and the settlement of Sahul," *Journal of Human Genetics* 65 (2020): 875–887; Gludhug Purnomo et al., "Mitogenomes reveal two major influxes of Papuan ancestry across Wallacea," *Genes* 12 (2021): 965.

48. Michael Bird et al., "Early human settlement of Sahul was not an accident," *Scientific Reports* 9 (2019): 8220.

49. Sue O'Connor et al., "Pelagic fishing at 42,000 years before the present," *Science* 334 (2011): 1117–1120; Peter Bellwood, ed., *The Spice Islands in Prehistory: Archaeology in the Northern Moluccas, Indonesia* (Australian National University Press, 2019).

50. In favor of human-caused extinctions, see Frédérick Saltré et al., "Climate change not to blame for late Quaternary megafauna extinctions in Australia," *Nature Communications* 7 (2015): 10511; Susan Rule et al., "The aftermath of megafaunal extinction: Ecosystem transformation in Pleistocene Australia," *Science* 335 (2012): 1483–14867. For an argument against the idea, see Julien Louys et al., "No evidence for widespread island extinctions after Pleistocene hominin arrival," *PNAS* 118 (2021): e2023005118.

51. Tim Flannery, *The Future Eaters: An Ecological History of the Australasian Lands and People* (Grove, 2002); Raquel Lopes dos Santos et al., "Abrupt vegetation change after the Late Quaternary megafaunal extinction in southeastern Australia," *Nature Geoscience* 6 (2013): 627–631.

52. Peter Bellwood, *First Migrants: Ancient Migration in Global Perspective* (Wiley Blackwell, 2013), 74.

53. Corey Bradshaw et al., "Minimum founding populations for the first peopling of Sahul," *Nature Ecology and Evolution* 3 (2019): 1057–1063; Jim Allen and James O'Connell, "A different paradigm for the initial colonization of Sahul," *Archaeology in Oceania* 55 (2020): 1–14.

54. O'Connell et al., "When did *Homo sapiens*"; Michael Bird et al., "Palaeogeography and voyage modeling indicates early human colonization of Australia was likely from Timor-Roti," *QSR* 191 (2018): 431–439.

55. Norma McArthur et al., "Small population isolates: A micro-simulation study," *Journal of the Polynesian Society* 85 (1976): 307–326.

56. The murderous story of the early years of the Pitcairn colony was reconstructed by Sir John Barrow in *The Mutiny of "The Bounty"* (1831; repr., Oxford University Press, 1960). See also John Terrell, *Prehistory in the Pacific Islands* (Cambridge University Press, 1986), 191; Bellwood, *First Migrants*, 75.

57. Joseph Birdsell, "Some population problems involving Pleistocene man," *Cold Spring Harbor Symposium on Quantitative Biology* 22 (1957): 47–69.

58. Many archaeologists argue strongly for survival of these early modern human populations into the modern human genetic and cultural landscape, even though the idea is resisted strongly by geneticists. I favored this viewpoint in my *First Migrants*. See also Nicole Boivin et al., "Human dispersal across diverse environments of Asia during the Upper Pleistocene," *Quaternary International* 300 (2013): 32–47; Hugo Reyes-Centeno et al., "Genomic and cranial phenotype data support multiple modern human dispersals from Africa and a southern route into Asia," *PNAS* 111 (2014): 7248–7253; Huw Groucutt et al., "Skhul lithic technology and the dispersal of *Homo sapiens* into Southwest Asia," *Quaternary International* 515 (2019): 30–52; Ryan Rabett, "The success of failed *Homo sapiens* dispersals out of Africa and into Asia," *Nature Ecology and Evolution* 2 (2018): 212–219.

59. Philip Habgood and Natalie Franklin, "The revolution that didn't arrive: A review of Pleistocene Sahul," *JHE* 55 (2008): 187–222; Maxime Aubert et al., "Palaeolithic cave art in Borneo," *Nature* 564 (2018): 254–257; Michelle Langley et al., "Symbolic expression in Pleistocene Sahul, Sunda and Wallacea," *QSR* 2019 (2019): 105883. At present, the oldest rock art in

Australia is about 17,000 years old; see Damien Finch et al., "Ages for Australia's oldest rock paintings," *Nature Human Behaviour* 5 (2021): 310–318.

60. Tingkayu is discussed, with references, in Peter Bellwood, *First Islanders: Prehistory and Human Migration in Island Southeast Asia* (Wiley Blackwell, 2017), 143–145.

Chapter 6

1. Vladimir Pitulko et al., "Early human presence in the Arctic: Evidence from 45,000-year-old mammoth remains," *Science* 351 (2016): 260–263.

2. Nicolas Zwyns et al., "The northern route for human dispersal in central and northeast Asia," *Scientific Reports* 9 (2019): 11759.

3. Yusuke Yokoyama et al., "Rapid glaciation and a two-step sea level plunge into the Last Glacial Maximum," *Nature* 559 (2018): 603–607.

4. Melinda Yang et al., "40,000-year-old individual from Asia provides insight into early population structure in Eurasia," *Current Biology* 27 (2017): 3202–3208.

5. Xiaowei Mao et al., "The deep population history of northern East Asia," *Cell* 184, no. 12 (2021): 3256–3266.e13.

6. X. Zhang et al., "The earliest human occupation of the high-altitude Tibetan Plateau 40 thousand to 30 thousand years ago," *Science* 362 (2018): 1049–1051; Emilia Huerta-Sanchez et al., "Altitude adaptation in Tibetans caused by introgression of Denisovan-like DNA," *Nature* 512 (2014): 194–197.

7. Vladimir Pitulko et al., "The oldest art of the Eurasian Arctic," *Antiquity* 86 (2012): 642–659; Pavel Nikolskiy and Vladimir Pitulko, "Evidence from the Yana Palaeolithic site, Arctic Siberia, yields clues to the riddle of mammoth hunting," *JAS* 40 (2013): 4189–4197; Martin Sikora, "The population history of northeastern Siberia since the Pleistocene," *Nature* 570 (2019): 182–188.

8. Sikora et al., "Population history of northeastern Siberia"; Naoki Osada and Yosuke Kawai, "Exploring models of human migration to the Japanese archipelago," *Anthropological Science* 129 (2021): 45–58.

9. Robin Dennell, *From Arabia to the Pacific: How Our Species Colonised Asia* (Taylor and Francis, 2020), 318.

10. John Hoffecker et al., "Beringia and the global dispersal of modern humans," *Evolutionary Anthropology* 25 (2016): 64–78. Horse and caribou bones with cut marks potentially made by humans are dated to 24,000 years ago in Bluefish Caves in Yukon Territory, but without any definite artifacts. See Lauriane Bourgeon et al., "Earliest human presence in North America dated to the Last Glacial Maximum," *PLoS One* 12 (2017): 0169486.

11. Ted Goebel and Ben Potter, "First traces: Late Pleistocene human settlement of the Arctic," in T. M. Friesen and O. Mason, eds., *The Oxford Handbook of the Prehistoric Arctic* (Oxford Handbooks Online, 2016), 223–252.

12. Michael Waters, "Late Pleistocene exploration and settlement of the Americas by modern humans," *Science* 365 (2019): eaat5447.

13. Yousuke Kaifu et al., "Palaeolithic seafaring in East Asia," *Antiquity* 93 (2019): 1424–1441; Dennis Normile, "Update: Explorers successfully voyage to Japan," *Science News*, July 20, 2019, https://www.sciencemag.org/news/2019/07/explorers-voyage-japan-primitive-boat-hopes

-unlocking-ancient-mystery; Yousuke Kaifu et al., "Palaeolithic voyage for invisible islands beyond the horizon," *Scientific Reports* 10 (2020): 19785.

14. On Japan and Sakhalin as potential Last Glacial Maximum refuge regions from Last Glacial Maximum cold, see Kelly Graf, "The good, the bad, and the ugly," *JAS* 36 (2009): 694–707; Peter Bellwood, *First Migrants: Ancient Migration in Global Perspective* (Wiley Blackwell, 2013), 81–83.

15. Kazuki Morisaki and Hiroyuki Sato, "Lithic technological and human behavioral diversity before and during the Late Glacial: A Japanese case study," *Quaternary International* 347 (2014): 200–210; Masami Izuho and Yousuke Kaifu, "The appearance and characteristics of the early Upper Paleolithic in the Japanese Archipelago," in Y. Kaifu et al., eds., *Emergence and Diversity of Modern Human Behavior in Paleolithic Asia* (Texas A&M University Press, 2015), 289–313.

16. Yousuke Kaifu et al., "Pleistocene seafaring and colonization of the Ryukyu Islands, southwestern Japan," in Kaifu et al., *Emergence and Diversity*, 345–361; Fuzuki Mizuno et al., "Population dynamics in the Japanese Archipelago since the Pleistocene," *Scientific Reports* 11 (2021): 12018.

17. Masaki Fujita et al., "Advanced maritime adaptation in the Western Pacific coastal region extends back to 35,000–30,000 years before present," *PNAS* 113 (2016): 11184–11189; Sue O'Connor, "Crossing the Wallace Line," in Kaifu et al., *Emergence and Diversity*, 214–224; Harumi Fujita, "Early Holocene pearl oyster circular fishhooks and ornaments on Espiritu Santo Island, Baja California Sur," *Monographs of the Western North American Naturalist* 7 (2014): 129–134; Matthew Des Lauriers et al., "The earliest shell fishhooks from the Americas," *American Antiquity* 82 (2017): 498–516.

18. See Yoshitake Tanomata and Andrey Tabarev, "A newly discovered cache of large biface lithics from northern Honshu, Japan," *Antiquity* 94, no. 374 (2020): E8, for a comparison between a cache of eight bifaces found in Initial Jomon Tohoku and similar caches found in Clovis North America.

19. Fuzuki Mizuno et al., "Population dynamics in the Japanese Archipelago since the Pleistocene," *Scientific Reports* 11 (2021): 12018.

20. Jon Erlandson et al., "The Kelp Highway Hypothesis," *JICA* 2 (2007): 161–174.

21. Eske Willerslev and David Meltzer, "Peopling of the Americas as inferred from ancient genomics," *Nature* 594 (2021): 356–364.

22. Bastien Llamas et al., "Ancient mitochondrial DNA provides high-resolution time scale of the peopling of the Americas," *Science Advances* 2, no. 4 (2016): e1501385.

23. Anders Bergström et al., "Insights into human genetic variation and population history," *Science* 367 (2020): eaay5012.

24. Lorena Becerra-Valdivia and Thomas Higham, "The timing and effect of the earliest human arrivals in North America," *Nature* 584 (2020): 93–97; Amy Goldberg et al., "Post-invasion demography of prehistoric humans in South America," *Nature* 532 (2016): 232–235.

25. Waters, "Late Pleistocene exploration."

26. Ciprian Ardelean et al., "Evidence of human occupation in Mexico around the Last Glacial Maximum," *Nature* 584 (2020): 87–92.

27. James Chatters et al., "Evaluating the claims of early human occupation at Chiquihuite Cave, Mexico," *PaleoAmerica* 8 (2022): 1–16; Ben Potter et al., "Current understanding of the earliest human occupations in the Americas," *PaleoAmerica* 8 (2022): 62–76.

28. For Diuktai Cave, see Yan Coutouly, "Migrations and interactions in prehistoric Beringia," *Antiquity* 90 (2016): 9–31. For details of Japanese stone tool sequences during the Late Pleistocene, see Kazuki Morisaki et al., "Human adaptive responses to environmental change during the Pleistocene-Holocene transition in the Japanese archipelago," in E. Robinson and F. Sellet, eds., *Lithic Technological Organization and Paleoenvironmental Change* (Springer, 2018), 91–122.

29. For surveys of the archaeological record discussed, see J. M. Adovasio and David Pedler, *Strangers in a New Land* (Firefly Books. 2016); David Meltzer, "The origins, antiquity, and dispersal of the first Americans," in Chris Scarre, ed., *The Human Past: World Prehistory and the Development of Human Societies*, 4th ed. (Thames and Hudson, 2018), 149–171; Waters, "Late Pleistocene exploration"; Loren Davis et al., "Late Upper Paleolithic occupation at Cooper's Ferry, Idaho, USA ~16,000 years ago," *Science* 365 (2019): 891–897. I also acknowledge Chris Gillam for discussion on the topic of a Japanese origin for First Americans; see his section in J. Uchiyama, et al., "Population dynamics in northern Eurasian forests," *Evolutionary Human Sciences* 2 (2020): E16.

30. Todd Braje et al., "Fladmark + 40: What have we learned about a potential Pacific coast peopling of the Americas?," *American Antiquity* 85 (2019): 1–21.

31. Joseph Greenberg, *Language in the Americas* (Stanford University Press, 1987); Joseph Greenberg et al., "The settlement of the Americas: A comparison of the linguistic, dental, and genetic evidence," *Current Anthropology* 27 (1986): 477–497.

32. In support of Greenberg, see Merritt Ruhlen, *The Origin of Language: Tracing the Evolution of the Mother Tongue* (Stanford University Press, 1994).

33. Morten Rasmussen et al., "The genome of a Late Pleistocene human from a Clovis burial site in western Montana," *Nature* 506 (2014): 225–229.

34. C. Scheib et al., "Ancient human parallel lineages within North America contributed to a coastal expansion," *Science* 360 (2018): 1024–1027; Cosimo Posth et al., "Reconstructing the deep population history of Central and South America," *Cell* 175 (2018): 1185–1197; J. Victor Moreno-Mayar et al., "Early human dispersals within the Americas," *Science* 362 (2018): eaav2621.

35. He Yu et al., "Paleolithic to Bronze Age Siberians reveal connections with First Americans and across Eurasia," *Cell* 181 (2020): 1232–1245.

36. Chao Ning et al., "The genomic formation of first American ancestors in east and North East Asia," *bioRxiv* (2020), https://doi.org/10.1101/2020.10.12.336628.

37. On the genetics of the Beringian standstill hypothesis, see Maanasa Raghavan et al., "Genomic evidence for the Pleistocene and recent population history of Native Americans," *Science* 349 (2015): aab3884; Llamas et al., "Ancient mitochondrial DNA"; J. Victor Moreno-Mayar, "Terminal Pleistocene Alaskan genome reveals first founding population of Native Americans," *Nature* 553 (2018): 203–207; Martin Sikora et al., "The population history of northeastern Siberia since the Pleistocene," *Nature* 570 (2019): 182–188.

38. On the dating of Ushki, see Ted Goebel et al., "New dates from Ushki-1, Kamchatka," *JAS* 37 (2010): 2640–2649.

39. Noreen von Cramon-Taubadel et al., "Evolutionary population history of early Paleoamerican cranial morphology," *Science Advances* 3 (2017): e1602289.

40. Moreno-Mayar et al., "Early human dispersals"; Ning et al., "Genomic formation."

41. André Strauss et al., "Early Holocene ritual complexity in South America: the archaeological record of Lapa do Santo," *Antiquity* 90 (2016): 1454–1473; Osamu Kondo et al., "A female human skeleton from the initial Jomon period," *Anthropological Science* 126 (2018): 151–164.

42. After Pontus Skoglund et al., "Genetic evidence for two founding populations of the Americas," *Nature* 525 (2015): 104–108; see also Raghavan, "Genomic evidence." The most recent genetic statement in favor of the existence of Population Y is by Marcos Castro e Silva et al., "Deep genetic affinity between coastal Pacific and Amazonian natives evidenced by Australasian ancestry," *PNAS* 118 (2021): e2025739118.

43. For instance, clear traces of Population Y ancestry appear to be absent in the Caribbean Islands according to Daniel Fernandes, "A genetic history of the pre-contact Caribbean," *Nature* 590, no. 7844 (2021): 103–110.

44. Melinda Yang et al., "40,000-year-old individual from Asia provides insight into early population structure in Eurasia," *Current Biology* 27 (2017): 3206.

45. Xiaoming Zhang et al., "Ancient genome of hominin cranium reveals diverse population lineages in southern East Asia during Late Paleolithic," submitted to *Cell*.

46. Davis et al., "Late Upper Paleolithic occupation."

47. Gustavo Politis and Luciano Prates, "Clocking the arrival of *Homo sapiens* in the southern cone of South America," in K. Harvati et al., eds., *New Perspectives on the Peopling of the Americas* (Kerns Verlag, 2018), 79–106.

48. Jon Erlandson et al., "Paleoindian seafaring, maritime technologies, and coastal foraging on California's Channel Islands," *Science* 331 (2011): 1181–1184.

49. Kurt Rademaker et al., "Paleoindian settlement of the high-altitude Peruvian Andes," *Science* 346 (2014): 466–469.

50. Angela Perri et al., "Dog domestication and the dual dispersal of people and dogs into the Americas," *PNAS* 118 (2021): e2010083118.

51. There is a large literature on Arctic prehistory in North America, from many disciplines. See the articles by Michael Fortescue and Max Friesen in P. Bellwood, ed., *The Global Prehistory of Human Migration* (Wiley, 2015), and many excellent articles in T. M. Friesen and O. Mason, eds., *The Oxford Handbook of the Prehistoric Arctic* (Oxford University Press, 2016).

52. Pavel Flegontov et al., "Palaeo-Eskimo genetic ancestry and the peopling of Chukotka and North America," *Nature* 570 (2019): 236–240; Sikora et al., "Population history of northeastern Siberia."

53. Mikkel-Holger Sinding et al., "Arctic-adapted dogs emerged at the Pleistocene-Holocene transition," *Science* 368 (2020): 1495–1499.

Chapter 7

1. *Note*: In this chapter and onward I switch to the use of BCE/CE (Common Era) terminology for absolute dates.

2. Abigail Page et al., "Reproductive trade-offs in extant hunter-gatherers suggest adaptive mechanism for the Neolithic expansion," *PNAS* 113 (2016): 4694–4699.

3. Jean-Pierre Bocquet-Appel, "When the world's population took off," *Science* 333 (2011): 560–561.

4. For a discussion of hunter-gatherer birth spacing, see Nicholas Blurton-Jones, "Bushman birth spacing: A test for optimal interbirth intervals," *Ecology and Sociobiology* 7 (1986): 91–105.

5. Richard Lee, *The !Kung San: Men, Women, and Work in a Foraging Society* (Cambridge University Press, 1979), 312.

6. Ian Keen, *Aboriginal Society and Economy: Australia at the Threshold of Colonisation* (Oxford University Press, 2004), 381. Fekri Hassan, *Demographic Archaeology* (Academic, 1981), 197, gives one person per 3.9 square kilometers for the Central Valley of California.

7. Nathan Wolfe et al., "Origins of major human infectious diseases," *Nature* 447 (2007): 279–283.

8. Anders Bergström et al., "Origins and genetic legacy of prehistoric dogs," *Science* 370 (2020): 557–564.

9. Shahal Abbo and Avi Gopher, "Plant domestication in the Neolithic Near East: The humans–plants liaison," *QSR* 242 (2020): 106412. See also Shahal Abbo et al., *Plant Domestication and the Origins of Agriculture in the Ancient Near East* (Cambridge University Press, 2022).

10. Dorian Fuller, "Contrasting patterns in crop domestication and domestication rates," *Annals of Botany* 100 (2007): 903–924; George Willcox, "Pre-domestic cultivation during the late Pleistocene and early Holocene in the northern Levant," in Paul Gepts et al., eds., *Biodiversity in Agriculture: Domestication, Evolution, and Sustainability* (Cambridge University Press, 2012), 92–109; Dorian Fuller et al., "From intermediate economies to agriculture: Trends in wild food use, domestication and cultivation among early villages in Southwest Asia," *Paléorient* 44 (2018): 59–74.

11. See, for instance, the rapid domestication of Fertile Crescent cereals and legumes that occurred around 8500 BCE at the site of Gusir Höyük in central Turkey, discussed by Ceren Kabukcu et al., "Pathways to domestication in Southeast Anatolia," *Scientific Reports* 11 (2021), Article 2112.

12. M. Kat Anderson, *Tending the Wild: Native American Knowledge and the Management of California's Natural Resources* (University of California Press, 2005); M. Kat Anderson and Eric Wohlgemuth, "California Indian proto-agriculture: Its characterization and legacy," in Gepts et al., *Biodiversity in Agriculture*, 190–224; Harry Allen, "The Bagundji of the Darling Basin: Cereal gatherers in an uncertain environment," *World Archaeology* 5 (1974): 309–322.

13. Peter Richerson et al., "Was agriculture impossible during the Pleistocene but mandatory during the Holocene?," *American Antiquity* 66 (2001): 387–411.

14. Joan Feynman and Alexander Ruzmaikin, "Climate stability and the development of agricultural societies," *Climate Change* 84 (2007): 295–311.

15. David Smith et al., "The early Holocene sea level rise," *QSR* 30 (2011): 1846–1860; Kurt Lambeck et al., "Sea level and global ice volumes from the Last Glacial Maximum to the Holocene," *PNAS* 111 (2014): 15296–15303; A. Dutton et al., "Sea-level rise due to polar ice-sheet mass loss during warm wet periods," *Science* 349 (2015): aaa4019.

16. As also suggested for the Middle East by Stephen Shennan, *The First Farmers of Europe: An Evolutionary Perspective* (Cambridge University Press, 2018).

17. Jared Diamond, "Evolution, consequences and future of plant and animal domestication," *Nature* 418 (2002): 34–41.

18. For instance, Mark Nathan Cohen, *The Food Crisis in Prehistory: Overpopulation and the Origins of Agriculture* (Yale University Press, 1977); Allen Johnson and Timothy Earle, *The Evolution of Human Societies: From Foraging Group to Agrarian State* (Stanford University Press, 2000).

19. Ian Gilligan, *Climate, Clothing, and Agriculture in Prehistory: Linking Evidence, Causes, and Effects* (Cambridge University Press, 2019).

Chapter 8

1. On the plants, see Ehud Weiss and Daniel Zohary, "The Neolithic Southwest Asian founder crops," *Current Anthropology* 52, suppl. 4 (2011): 237–254; Daniel Zohary et al., *Domestication of Plants in the Old World*, 4th ed. (Oxford University Press, 2012). Some archaeologists think that Fertile Crescent cereals and legumes were domesticated in more than one place; see, for instance, Eleni Asouti and Dorian Fuller, "A contextual approach to the emergence of agriculture in Southwest Asia," *Current Anthropology* 54 (2013): 299–345. On animals, see Melinda Zeder, "Out of the Fertile Crescent: The dispersal of domestic livestock through Europe and Africa," in Nicole Boivin et al., eds., *Human Dispersal and Species Movement: From Prehistory to Present* (Cambridge University Press, 2017), 261–303.

2. Dorian Fuller and Chris Stevens, "Between domestication and civilization," *Vegetation History and Archaeobotany* 28 (2019): 263–282.

3. W. J. Perry, *The Growth of Civilization* (1924; repr., Pelican, 1937), 46; V. Gordon Childe, *New Light on the Most Ancient East* (1928; repr., Routledge and Kegan Paul, 1958), 32.

4. Daniel Stanley and Andrew Warne, "Sea level and initiation of Predynastic culture in the Nile Delta," *Nature* 363 (1993): 425–428.

5. Robert Braidwood and Bruce Howe, *Prehistoric Investigations in Iraqi Kurdistan* (Oriental Institute of the University of Chicago, 1960), 1.

6. Amaia Arranz-Otaegui et al., "Archaeobotanical evidence reveals the origins of bread 14,400 years ago in northeastern Jordan," *PNAS* 115 (2018): 7925–7930.

7. Stephen Shennan, *The First Farmers of Europe: An Evolutionary Perspective* (Cambridge University Press, 2018). The reasoning here is that numbers of radiocarbon dates can be a proxy for the intensity of human activity. This may or may not have been true in specific prehistoric situations, but as a general indicator of past demography the technique has a certain validity.

8. As suggested by Andrew Moore and Gordon Hillman, "The Pleistocene to Holocene transition and human economy in Southwest Asia," *American Antiquity* 57 (1992): 482–494.

9. Eleni Asouti and Dorian Fuller, "A contextual approach to the emergence of agriculture in Southwest Asia," *Current Anthropology* 54 (2013): 299–345; George Willcox, "Pre-domestic cultivation during the late Pleistocene and early Holocene in the northern Levant," in Paul Gepts et al., eds., *Biodiversity in Agriculture: Domestication, Evolution, and Sustainability* (Cambridge University Press, 2012), 92–109.

10. Kathleen Kenyon, *Archaeology in the Holy Land* (Benn, 1960). Jericho lies almost 300 meters below sea level.

11. Harald Hauptmann, "Les sanctuaires mégalithiques de Haute-Mésopotamie," in Jean-Paul Demoule, ed., *La révolution néolithique dans le monde* (CNRS Editions, 2009), 359–382; Oliver Dietrich et al., "The role of cult and feasting in the emergence of Neolithic communities," *Antiquity* 86 (2012): 674–695.

12. Klaus Schmidt, *Göbekli Tepe* (Ex Oriente, 2012). I had the pleasure of meeting the late Klaus Schmidt at the Shanghai Archaeological Forum and discussing the site with him in 2013.

13. Andrew Curry, "The ancient carb revolution," *Nature* 594 (2021): 488–491.

14. Megan Gannon, "Archaeology in a divided land," *Science* 358 (2017): 28–30.

15. Sturt Manning et al., "The earlier Neolithic in Cyprus," *Antiquity* 84 (2010): 693–706.

16. Jean-Denis Vigne et al., "The early process of mammal domestication in the Near East," *Current Anthropology* 52, suppl. 4 (2011): 255–271.

17. Leilani Lucas and Dorian Fuller, "Against the grain: Long-term patterns in agricultural production in prehistoric Cyprus," *Journal of World Prehistory* 33 (2020): 233–266.

18. Alessio Palmisano et al., "Holocene regional population dynamics and climatic trends in the Near East," *QSR* 252 (2021): 106739.

19. Ian Hodder, *Çatalhöyük: The Leopard's Tale* (Thames and Hudson, 2006). An excellent artist's reconstruction of the Çatalhöyük settlement can be seen in Ann Gibbons, "How farming shaped Europeans' immunity," *Science* 373 (2021): 1186.

20. Alessio Palmisano et al., "Holocene landscape dynamics and long-term population trends in the Levant," *The Holocene* 29 (2019): 708–727.

21. David Friesem et al., "Lime plaster cover of the dead 12,000 years ago," *Evolutionary Human Sciences* 1 (2020): E9.

22. There are many accounts of the Chinese Neolithic, but for recent developments, see David Cohen, "The beginnings of agriculture in China," *Current Anthropology* 52, suppl. 4 (2011): 273–306; Gideon Shelach-Lavi, "Main issues in the study of the Chinese Neolithic," in P. Goldin, ed., *Routledge Handbook of Early Chinese History* (Routledge, 2018), 15–38.

23. Gideon Shelach-Lavi et al., "Sedentism and plant agriculture in northeast China emerged under affluent conditions," *PLoS One* 14, no. 7 (2019): e0218751.

24. Yunfei Zheng et al., "Rice domestication revealed," *Scientific Reports* 6 (2016): 28136; Xiujia Huan et al., "Spatial and temporal pattern of rice domestication during the early Holocene," *The Holocene* 31 (2021): 1366–1375.

25. Christian Peterson and Gideon Shelach, "The evolution of early Yangshao village organization," in Matthew Bandy and Jake Fox, eds., *Becoming Villagers: Comparing Early Village Societies* (University of Arizona Press, 2010), 246–275.

26. For detailed studies of rice domestication in China, see Fabio Silva et al., "Modelling the geographical origin of rice cultivation in Asia," *PLoS One* 10 (2015): e0137024; Dorian Fuller et al., "Pathways of rice diversification across Asia," *Archaeology International* 19 (2016): 84–96; Yongchao Ma et al., "Multiple indicators of rice remains and the process of rice domestication," *PLoS One* 13 (2018): e0208104.

27. Patrick McGovern et al., "Fermented beverages of pre- and proto-historic China," *PNAS* 101 (2004): 17593–17598; Ningning Dong and Jing Yuan, "Rethinking pig domestication in China," *Antiquity* 94 (2020): 864–879. For a new genetic analysis of pig domestication in China, see Ming Zhang et al., "Ancient DNA reveals the maternal genetic history of East Asian domestic pigs," *Journal of Genetics and Genomics* (pre-proof), doi.org/10.1016/j.jgg.2021.11.014.

28. Li Liu et al., "The brewing function of the first amphorae in the Neolithic Yangshao culture, North China," *AAS* 12 (2020): 28.

29. Bin Liu et al., "Earliest hydraulic enterprise in China, 5100 years ago," *PNAS* 114 (2017): 13637–13642; Colin Renfrew and Bin Liu, "The emergence of complex society in China: The case of Liangzhu," *Antiquity* 92 (2018): 975–990.

30. Yanyan Yu et al., "Spatial and temporal changes of prehistoric human land use in the Wei River valley, northern China," *The Holocene* 26 (2016): 1788–1901.

31. Melinda Zeder, "Out of the Fertile Crescent: The dispersal of domestic livestock through Europe and Africa," in Boivin et al., *Human Dispersal and Species Movement*, 261–303.

32. On the Saharan humid phase, see Rudolph Kuper and Stefan Kröper, "Climate-controlled Holocene occupation in the Sahara," *Science* 313 (2006): 803–807; Wim van Neer et al., "Aquatic fauna from the Takarkori rock shelter," *PLoS One* 15 (2020): e0228588.

33. Noriyuki Shirai, "Resisters, vacillators or laggards? Reconsidering the first farmer-herders in prehistoric Egypt," *Journal of World Prehistory* 33 (2020): 457–512.

34. Marieke van de Loosdrecht et al., "Pleistocene North African genomes link Near Eastern and Sub-Saharan African human populations," *Science* 360 (2018): 548–552.

35. Julie Dunne et al., "First dairying in green Saharan Africa in the fifth millennium BCE," *Nature* 486 (2012): 390–394.

36. Frank Winchell et al., "On the origins and dissemination of domesticated sorghum and pearl millet across Africa and into India," *African Archaeological Review* 35 (2018): 483–505; N. Scarcelli et al., "Yam genomics supports West Africa as a major cradle of crop domestication," *Science Advances* 5 (2019): eaaw1947.

37. Aleese Barron et al., "Snapshots in time," *JAS* 123 (2020): 105259; Dorian Fuller et al., "Transition from wild to domesticated pearl millet (*Pennisetum glaucum*)," *African Archaeological Review* 38 (2021): 211–230.

38. Andrea Kay et al., "Diversification, intensification and specialization: Changing land use in western Africa from 1800 BC to AD 1500," *Journal of World Prehistory* 32 (2019): 179–228.

39. Robert Power et al., "Asian crop dispersal in Africa and late Holocene human adaptation to tropical environments," *Journal of World Prehistory* 32 (2019): 353–392; Alison Crowther et al., "Subsistence mosaics, forager-farmer interactions, and the transition to food production in eastern Africa," *Quaternary International* 489 (2018): 101–120.

40. Jack Golson et al., eds., *10,000 Years of Cultivation at Kuk Swamp* (Australian National University Press, 2017).

41. Tim Denham, *Tracing Early Agriculture in the Highlands of New Guinea* (Routledge, 2018).

42. Ibrar Ahmed et al., "Evolutionary origins of taro (*Colocasia esculenta*) in Southeast Asia," *Ecology and Evolution* 10, no. 23 (2020): 13530–13543.

43. Ben Shaw, "Emergence of a Neolithic in Highland New Guinea by 5000 to 4000 years ago," *Science Advances* 6 (2020): eaay4573.

44. For an example see Philip Guddemi, "When horticulturalists are like hunter-gatherers: The Sawiyano of Papua New Guinea," *Ethnology* 31 (1992): 303–314.

45. Nicole Pedro et al., "Papuan mitochondrial genomes and the settlement of Sahul," *Journal of Human Genetics* 65 (2020): 875–887.

46. Glenn Summerhayes, "Austronesian expansions and the role of mainland New Guinea," *Asian Perspectives* 58 (2019): 250–260.

47. Tim Denham et al., "Horticultural experimentation in northern Australia reconsidered," *Antiquity* 83 (2009): 634–648.

48. On the spread of maize, see Logan Kistler et al., "Multiproxy evidence highlights a complex evolutionary legacy of maize in South America," *Science* 362 (2018): 1309–1312. Duccio Bonavia thinks that maize was spread from Mexico to South America by birds: see *Maize: Origin, Domestication, and Its Role in the Development of Culture* (Cambridge University Press, 2013). On Mesoamerica–Ecuador contacts, see Patricia Anawalt, "Traders of the Ecuadorian littoral," *Archaeology* 50, no. 6 (1997): 48–52.

49. James Ford, *A Comparison of Formative Cultures in the Americas: Diffusion or the Psychic Unity of Man* (Smithsonian Institution Press, 1969).

50. On the issue of contacts between the Old and New Worlds in prehistory, see Peter Watson, *The Great Divide: History and Human Nature in the Old World and the New* (Weidenfeld and Nicholson, 2012); Stephen Jett, *Ancient Ocean Crossings: Reconsidering the Case for Contacts with the Pre-Columbian Americas* (University of Alabama Press, 2017); and my review of Jett's excellent book in *Journal of Anthropological Research* 74 (2018): 281–284.

51. David Malakoff, "Great Lakes people amongst first coppersmiths," *Science* 371 (2021): 1299.

52. John Smalley and Michael Blake, "Sweet beginnings: Stalk sugar and the domestication of maize," *Current Anthropology* 44 (2003): 675–704; David Webster et al., "Backward bottlenecks," *Current Anthropology* 52 (2011): 77–104; Robert Kruger, "Getting to the grain," in Basil Reid, ed., *The Archaeology of Caribbean and Circum-Caribbean Farmers 6000 BC–AD 1500* (Routledge, 2018), 353–369; Tiffany Tung et al., "Early specialized maritime and maize economies on the North Coast of Peru," *PNAS* 117, no. 51 (2020): 32308–32319.

53. Jazmin Ramos-Madrigal et al., "Genome sequence of a 5,310-year-old maize cob provides insights into the early stages of maize domestication," *Current Biology* 26 (2016): 3195–3201; Miguel Vallebueno-Estrada et al., "The earliest maize from San Marcos Tehuacán is a partial domesticate with genomic evidence of inbreeding," *PNAS* 113 (2016): 14151–14156.

54. Douglas Kennett et al., "High-precision chronology for Central American maize diversification from El Gigante rockshelter, Honduras," *PNAS* 114 (2017): 9026–9031.

55. Douglas Kennett et al., "Early isotopic evidence for maize as a staple grain in the Americas," *Science Advances* 6 (2020): eaba3245.

56. Tom D. Dillehay, ed., *From Foraging to Farming in the Andes: New Perspectives on Food Production and Social Organization* (Cambridge University Press, 2011).

57. John Clark et al., "First towns in the Americas," in Bandy and Fox, *Becoming Villagers*, 205–245; Deborah Pearsall et al., "Food and society at Real Alto, an Early Formative community in Southwest coastal Ecuador," *Latin American Antiquity* 31 (2020): 122–142.

58. Dolores Piperno, "The origins of plant cultivation and domestication in the New World tropics," *Current Anthropology* 52, suppl. 4 (2011): 453–470. See also Kennett et al., "Early isotopic evidence," for possible consequences.

59. Tung et al., "Early specialized maritime," 32308–32319.

60. Jennifer Watling, "Direct evidence for southwestern Amazonia as an early plant domestication and food production centre," *PLoS One* 13 (2018): e0199868; S. Yoshi Maezumi et al., "The legacy of 4500 years of polyculture agroforestry in the eastern Amazon," *Nature Plants* 4 (2018): 540–547; Umberto Lombardo et al., "Early Holocene crop cultivation and landscape modification in Amazonia," *Nature* 581 (2020): 190–193; Jose Iriarte et al., "The origins of Amazonian landscapes," *QSR* 248 (2020): 106582.

61. Michael Moseley and Michael Heckenberger, "From village to empire in South America," in Chris Scarre, ed., *The Human Past: World Prehistory and the Development of Human Societies*, 4th ed. (Thames and Hudson, 2018), 636–669.

62. Mark Nathan Cohen, "Population pressure and the origins of agriculture," in C. Reed, ed., *Origins of Agriculture* (Mouton, 1977), 135–178.

63. Amy Goldberg et al., "Post-invasion demography of prehistoric humans in South America," *Nature* 532 (2016): 232–235.

64. Kistler et al., "Multiproxy evidence," 1309–1312; Keith M. Prufer et al., "Terminal Pleistocene through Middle Holocene occupations in southeastern Mesoamerica," *Ancient Mesoamerica* 32 (2021): 439–460.

65. Kent Flannery and Joyce Marcus, *The Creation of Inequality: How Our Prehistoric Ancestors Set the Stage for Monarchy, Slavery, and Empire* (Harvard University Press, 2012), 299; William Sanders and Carson Murdy, "Cultural evolution and ecological succession in the Valley of Guatemala 1500 B.C.–A.D. 1524," in Kent Flannery, ed., *Maya Subsistence: Studies in Memory of Dennis E. Puleston* (Academic Press, 1982), 19–63. See also Richard Lesure, "The Neolithic demographic transition in Mesoamerica," *Current Anthropology* 55 (2014): 654–664, although he does not consider burial evidence from the Early Formative.

66. See the recent analysis of population growth in early agricultural Mesoamerica and Arizona by Richard Lesure et al., "Large scale patterns in the Agricultural Demographic Transition of Mesoamerica and southwestern North America," *American Antiquity* 86 (2021): 593–612.

67. Bruce Smith, "The cultural context of plant domestication in eastern North America," *Current Anthropology* 52, suppl. 4 (2011): 471–484.

Chapter 9

1. Colin Renfrew, *Archaeology and Language: The Puzzle of Indo-European Origins* (Jonathan Cape, 1987); Peter Bellwood, "The great Pacific migration," in *1984 Britannica Yearbook of Science and the Future* (Encyclopaedia Britannica, 1984), 80–93; Robert Blust, "The Austronesian homeland: A linguistic perspective," *Asian Perspectives* 26 (1984–1985): 45–67.

2. Peter Bellwood and Colin Renfrew, eds., *Examining the Farming/Language Dispersal Hypothesis* (McDonald Institute for Archaeological Research, 2002); Peter Bellwood, *First Farmers: The Origins of Agricultural Societies* (Blackwell, 2005).

3. Jared Diamond and Peter Bellwood, "Farmers and their languages: The first expansions," *Science* 300 (2003): 597–603.

4. Bellwood, *First Farmers*; Peter Bellwood, *First Migrants: Ancient Migration in Global Perspective* (Wiley Blackwell, 2013).

5. David Reich, *Who We Are and How We Got Here: Ancient DNA and the New Science of the Human Past* (Oxford University Press, 2018), xv.

6. See Bellwood, *First Migrants*, 8–9.

7. Charles Darwin, *The Descent of Man, and Selection in Relation to Sex* (John Murray, 1871), 175, 113.

8. Marianne Mithun, *The Native Languages of North America* (Cambridge University Press, 1999), 2.

9. Alfred Crosby, *Ecological Imperialism: The Biological Expansion of Europe, 900–1900* (Cambridge University Press, 1986).

10. Bernal Diaz, *The Conquest of New Spain* (Penguin, 1963); Jared Diamond, *Guns, Germs, and Steel: The Fates of Human Societies* (Jonathan Cape, 1997).

11. Alexander Koch et al., "Earth system impacts of the European arrival and Great Dying in the Americas after 1492," *QSR* 207 (2019): 13–36; Linda Ongaro et al., "The genomic impact of European colonization of the Americas," *Current Biology* 29 (2020): 3974–3986.

12. S. Heath and R. Laprade, "Castilian colonization and indigenous languages," in Robert Cooper, ed., *Language Spread: Studies in Diffusion and Social Change* (Indiana University Press, 1982), 137.

13. Translator Jona Lendering, Livius.org.

14. Nicholas Ostler, *Empires of the Word: A Language History of the World* (HarperPerennial, 2005), 275, 525, 534, 536.

Chapter 10

1. Kurt Gron et al., "Cattle management for dairying in Scandinavia's earliest Neolithic," *PLoS One* 10, no. 7 (2015): e0131267; Laure Ségurel and Celine Bon, "On the evolution of lactase persistence in humans," *Annual Review of Genomics and Human Genetics* 18 (2017): 297–319; Sophy Charlton et al., "Neolithic insights into milk consumption through proteomic analysis of dental calculus," *AAS* 11 (2019): 6183–6196.

2. Maria Bodnar, "Prehistoric innovations: Wheels and wheeled vehicles," *Acta Archaeologica Academiae Scientarum Hungaricae* 69 (2018): 271–298.

3. Kurt Lambeck et al., "Sea level and global ice volumes from the Last Glacial Maximum to the Holocene," *PNAS* 111 (2014): 15296–15303; Stephen Shennan, *The First Farmers of Europe: An Evolutionary Perspective* (Cambridge University Press, 2018), 27–28.

4. Gary Rollefson and Ilse Kohler-Rollefson, "PPNC adaptations in the first half of the 6th millennium BC," *Paléorient* 19 (1993): 33–42.

5. Clark Larsen et al., "Bioarchaeology of Neolithic Çatalhöyük reveals fundamental transitions in health, mobility, and lifestyle in early farmers," *PNAS* 116 (2019): 12615–12623.

6. As discussed by Nigel Goring-Morris and Anna Belfer-Cohen, "'Great Expectations,' or the inevitable collapse of the early Neolithic in the Near East," in Matthew S. Bandy and Jake R. Fox, eds., *Becoming Villagers: Comparing Early Village Societies* (University of Arizona Press, 2010), 62–77.

7. Nicolas Rascovan et al., "Emergence and spread of basal lineages of *Yersinia pestis* during the Neolithic decline," *Cell* 176 (2019): 295–305. Julian Susat et al., "A 5,000-year-old hunter-gatherer already plagued by *Yersinia pestis*," *Cell Reports* 35 (2021): 1092678, give a date of circa 7,000 years ago for the Neolithic origin of this plague bacterium.

8. Arkadiusz Marciniak, "Çatalhöyük and the emergence of the late Neolithic network," in Maxime Brami and Barbara Horejs, eds., *The Central/Western Anatolian Farming Frontier: Proceedings of the Neolithic Workshop Held at 10th ICAANE in Vienna, April 2016* (Austrian Academy of Sciences, 2019), 127–142; Ian Hodder, seminar given at the Australian National University, May 21, 2021.

9. Katerina Douka et al., "Dating Knossos and the arrival of the earliest Neolithic in the southern Aegean," *Antiquity* 91 (2017): 304–321.

10. On the expansion of the Anatolian Neolithic, see Mehmet Ozdogan, "Archaeological evidence on the westward expansion of farming communities from eastern Anatolia to the Aegean and the Balkans," *Current Anthropology* 52, suppl. 4 (2011): 397–413; Douglas Baird et al., "Agricultural origins on the Anatolian Plateau," *PNAS* 115 (2018): E3077–E3086; and the chapters in Brami and Horejs, *Central/Western Anatolian Farming Frontier.*

11. Marko Porcic et al., "Expansion of the Neolithic in southeastern Europe," *AAS* 13 (2021): 77.

12. Jerome Dubouloz, "Impacts of the Neolithic demographic transition on Linear Pottery Culture settlement," in Jean-Pierre Bocquet-Appel and Ofer Bar-Yosef, eds., *The Neolithic Demographic Transition and Its Consequences* (Springer, 2008), 208. On evidence for massacres, see Mark Golitko and Lawrence Keeley, "Beating ploughshares back into swords: Warfare in the Linearbandkeramik," *Antiquity* 81 (2007): 332–343.

13. Stephen Shennan et al., "Regional population collapse followed initial agriculture booms in mid-Holocene Europe," *Nature Communications* 4 (2013): 2486; Andrew Bevan et al., "Holocene fluctuations in human population demonstrate repeated links to food production and climate," *PNAS* 114 (2017): E10524–10531. These declines are also discussed by Shennan, *First Farmers of Europe*.

14. Wolfgang Haak et al., "Massive migration from the steppe was a source for Indo-European languages in Europe," *Nature* 522 (2015): 207–211.

15. Krisztian Oross et al., "'It's still the same old story': The current southern Transdanubian approach to the Neolithisation process of central Europe," *Quaternary International* 560-561 (2020): 154–178.

16. Selina Brace et al., "Ancient genomes indicate population replacement in early Neolithic Britain," *Nature Ecology and Evolution* 3 (2019): 765–771.

17. Lia Betti et al., "Climate shaped how Neolithic farmers and European hunter-gatherers interacted," *Nature Human Behaviour* 4 (2020): 1004–1010.

18. Gulsah Kilinç et al., "The demographic development of the first farmers in Anatolia," *Current Biology* 26 (2016): 1–8; Mark Lipson et al., "Parallel palaeogenomic transects reveal complex genetic history of early European farmers," *Nature* 551 (2017): 369–372; Michal Feldman et al., "Late Pleistocene human genome suggests a local origin for the first farmers of central Anatolia," *Nature Communications* 10 (2019): 1218.

19. Iosif Lazaridis et al., "Genomic insights into the origin of farming in the ancient Near East," *Nature* 536 (2016): 419–424.

20. Gordon Hillman, "Late Pleistocene changes in wild plant-foods available to hunter-gatherers of the northern Fertile Crescent: Possible preludes to cereal cultivation," in D. Harris, ed., *The Origins and Spread of Agriculture and Pastoralism in Eurasia* (UCL Press, 1996), 159–203.

21. For instance, at Ganj Dareh in Kermanshah Province. See Philip Smith, "Architectural innovation and experimentation at Ganj Dareh, Iran," *World Archaeology* 21 (1990): 323–335.

22. Jean-François Jarrige, "Mehrgarh Neolithic," *Pragdhara* (Lucknow) 18 (2007–2008): 135–154.

23. Farnaz Broushaki et al., "Early Neolithic genomes from the eastern Fertile Crescent," *Science* 353 (2016): 499–503; Iain Mathieson et al., "The genomic history of southeastern Europe," *Nature* 555 (2018): 197–203.

24. Lazaridis et al., "Genomic insights"; C. Wang et al., "Ancient human genome-wide data from a 3000-year interval in the Caucasus," *Nature Communications* 10 (2019): 590.

25. Reyhan Yaka et al., "Variable kinship patterns in Neolithic Anatolia revealed by ancient genomes," *Current Biology* 11 (2021): 244–268.

26. Robin Coningham and Ruth Young, *The Archaeology of South Asia: From the Indus to Asoka, c. 6500 BCE–200 CE* (Cambridge University Press, 2015).

27. Dorian Fuller, "Finding plant domestication in the Indian subcontinent," *Current Anthropology* 52, suppl. 4 (2011): 347–362.

28. Vasant Shinde et al., "Ancient Harappan genome lacks ancestry from steppe pastoralists or Iranian farmers," *Cell* 179 (2019): 729–735. For Gonur, see Megan Gannon, "An oasis civilization rediscovered," *Archaeology* 74, no. 1 (2021): 40–47.

29. Gordon Childe, *The Aryans* (K. Paul, Trench, Trubner & Co., 1926); Marija Gimbutas, "Primary and secondary homelands of the Indo-Europeans," *Journal of Indo-European Studies* 13 (1985): 185–202; David Anthony, *The Horse, the Wheel, and Language: How Bronze-Age Riders from the Eurasian Steppes Shaped the Modern World* (Princeton University Press, 2007).

30. David Anthony, "Ancient DNA, mating networks, and the Anatolian split," in Matilde Serangeli and Thomas Olander, eds., *Dispersals and Diversification: Linguistic and Archaeological Perspectives on the Early Stages of Indo-European* (Brill, 2020), 21–53.

31. Philip L. Kohl, *The Making of Bronze Age Eurasia* (Cambridge University Press, 2007).

32. Pablo Librado et al., "The origins and spread of domestic horses from the western Eurasian steppes," *Nature* 598 (2021): 634–640.

33. Jean Manco, *Ancestral Journeys: The Peopling of Europe from the First Venturers to the Vikings* (Thames and Hudson, 2013); Kristian Kristiansen et al., "Re-theorising mobility and the formation of culture and language among the Corded Ware culture in Europe," *Antiquity* 91 (2017): 334–347.

34. Hannes Schroeder et al., "Unravelling ancestry, kinship, and violence in a late Neolithic mass grave," *PNAS* 116 (2019): 10705–10710.

35. Iosif Lazaridis et al., "Genetic origins of the Minoans and Mycenaeans," *Nature* 548 (2017): 214–218; Mathieson et al., "Genomic history," 197–203; Florian Clemente et al., "The genomic history of the Aegean palatial civilizations," *Cell* 184, no. 10 (2021): 2565–2586.e21.

36. Cristina Valdiosera et al., "Four millennia of Iberian biomolecular prehistory," *PNAS* 115 (2018): 3428–3433; Daniel Fernandes et al., "The spread of steppe and Iranian-related ancestry in the islands of the western Mediterranean," *Nature Ecology and Evolution* 4 (2020): 334–345; Fernando Racimo et al., "The spatiotemporal spread of human migrations during the European Holocene," *PNAS* 117 (2020): 8989–9000.

37. Andaine Seguin-Orlando et al., "Heterogeneous hunter-gatherer and steppe-related ancestries in Late Neolithic and Bell Beaker genomes from present-day France," *Current Biology* 31 (2021): 1072–1083.

38. Helene Malström et al., "The genomic ancestry of the Scandinavian Battle Axe culture people," *Proceedings of the Royal Society of London, Series B: Biological Sciences* 286 (2019): 20191528.

39. Iñigo Olalde et al., "The Beaker phenomenon and the genomic transformation of northwest Europe," *Nature* 555 (2018): 190–196; Thomas Booth et al., "Tales from the supplementary information: Ancestry change in Chalcolithic–Early Bronze Age Britain was gradual with varied kinship organization," *Cambridge Archaeological Journal* 31, no. 3 (2021): 379–400.

40. Satya Pachori, *Sir William Jones: A Reader* (Oxford University Press, 1993), 175.

41. Nicholas Thomas et al., eds., *Observations Made during a Voyage around the World* (University of Hawai'i Press, 1996), 185.

42. For a general presentation of the Indo-European languages, see Benjamin Fortson, *Indo-European Language and Culture: An Introduction* (Wiley Blackwell, 2011).

43. J. P. Mallory and Victor Mair, *The Tarim Mummies: Ancient China and the Mystery of the Earliest Peoples from the West* (Thames and Hudson, 2000).

44. Harry Hoenigswald, "Our own family of languages," in A. Hill, ed., *Linguistics* (US Information Service, 1969), 67–80.

45. This is partly because the features of Indo-European languages that are studied by linguists in order to define subgroups do not have correlated and identical distributions. For instance, gender distinctions in nouns (masculine, feminine, neuter), ways of creating verbal tenses, pronunciation of words for 100 (as in Latin *centum* versus Old Iranian *satem*), and specific items of cultural vocabulary cross-cut each other in distribution, across different subgroups, in ways so complex that all recent classifications of the Indo-European family tree, no matter what statistics are used, come up with remarkably different internal relationships.

46. Remco Bouckaert et al., "Mapping the origins and expansion of the Indo-European language family," *Science* 337, no. 6097 (2012): 957–960 (see also figure 10.4 in this chapter). They used statistical calculations based on cognate vocabulary to place Proto-Indo-European in Anatolia around 6500 BCE, which was later corrected to 5500 BCE, in Remco Bouckaert et al., "Corrections and clarifications," *Science* 342 (2013): 1446.

47. Colin Renfrew, *Archaeology and Language: The Puzzle of Indo-European Origins* (Jonathan Cape, 1987).

48. In favor of substrate Indo-European languages in Europe, see Bernard Mees, "A genealogy of stratigraphy theories from the Indo-European West," in Henning Andersen, ed., *Language Contacts in Prehistory: Language Contacts in Prehistory: Studies in Stratigraphy* (John Benjamins, 2003), 11–44.

49. For useful discussion of these issues, see the two chapters by Paul Heggarty, "Why Indo-European?," and "Indo-European and the ancient DNA revolution," in Guus Kroonen et al., eds., *Talking Neolithic: Proceedings of the Workshop on Indo-European Origins Held at the Max Planck Institute for Evolutionary Anthropology, Leipzig, December 2–3, 2013* (Institute for the Study of Man, 2018), 69–119, 120–173. For an exercise in genetic simulation that indicates the likely absence of a single large migration from the Pontic Steppes into central Europe, rather than a more continuous trickle, see Jérémy Rio et al., "Spatially explicit paleogenomic simulations support cohabitation with limited admixture between Bronze Age Central European populations," *Communications Biology* 4 (2021): 1163. For an ancient DNA analysis of the hepatitis B virus that suggests a spread of this disease through Europe with Neolithic rather than Yamnaya populations, see Arthur Kocher et al., "Ten millennia of hepatitis B virus evolution," *Science* 374 (2021): 182–188.

50. Gimbutas, "Primary and secondary homelands."

51. Lara Cassidy et al., "A dynastic elite in monumental Neolithic society," *Nature* 582 (2020): 384–388.

52. On evidence for Yamnaya dairying from dental calculus, see Shevan Wilkin et al., "Dairying enabled Early Bronze Age Yamnaya steppe expansions," *Nature* 598 (2021): 629–633.

53. Rascovan et al., "Emergence and spread"; Susat et al., "5,000-year-old hunter-gatherer."

54. On Uralic prehistory, see Vaclav Blazek, "Northern Europe and Russia: Uralic linguistic history," in Peter Bellwood, ed., *The Global Prehistory of Human Migration* (Wiley-Blackwell, 2015), 178–183.

55. Thiseas Lamnidis et al., "Ancient Fennoscandian genomes reveal origin and spread of Siberian ancestry in Europe," *Nature Communications* 9 (2018), article 5018.

56. Peter Damgaard et al., "The first horse herders and the impact of early Bronze Age steppe expansions into South Asia," *Science* 360 (2018): eaar7711; Vagheesh Narasimhan et al., "The formation of human populations in South and Central Asia," *Science* 365 (2019): eaat7487.

57. Asko Parpola, *The Roots of Hinduism: The Early Aryans and the Indus Civilization* (Oxford University Press, 2015).

58. Nils Riedel et al., "Monsoon forced evolution of savanna and the spread of agro-pastoralism in peninsular India," *Scientific Reports* 11 (2021): 9032.

59. Dorian Fuller, "South Asia: Archaeology," in Bellwood, *Global Prehistory*, 245–253.

60. Franklin Southworth and David McAlpin, "South Asia: Dravidian linguistic history," in Bellwood, *Global Prehistory*, 235–244.

61. Vishnupriya Kolipakam et al., "A Bayesian phylogenetic study of the Dravidian language family," *Royal Society Open Science* 5 (2018): 171504.

62. Franklin Southworth, *Linguistic Archaeology of South Asia* (Routledge Curzon, 2005).

63. Guillermo Algaze, *The Uruk World System: The Dynamics of Expansion of Early Mesopotamian Civilization* (University of Chicago Press, 2004).

Chapter 11

1. See, for instance, the ancient DNA evidence for the existence of at least three separate late Paleolithic populations in southern China presented by Tianyi Wang et al., "Human population history at the crossroads of East and Southeast Asia since 11,000 years ago," *Cell* 184 (2021): 3829–3841.

2. Murray Cox, "The genetic history of human populations in island Southeast Asia during the Late Pleistocene and Holocene," in Peter Bellwood, *First Islanders: Prehistory and Human Migration in Island Southeast Asia*, 107–116 (Wiley Blackwell, 2017).

3. Hirofumi Matsumura et al., "Craniometrics reveal two layers of prehistoric human dispersal in eastern Eurasia," *Scientific Reports* 9 (2019): 1451; Hirofumi Matsumura et al., "Female craniometrics support the 'two layer model' of human dispersal in eastern Eurasia," *Scientific Reports* 11 (2021): 20830.

4. Hugh McColl et al., "The prehistoric peopling of Southeast Asia," *Science* 361 (2018): 88–92; Mark Lipson et al., "Ancient genomes document multiple waves of migration in Southeast Asian prehistory," *Science* 361 (2018): 92–95; Melinda Yang et al., "Ancient DNA indicates human population shifts and admixture in northern and southern China," *Science* 369, no. 6501 (2020): 282–288; Chuan-chao Wang et al., "Genomic insights into the formation of human populations in East Asia," *Nature* 591 (2021): 413–419; Selina Carlhoff et al., "Genome of a middle Holocene hunter-gatherer from Wallacea," *Nature* 596 (2021): 543–547.

5. Martine Robbeets et al., "Triangulation supports agricultural spread of the Transeurasian languages," *Nature* 599 (2021): 616–621; Peter Bellwood, "Tracking the origin of Transeurasian languages," *Nature* 599 (2021): 557–558.

6. Chao Ning et al., "Ancient genomes from northern China," *Nature Communications* 11 (2020): 2700. This situation of early separation followed by later genetic mixing parallels that described for the Fertile Crescent in chapter 8.

7. Martine Robbeets and Mark Hudson, "Archaeolinguistic evidence for the farming/language dispersal of Koreanic," *Evolutionary Human Sciences* 2 (2020): E52; Tao Li et al., "Millet agriculture dispersed from northeast China to the Russian Far East," *Archaeological Research in Asia* 22 (2020): 100177; Yating Qu et al., "Early interaction of agropastoralism in Eurasia," *AAS* 12 (2020): 195; Sarah Nelson et al., "Tracing population movements in ancient East Asia through the linguistics and archaeology of textile production," *Evolutionary Human Sciences* 2 (2020): E5.

8. Gary Crawford, "Advances in understanding early agriculture in Japan," *Current Anthropology* 52, suppl. 4 (2011): 331–345.

9. Rafal Gutaker et al., "Genomic history and ecology of the geographic spread of rice," *Nature Plants* 6 (2020): 492502.

10. Choongwon Jeong et al., "A dynamic 6,000-year genetic history of Eurasia's eastern steppes," *Cell* 183, no. 4 (2020): 890–904. Ancient Mongolians and Tungus carried an Amur Valley genetic signature; see Wang et al., "Genomic insights."

11. Chuan-chao Wang and Martine Robbeets, "The homeland of Proto-Tungusic inferred from contemporary words and ancient genomes," *Evolutionary Human Sciences* 2 (2020): E8.

12. Junzo Uchiyama et al., "Population dynamics in northern Eurasian forests," *Evolutionary Human Sciences* 2 (2020): E16.

13. Dominic Hosner et al., "Spatiotemporal distribution patterns of archaeological sites in China," *The Holocene* 26 (2016): 1576–1593.

14. Menghan Zhang et al., "Phylogenetic evidence for Sino-Tibetan origin in northern China in the late Neolithic," *Nature* 569 (2019): 112–115; Laurent Sagart et al., "Dated language phylogenies shed light on the ancestry of Sino-Tibetan," *PNAS* 116 (2019): 10317–10322; Hanzhi Zhang et al., "Dated phylogeny suggests early Neolithic origins of Sino-Tibetan languages," *Scientific Reports* 10 (2020): 20792.

15. Guiyun Jin et al., "The Beixin culture," *Antiquity* 94 (2020): 1426–1443.

16. Randy LaPolla, "The role of migration and language contact in the development of the Sino-Tibetan language family," in Alexandra Aikhenvald and Robert Dixon, eds., *Areal Diffusion and Genetic Inheritance: Problems in Comparative Linguistics* (Oxford University Press, 2001), 225–254.

17. Lele Ren et al., "Foraging and farming: Archaeobotanical and zooarchaeological evidence for Neolithic exchange on the Tibetan plateau," *Antiquity* 94: 637–652 (2020).

18. Ruo Li et al., "Spatio-temporal variation of cropping patterns in relation to climate change in Neolithic China," *Atmosphere* 11, no. 7 (2020): 677.

19. Ting Ma et al., "Holocene coastal evolution preceded the expansion of paddy field rice farming," *PNAS* 117 (2020): 24138–24143.

20. Zhang Chi and Hsiao-chun Hung, "Eastern Asia: Archaeology," in Peter Bellwood, ed., *The Global Prehistory of Human Migration* (Wiley, 2015), 209–216.

21. Matsumura et al., "Craniometrics reveal two layers."

22. Marc Oxenham et al., "Between foraging and farming," *Antiquity* 92 (2018): 940–957; Hsiao-chun Hung, "Prosperity and complexity without farming: The South China coast, c. 5000–3000 BC," *Antiquity* 93 (2019): 325–341.

23. For Gaomiao, see Hirofumi Matsumura et al., "Mid-Holocene hunter-gatherers 'Gaomiao' in Hunan, China," in Philip Piper et al., eds., *New Perspectives in Southeast Asia and Pacific*

Prehistory (Australian National University Press, 2017), 61–78. For Man Bac, see Marc Oxenham et al., *Man Bac: The Excavation of a Neolithic Site in Northern Vietnam* (Australian National University Press, 2011); Lipson et al., "Ancient genomes."

24. Tim Denham et al., "Is there a centre of early agriculture and plant domestication in southern China?," *Antiquity* 92 (2018): 1165–1179.

25. Wang et al., "Human population history"; Yang et al., "Ancient DNA."

26. Charles Higham, *Early Mainland Southeast Asia: From First Humans to Angkor* (River Books, 2014); Philip Piper et al., "The Neolithic of Vietnam," in Charles Higham and Nam Kim, eds., *The Oxford Handbook of Early Southeast Asia* (Oxford University Press, 2021), 194–215.

27. Zhenhua Deng et al., "Bridging the gap on the southward dispersal of agriculture in China," *AAS* 12 (2020): 151.

28. Ming-Shan Wang et al., "863 Genomes reveal the origin and domestication of chicken," *Cell Research* 30 (2020): 693–701.

29. Jade d'Alpoim Guedes et al., "3000 Years of farming strategies in central Thailand," *Antiquity* 94 (2020): 966–982.

30. Fiorella Rispoli, "The incised and impressed pottery of mainland Southeast Asia," *East and West* 57 (2007): 235–304. Illustrations of this kind of pottery can be found in Bellwood, *First Islanders*, plates 6 and 7.

31. Weera Ostapirat, "Kra-Dai and Austronesian: Notes on phonological correspondences and vocabulary distribution," in Laurent Sagart et al., eds., *The Peopling of East Asia: Putting Together Archaeology, Linguistics and Genetics* (Routledge Curzon, 2005), 107–131; Jin Sun et al., "Shared paternal ancestry of Han, Tai-Kadai-speaking, and Austronesian-speaking populations," *AJPA* 174 (2020): 686–700.

32. Wang et al., "Genomic insights"; S. Wen et al., "Y-chromosome-based genetic pattern in East Asia affected by Neolithic transition," *Quaternary International* 426 (2016): 50–55.

33. Jim Goodman, *Delta to Delta: The Vietnamese Move South* (The Gioi, 2015).

34. Felix Rau and Paul Sidwell, "The Munda maritime hypothesis," *Journal of the Linguistic Society of Southeast Asia* 12 (2019): 35–57.

35. Kai Tätte et al., "The genetic legacy of continental scale admixture in Indian Austro-Asiatic speakers," *Scientific Reports* 9 (2019): 3818.

36. For recent overviews, see Bellwood, *First Islanders*; Patrick Kirch, *On the Road of the Winds: An Archaeological History of the Pacific Islands before European Contact*, rev. ed. (University of California Press [Berkeley], 2017); Mike Carson, *Archaeology of Pacific Oceania: Inhabiting a Sea of Islands* (Routledge, 2018); Peter Bellwood and Peter Hiscock, "Australia and the Pacific Islands," in Chris Scarre, ed., *The Human Past: World Prehistory and the Development of Human Societies*, 5th ed. (Thames and Hudson, 2022, in production).

37. For Austronesian linguistic history, see Robert Blust, "The Austronesian homeland and dispersal," *Annual Review of Linguistics* 5 (2019): 417–434.

38. Kuo-Fang Chung, "Paper mulberry DNA attests Taiwan as Austronesian ancestral homeland," in *The Origins of the Austronesians* (Council of Indigenous Peoples, Taiwan, 2021), 157–197.

39. Christopher Buckley, "Looms, weaving and the Austronesian expansion," in A. Acri et al., eds., *Spirits and Ships: Cultural Transfers in Early Monsoon Asia* (Institute of Southeast Asian Studies, Singapore, 2017), 273–374.

40. Peter Bellwood, "Holocene population history in the Pacific region as a model for world-wide food producer dispersals," *Current Anthropology* 52, suppl. 4 (2011): 363–378; Peter Bell-wood et al., "Are cultures inherited?," in Benjamin Roberts and Marc Vander Linden, eds., *Investigating Archaeological Cultures: Material Culture, Variability, and Transmission* (Springer, 2011), 321–354; Bellwood, *First Islanders*.

41. Victoria Chen et al., "Is Malayo-Polynesian a primary branch of Austronesian? A view from morphosyntax," *Diachronica*, in press.

42. Kai Tätte et al., "The Ami and Yami aborigines of Taiwan and their genetic relationship to East Asian and Pacific populations," *European Journal of Human Genetics* 29 (2021): 1092–1102; Jeremy Choin et al., "Genomic insights into population history and biological adaptation in Oceania," *Nature* 592 (2021): 583–589.

43. Mike Carson, *Archaeology of Pacific Oceania* (Routledge, 2018); I. Pugach et al., "Ancient DNA from Guam and the peopling of the Pacific," *PNAS* 118 (2021): e2022112118.

44. Pontus Skoglund et al., "Genomic insights into the peopling of the southwest Pacific," *Nature* 538 (2016): 510–513.

45. Antoinette Schapper, "Farming and the Trans-New Guinea family," in Martine Robbeets and Alexander Savelyev, eds., *Language Dispersal beyond Farming* (John Benjamins, 2017), 155–182.

46. Mark Lipson et al., "Three phases of ancient migration shaped the ancestry of human populations in Vanuatu," *Current Biology* 30 (2020): 4846–4856.

47. Kirch, *On the Road of the Winds*; Alexander Ioannidis et al., "Paths and timings of the peopling of Polynesia inferred from genomic records," *Nature* 597 (2021): 522–526.

48. William Wilson, "The northern outliers–East Polynesian theory expanded," *Journal of the Polynesian Society* 127 (2018): 389–423.

49. Lipson et al., "Three phases of ancient migration."

50. Alexander Ioannidis et al., "Native Native American gene flow into Polynesia predating Easter Island settlement," *Nature* 583 (2020): 572–577; Thor Heyerdahl, *The Kon-Tiki Expedition: By Raft across the South Seas* (Allen and Unwin, 1950).

51. Cheng-hwa Tsang and Kuang-ti Li, *Archaeological Heritage in the Tainan Science Park of Taiwan* (National Museum of Prehistory, Taitung, 2015).

52. Zhenhua Deng et al., "Validating earliest rice farming in the Indonesian Archipelago," *Scientific Reports* 10 (2020): 10984.

53. Ornob Alam et al., "Genome analysis traces regional dispersal of rice in Taiwan and Southeast Asia," *Molecular Biology and Evolution*, 38 (2021): 4832–4846.

Chapter 12

1. Roger Blench, *Archaeology, Linguistics and the African Past* (Altamira, 2006).

2. Alexander Militarev, "The prehistory of a dispersal: The Proto-Afrasian (Afro-Asiatic) farming lexicon," in Peter Bellwood and Colin Renfrew, eds., *Examining the Farming/Language Dispersal Hypothesis* (McDonald Institute, Cambridge University, 2002), 135–150.

3. Vaclav Blazek, "Levant and North Africa: Afro-Asiatic linguistic history," in Peter Bell-wood, ed., *The Global Prehistory of Human Migration* (Wiley Blackwell, 2015), 125–132.

4. Aharon Dolgopolsky, "More about the Indo-European homeland problem," *Mediterranean Language Review* 6 (1993): 230–248.

5. Shyamalika Gopalan et al., "Hunter-gatherer genomes reveal diverse demographic trajectories following the rise of farming in East Africa," *bioRxiv* (2019), https://dx.doi.org/10.1101/517730 (see their figure 1); Carina Schlebusch, "Population migration and adaptation during the African Holocene," in Yonatan Sahle et al., eds., *Modern Human Origins and Dispersal* (Kerns Verlag, 2019), 261–283. A recent genetic classification of living populations places the Cushitic, Omotic, and Chadic speakers with contemporary Sub-Saharan African populations, whereas Berber and Semitic populations group with Middle Eastern and European populations. See Pavel Duda and Jan Zrzavy, "Towards the global phylogeny of human populations based on genetic and linguistic data," in Sahle et al., *Modern Human Origins*, 331–359.

6. Verena Schuenemann et al., "Ancient Egyptian mummy genomes," *Nature Communications* 8 (2017): 15694.

7. Rosa Fregel et al., "Ancient genomes from North Africa," *PNAS* 115 (2018): 6774–6779.

8. Fiona Marshall and Lior Weisbrod, "Domestication processes and morphological change: Through the lens of the donkey and African pastoralism," *Current Anthropology* 52, suppl. 4 (2011): 397–414.

9. Carina Schlebusch et al., "Khoe-San genomes reveal unique variation and confirm the deepest population divergence in *Homo sapiens*," *Molecular Biology and Evolution* 37 (2020): 2944–2954.

10. Schlebusch, "Population migration and adaptation"; Ke Wang et al., "Ancient genomes reveal complex patterns of population movement, interaction and replacement in sub-Saharan Africa," *Science Advances* 6 (2020): eaaz0183. The Khoekhoen were formerly known as Hottentots.

11. Christopher Ehret, "Sub-Saharan Africa: Linguistics," in Bellwood, *Global Prehistory*, 96–106.

12. Sen Li et al., "Genetic variation reveals large scale population expansion and migration during the expansion of Bantu-speaking peoples," *Proceedings of the Royal Society of London, Series B: Biological Sciences* 281 (2014): 20141448.

13. Ezequiel Koile et al., "Phylogeographic analysis of the Bantu language expansion supports a rainforest route," forthcoming in *PNAS*. See also Koen Bostoen et al., "Middle to late Holocene paleoclimatic change and the early Bantu expansion," *Current Anthropology* 56 (2015): 327–353; Rebecca Grollemund et al., "Bantu expansion shows that habitat alters the route and pace of human dispersals," *PNAS* 112 (2015): 13296–13301; Etienne Patin et al., "Dispersals and genetic adaptation of Bantu-speaking populations in Africa and North America," *Science* 356 (2017): 543–546.

14. Peter Robertshaw, "Sub-Saharan Africa: Archaeology," in Bellwood, *Global Prehistory*, 107–114; Dirk Seidensticker et al., "Population collapse in Congo rainforests from 400 CE urges reassessment of the Bantu expansion," *Science Advances* 7 (2021): eabd8352.

15. Armando Semo et al., "Along the Indian Ocean coast: Genomic variation in Mozambique provides new insights into the Bantu expansion," *Molecular Evolution and Biology* 37, no. 2 (2019): 406–416; Ananyo Choudhury et al., "High-depth African genomes inform human migration and health," *Nature* 586 (2020): 741–748.

16. James Webb, "Malaria and the peopling of early tropical Africa," *Journal of World History* 16 (2005): 269–291.

17. On Australian prehistory, see Peter Hiscock, *Archaeology of Ancient Australia* (Routledge, 2008). For the Toalian, see Peter Bellwood, *First Islanders: Prehistory and Human Migration in Island Southeast Asia* (Wiley Blackwell, 2017), 155–159.

18. For a suggested link between backed tool manufacture and El Niño droughts in southern Australia between 5,000 and 4,000 years ago, see Peter Hiscock, "Pattern and context in the Holocene proliferation of backed artifacts in Australia," in Robert Elston and Steven Kuhn, eds., *Thinking Small: Global Perspectives on Microlithization* (American Anthropological Association, 2002), 163–177.

19. Patrick McConvell, "Australia: Linguistic history," in Bellwood, *Global Prehistory*, 362–368; Remco Bouckaert et al., "The origin and expansion of Pama-Nyungan languages across Australia," *Nature Ecology and Evolution* 2 (2018): 741–749.

20. Peter Sutton, "Small language survival and large language expansion on a hunter-gatherer continent," in Tom Güldemann et al., eds., *The Language of Hunter-Gatherers* (Cambridge University Press, 2020), 356–391.

21. Arman Ardalan et al., "Narrow genetic basis for the Australian dingo," *Genetica* 140 (2012): 65–73. For the date of dingo arrival in Australia see Jane Balme et al., "New dates on dingo bones from Majura cave provide oldest firm evidence for arrival of the species in Australia," *Scientific Reports* 8 (2018): 9933.

22. Loukas Koungoulos and Melanie Fillios, "Hunting dogs down under?," *Journal of Anthropological Archaeology* 58 (2020): 101146.

23. Ray Wood, "*Wangga*," *Oceania* 88 (2018): 202–231.

24. Tim Denham et al., "Horticultural experimentation in northern Australia reconsidered," *Antiquity* 83 (2009): 634–648.

25. Peter Bellwood, *First Migrants: Ancient Migration in Global Perspective* (Wiley Blackwell, 2013).

26. Ray Tobler et al., "Aboriginal mitogenomes reveal 50,000 years of regionalism in Australia," *Nature* 544 (2017): 180–184.

27. Manfred Kayser et al., "Independent histories of human Y chromosomes from Melanesia and Australia," *American Journal of Human Genetics* 68 (2001): 173–190.

28. Anna-Sapfo Malaspinas et al., "A genomic history of Aboriginal Australia," *Nature* 538 (2016): 207–214.

29. Many Australian readers will be aware that there has been considerable debate in recent years about the question of whether or not indigenous Australians were engaged in practices akin to farming in prehistoric times. See Bill Gammage, *The Biggest Estate on Earth: How Aborigines Made Australia* (Allen & Unwin, 2011); Bruce Pascoe, *Dark Emu: Aboriginal Australia and the Birth of Agriculture* (Magabala Books, 2018); Peter Sutton and Keryn Walshe, *Farmers or Hunter-Gatherers? The Dark Emu Debate* (Melbourne University Press, 2021). I am aware of no evidence that would suggest a presence of food production in Australia according to the definition I use in this book (chapter 7). See my book review of Sutton and Walshe 2021 in *Oceania* 91 (2021): 375–376.

30. Sutton and Walshe, *Farmers or Hunter-Gatherers?*

31. Grover Krantz, "On the nonmigration of hunting peoples," *Northwestern Anthropology Research Notes* 10 (1976): 209–216.

32. Douglas Kennett et al., "South-to-north migration preceded the advent of intensive farming in the Maya Region," *Nature Communications*, in press (2021); Keith M. Prufer et al., "Terminal Pleistocene through Middle Holocene occupations in southeastern Mesoamerica," *Ancient Mesoamerica* 32 (2021): 439–460. I also thank David Reich and Keith Prufer for discussion of this important piece of research.

33. See the relevant sections, with references, in Peter Bellwood, *First Farmers: The Origins of Agricultural Societies* (Blackwell, 2005); Bellwood, *First Migrants*; and Bellwood, *Global Prehistory*. See also Robert Walker and Lincoln Ribeiro, "Bayesian phylogeography of the Arawak expansion in lowland South America," *Proceedings of the Royal Society of London, Series B: Biological Sciences* 278 (2011): 2562–2577; Jose Iriarte et al., "The origins of Amazonian landscapes," *QSR* 248 (2020): 106582.

34. Paul Heggarty and David Beresford-Jones, "Agriculture and language dispersals," *Current Anthropology* 51 (2010): 163–192.

35. Matthew Napolitano et al., "Reevaluating human colonization of the Caribbean," *Science Advances* 5 (2019): eaar7806; Daniel Fernandes et al., "A genetic history of the pre-contact Caribbean," *Nature* 590 (2021): 103–110.

36. Stuart Fiedel, "Are ancestors of contact period ethnic groups recognizable in the archaeological record of the Early Lake Woodland?," *Archaeology of Eastern North America* 41 (2013): 221–229.

37. Jane Hill, "Proto-Uto-Aztecan as a Mesoamerican language," *Ancient Mesoamerica* 23 (2012): 57–68.

38. Volume 23, no. 1 of *Archaeology Southwest Magazine* (Center for Desert Archaeology, 2009, https://www.archaeologysouthwest.org/pdf/arch-sw-v23-no1.pdf) has good accounts of these finds. See also James Vint and Jonathan Mabry, "The Early Agricultural period," in Barbara Mills and Severin Fowles, eds., *The Oxford Handbook of Southwest Archaeology* (Oxford University Press, 2017), 247–264.

39. Rute Da Fonseca et al., "The origin and evolution of maize in the Southwestern United States," *Nature Plants* 1 (2015): 14003.

40. Victor Moreno-Mayar et al., "Early human dispersals within the Americas," *Science* 362 (2018): eaav2621; Jane Hill, "Uto-Aztecan hunter-gatherers," in Güldemann et al., *Language of Hunter-Gatherers*, 577–604.

INDEX

Note: Page numbers in *italics* indicate figures and tables.

First Australians, 107–108
First Japanese, 120
fishhooks, 106, 121, 132, 291
Flannery, Tim, 100
Flores Island (Indonesia), 53–56
food production: about, 7–12, 102, 136–151, 310–312; advantages of, 138–140; in Africa, 174–180, 278–285; in the Americas, 184–191, 293–304; in Asia, 167–174, 251–277; in Australia, 106, 290; and climate changes, 150; in Cyprus, 163–164; early farming dispersal hypothesis, 197–199; in Egypt, 155–156, 278–279; in Europe, 224–229; expansion of, 220–221, 224; Fertile Crescent, 153–157, 163–164, 221–224, 243, 280; homelands, 152–196; and hunter-gatherers, 146–147, 312; irrigation, 192; and language, 197–219; main developments in, *195*; in Mesoamerica, 193; in Mesopotamian lowlands, 156; and migration, 12; New Guinea, 99, 180–184; origins of, 141–151, 156; and population growth, 139–140, 310; South Asia, 231–234, 245–249
Ford, James, 185
Forster, Johann Reinhold, 238, 271
FOXP2 (gene), 80

Ganges Basin, 248
Gangetic Plain, 233–234
Gaomiao (China), 262
Garrod, Dorothy, 157
gazelles, 150, 157, 162, 176
Gibbons, Ann, 79
Gimbutas, Marija, 235, 244
glacial-interglacial cycles, 44–46
global warming, 10
Globular Amphora (culture), 236
Göbekli Tepe (Turkey), 159–162, 177, 226
Golson, Jack, 181
Gona, 31
Gonur (Turkmenistan), 234
Gopher, Avi, 146
gorillas, 18
Gowlett, John, 38

Gran Dolina (Spain), 66–67, 69, 71
Greek (language), 217
Greenberg, Joseph, 127
Groves, Colin, 87
Guatemala, 193

Hadza hunter-gatherers, 91, 92
hair, 3, 23
Halafians, 249
hand axes, 69–70, 75, 87
hand grip (in hominins), 18
Harappan phase, 233–234
Harar, 20
"Harbin human group" (species), 33, 59–60, 72, 78–79
Heidelberg (Germany), 67
Heilongjiang (China), 78, 116
Helwan Point, 177
Hemudu (China), 172
Herodotus, xxii
Heyerdahl, Thor, 275
Hill, Jane, 301
Hinduism, 247
Hittites, 238
Hoabinhian, 261–262, 266
Hoenigswald, Harry, 239–240
Holocene (era): climates, 148; and food production, 23, 102; as an interglacial, 12, 43; Japan during, 258; migrations in the Americas, 293–294; population movements during, *283*; settlement during, 133–135; stone tool industries, 287
hominins, xviii, 3; about, 18–23; in Africa, 27–30, 33; ape ancestors of, 18; versus apes, 18–24; bipedality of, 19–20, 58; brains of, 36, 37; in China, 51; in Eurasia, 33; in Java, 52; populations, 7; Sima de Los Huesos, 74–75; species, 42. *See also* panins
Homo antecessor, 60, 66–67, 69, 79
Homo erectus, xx, 3; and the Arctic Circle, 114; departure from Africa, 9, 47–51; discovery of, 23–24; *floresiensis*, 55; in Java, 51–52; Nariokotome Boy, 36; in

A NOTE ON THE TYPE

This book has been composed in Arno, an Old-style serif typeface in the classic Venetian tradition, designed by Robert Slimbach at Adobe.